2020 36th Semiconductor Thermal Measurement, Modeling & Management Symposium (SEMI-THERM 2020)

San Jose, California, USA
16-20 March 2020

IEEE Catalog Number: CFP20SEM-POD
ISBN: 978-1-7281-9588-9

Copyright © 2020, SEMI-THERM Educational Foundation (STEF)
All Rights Reserved

*** This is a print representation of what appears in the IEEE Digital Library. Some format issues inherent in the e-media version may also appear in this print version.

IEEE Catalog Number: CFP20SEM-POD
ISBN (Print-On-Demand):   978-1-7281-9588-9
ISBN (Online):   978-0-578-43862-7

Additional Copies of This Publication Are Available From:

Curran Associates, Inc
57 Morehouse Lane
Red Hook, NY 12571 USA
Phone: (845) 758-0400
Fax: (845) 758-2633
E-mail: curran@proceedings.com
Web: www.proceedings.com

Thirty-Sixth Annual

SEMICONDUCTOR THERMAL MEASUREMENT, MODELING AND MANAGEMENT SYMPOSIUM

PROCEEDINGS 2020

San Jose, CA USA
March 16-20, 2020

Thirty Sixth Annual

SEMICONDUCTOR THERMAL MEASUREMENT, MODELING AND MANAGEMENT SYMPOSIUM

PROCEEDINGS 2020

San Jose, CA   USA
March 16-20, 2020

The 2020 Semiconductor Thermal Measurement, Modeling and Management (SEMI-THERM) Symposium is an annual international forum for the presentation of new developments in and applications relating to generation and removal of heat within semiconductor devices, and measurement of junction temperatures under various application and environmental conditions.

Attendance at the Symposium is limited, to preserve the close interaction among attendees and presenters. The format of the symposium this year couples eight sessions of selected technical papers, a more-intimate Poster Session for one-on-one discussion of results, a luncheon talk, and a series of Workshops focused on products and techniques and an embedded tutorial.

This year, the Symposium is preceded by seven Short Courses: **"Introduction to Electronics Cooling," "Introduction to Thermal Modeling with OpenFOAM," "Design and Optimization of Heat Sinks," "Thermal Management of Li-Ion Battery Packs," "Air Movers and Aeroacoustics for Electronics Cooling," "Micro-Two-Phase Electronics Cooling...Getting it on its Way,"** and **"Let's Work Together: How Co-Design Leads to Better Solutions in Thermal Management."** In addition, an exhibits area offers displays of equipment, software, and other resources within the thermal measurements field.

We trust you will take advantage of the rich array of information and experiences developed by this year's Steering and Program Committees, and consider submitting an abstract for next year's SEMI-THERM.

General Chair
Pablo Hidalgo,
    Aavid

Program Chair
Marcelo del Valle,
    Intel

Vice Program Chair
Joshua Gess,
    Oregon State University

International Liaisons
John Parry,
    Mentor Graphics
Sobo Sun,
    Celsia Inc.
Winston Zhang,
    Novark

Symposium Marketing
Denise Rael,
    +1- 480 839-8988,
    drael@semi-therm.org

Symposium Management
Bonnie Crystall,
C/S Communications, Inc

Social Media
Robin Bornoff,
    Mentor

Proceedings
Paul Wesling

SEMI-THERM 36

Welcome!

Pablo Hidalgo, AMD

Dear Colleagues,

It is my pleasure to welcome you to SEMI-THERM 36 annual symposium. The entire organizing team and I want to thank you for your presence in this year's event and we hope you find the symposium extremely helpful and beneficial for your career and your business. The SEMI-THERM organization is dedicated to providing a platform for discussion on the latest advancements in thermal management for both industry professionals as well as members of academia so during the thermal week, all participants can share their knowledge and promote collaboration between entities.

In this edition, the Program Chair, Dr. Marcelo del Valle, the Program Vice-Chair, Prof. Joshua Gess and the entire program committee has assembled a superb list of presentations and papers with a particular emphasis on thermal management from die level to infrastructure level for data centers, as well as measurement techniques of semiconductors and components. The program consists of seven short courses on Monday that will span from the morning till the afternoon. Technical presentations will be given every morning from Tuesday through Thursday. This year's Keynote speaker is Dr. Andy Delano from Microsoft and he will be talking about "Innovations in Thermal Management of Electronic Devices". The symposium will also have two luncheon speakers, one on Tuesday and one on Wednesday, a panel session on data center cooling technologies, one embedded tutorial and one evening tutorial. Tuesday and Wednesday afternoon, you will have the opportunity to enjoy the exhibits and vendor workshops, that you can use for networking with other fellow engineers and attend product presentations provided by vendors.

During the symposium, we will also have several awards and I would like to take this opportunity to congratulate every single one of the recipients for their achievements. The THERMI award, which this year is sponsored by The University of Texas in Arlington, recognizes the recipient's contributions to thermal issues and overcome the challenges presented in electronics cooling. I am very pleased to announce that the 2020 THERMI award recipient is for Dr. Ross Wilcoxon from Collins Aerospace. SEMI-THERM also instituted the Thermal Hall of Fame Lifetime Achievement Award, which this year is sponsored by Celsia, recognizes an individual in the thermal management field that has made significant contributions to the development and commercialization of thermal management technologies during their career. I am also very honored to announce that this

SEMI-THERM 36

distinguished award will be given to Prof. Dereje Agonafer from the University of Texas in Arlington. Last but not least, the 2019 Harvey Rosten Award, which is sponsored by Mentor a Siemens business, for outstanding work recently published or in the public domain, for advances in analysis or modeling of thermal effects in electronic equipment or components, will be given to Baver Ozceylan, Prof. Boudenwijn R. Haverkort, Dr. Maurits de Graaf and Dr. Marco E.T. Gerards.

SEMI-THERM won't be as successful as it is without all the individuals that devote their own personal time to help in the preparation and organization of this symposium. From reviewers to session chairs, organizing, steering and technical committee members, the unconditional guidance and support from Bonnie Crystall, Denise Rael and Robert Schuch and I cannot leave behind the continuous mentorship and influence by George Meyer and SEMI-THERM's co-founder Bernie Siegal. I would like to take this opportunity to acknowledge them and give thanks for all your support during the last year.

I hope that SEMI-THERM 36 leaves a positive impact on everyone that participates this year and that everyone takes away the main reason of the symposium existence, education on the best and newest thermal technologies and the ability to build new connections with leaders in the thermal management community.

Please join me in congratulating the Program Chair, Dr. Marcelo del Valle, the Program Vice-Chair, Prof. Joshua Gess for putting together such an exceptional program and I will be looking forward to enjoy with everyone of you a wonderful thermal week.

Sincerely,

Pablo Hidalgo
Symposium General Chair

SEMI-THERM 36

SEMI-THERM 36 SYMPOSIUM PERSONNEL

General Chair:
Pablo Hidalgo, AMD   pablo.hidalgoardana@amd.com

Program Chair:
Marcelo del Valle, Infinera Corporation   mvalle@infinera.com

Program Vice Chair:
Joshua Gess   joshua.gess@oregonstate.edu

International Liaisons:
John Parry, Mentor, a Siemens business
    john_parry@mentor.com

Sobo Sun, Celsia Inc.   ssun@celsiainc.com

Winston Zhang, Novark, China   winstonzhang@novark.com.cn

Social Media:
Robin Bornoff Mentor, a Siemens business
    robin_bornoff@mentor.com

Symposium Management:
SEMI-THERM Symposium Manager
Bonnie Crystall, C/S Communications, Inc.
    cscomm@earthlink.net

Proceedings IEEE Region 6:
Paul Wesling   p.wesling@ieee.org

SEMI-THERM 36 Steering/Technical Committee

Chair
George Meyer   gmeyer@celsiainc.com

Technical Chair
Ross Wilcoxon   ross.wilcoxon@collins.com

Finance Chair
Jim Wilson   jsw@raytheon.com

Steering/Technical Committee
Dereje Agonafer   agonafer@uta.edu
Herman Chu
Bruce Guenin   bguenin@usa.net
Genevieve Martin   genevieve.martin@signify.com
Bill Maltz   wmaltz@ecooling.com
Veerendra Mulay   vmulay@fb.com
Alfonso Ortega   alfonso.ortega@villanova.edu
John Parry   john_parry@mentor.com
Adrianna Rangel   adromero@cisco.com
Dave Saums   dsaums@dsa-thermal.com
Bernie Siegal   bsiegal@thermengr.net
Tom Tarter   ttarter@pkgscience.com
Winston Zhang   winstonzhang@novark.com.cn

SEMI-THERM Exhibits/Registration:
Bob Schuch   rschuch@semi-therm.org

SEMI-THERM Marketing
Denise Rael   drael@semi-therm.org

Graphic Design:
William Schuch   bill.schuch@semi-therm.org

SEMI-THERM 36 TOPIC CHAMPIONS, SESSION CHAIRS AND PROGRAM/REVIEW COMMITTEE

Cathy Biber	Intel	George Meyer	Celsia
Robin Bornoff	Mentor, a Siemens business	Koroush Nemati	Future Facilities
Mark Carbone	Intel	Alex Ockfen	Facebook
Marcelo del Valle	Infinera Corporation	Pritish Parida	IBM
Roger Dickinson	Boyd Corp.	Devin Pellicone	ACT
Valerie Eveloy	Khalifa University, Abu Dhabi, UAE	Adriana Rangel	Cisco Systems
Pablo Hidalgo	AMD	Peter Rodgers	Khalifa University, Abu Dhabi, UAE
Shailesh Joshi	Toyota	David Saums	DS&A LLC
Taravat Khadivi	Qualcomm	Mohammad Reza Shaeri	STERIS Endoscopy
Wendy Luiten	WLC Wendy Luiten Consultancy	Tim Shedd	ZutaCore
Bonnie Mack	Ciena	Jason Strader	Laird
Genevieve Martin	Signify	Ross Wilcoxon	Collins Aerospace
		Jim Wilson	Raytheon

www.semi-therm.org

SEMI-THERM 36

Short Courses Monday, March 16, 2020

Short Course 1   Morning   8:00 a.m. – 12:00 p.m.
Introduction to Electronics Cooling
Patrick Loney, Northrop Grumman Mission Systems

As electronic packages get smaller and the power dissipations increase, performing robust thermal analyses is an increasingly important step in the electronics packaging design process. This course will focus on the component level of the electronics assembly. Thermal management, proper cooling techniques, component attachment, and analytical modeling methods will be presented. How to decipher vendor datasheets will be discussed as well as the basics of how to model custom components. Best practices for steady state and transient operational modes are included. Process development will also be presented along with discussions on requirements compliance. Students will finish the course with an understanding of how to determine the limits and requirements of an electronics component, assess the thermal performance, how to integrate the performance model into a Next Higher Assembly (NHA) thermal model, and most importantly, how to communicate this information to their internal and external customers who are dependent on this data.

Patrick Loney recently celebrated his 30 th anniversary with Northrop Grumman Corporation. He has over 35 years of experience in the thermal engineering/electronics cooling industry. He received his Batchelor of Sciences degree in Nuclear Engineering from the University of Illinois and his Masters of Sciences degree in Mechanical Engineering from Cleveland State University. He holds several US Patents and Trade Secrets, mostly dealing with thermal management and electronics cooling techniques. He has presented similar courses to internal customers as well as the 2019 IPC AMEX Expo.

Short Course 2   Morning   8:00 a.m. – 12:00 p.m.
8:00 a.m. Short Course 2: Introduction to Thermal Modeling with OpenFOAM
John F. Maddox, University of Kentucky

OpenFOAM is the leading free, open source software for computational fluid dynamics (CFD). This course is an introduction to thermal modeling using OpenFOAM for users familiar with CFD and heat transfer, however, no prior experience with OpenFOAM is required. Attendees will be introduced to the OpenFOAM environment through hands-on tutorials covering meshing, solving, and post-processing with a focus on conjugate heat transfer. Attendees wishing to participate in the hand-on tutorials will need to bring a laptop with a 64-bit operating system (Window, Mac, or Linux) and Oracle VM VirtualBox installed. All the software required for this course will be free and open source.

Dr. John F. Maddox is an Assistant Professor of Mechanical Engineering at the University of Kentucky, Paducah Campus. He received his Ph.D. in mechanical engineering from Auburn University in 2015. His primary research areas are thermal management of high power electronics through jet impingement and thermal characterization of advanced materials used in aerospace and electronics cooling applications.

SEMI-THERM 36

Short Courses Monday, March 16, 2020

Short Course 3   Morning   8:00 a.m. – 12:00 p.m.

Design and Optimization of Heat Sinks

Marc Hodes and Georgios Karamanis, Transport Phenomena Technologies, LLC

This course provides the audience with an understanding of heat sink design and optimization in the context of the thermal management of electronics. The course has two parts. The first part begins with an overview of common methods to manufacture heat sinks such as extrusion, die casting and forging, and discusses their advantages and disadvantages with respect to cost and fin geometry. Attention then shifts to the theory of spreading resistance and how it can be calculated in order to properly size the thicknesses of the bases of heat sinks. Next, the theory of the operation of heat pipes in tubular and flat (vapor chamber) configurations is presented along with their roles in smoothing out temperature gradients in the fins and bases of heat sinks. In the second part of the course, single-phase conjugate heat transfer, where conduction in the heat sink is coupled to convection in the coolant, i.e., air or water, flowing through the heat sink is highlighted. We discuss why the constant heat transfer coefficient assumption tends to be an invalid one in real heat sinks by using specific examples. Then, the use of computational fluid dynamics (CFD) to compute conjugate Nusselt numbers is considered. The course concludes with a discussion of how to embed pre-computed results for conjugate Nusselt numbers and dimensionless flow resistances for heat sinks in flow network models (FNMs) of circuit packs such as blade servers. Finally, how to use a multi-variable optimization scheme to optimize the geometry (fin thickness, spacing, height, length, say) of an array of heat sinks in a circuit pack represented by an FNM model with embedded tabulations of CFD results is discussed.

Marc Hodes is a Professor of Mechanical Engineering at Tufts University and the CTO of Transport Phenomena Technologies, LLC. He received his B.S., M.S. and Ph.D. degrees in Mechanical Engineering, the latter from MIT in 1998. He held a succession of appointments at Alcatel-Lucent's (now Nokia's) Bell Laboratories from Postdoctoral Scientist to Manager of a Thermal Management Research Group between 1998 and 2008, when he joined Tufts University.

Georgios (George) Karamanis is a Co-Founder and Senior Engineer in Transport Phenomena Technologies, LLC. He received his Ph.D. and M.S. in Mechanical Engineering from Tufts University. He has expertise in analytical, numerical and experimental techniques relevant to convective transport. He is the PI in a NSF Phase I SBIR awarded to Transport Phenomena Technologies, LLC, to develop specialized thermal modeling software and hardware for Data/Telco centers.

SEMI-THERM 36

www.semi-therm.org

Short Courses Monday, March 16, 2020

Short Course 4   Morning   8:00 a.m. – 12:00 p.m.
Thermal Management of Li-Ion Battery Packs
Azita Soleymani Ph.D., Electronic Cooling Solutions, Inc.
The thermal management of li-ion battery packs is crucial, as the cooling is directly related to the safety, reliability, performance and durability of battery packs. In this course, the specific thermal requirements of li-ion battery cells and battery packs will be discussed. The empirical tests necessary to characterize the thermal performance of cells will be presented; and it will be shown how the test results can be utilized to estimate real-time heat generation rates of cells at different state-of-charge, current and temperature.
Practical considerations in the design of thermal management system of battery packs will be provided in detail. The course will cover comprehensively the use of different simulation approaches such as Computational Fluid Dynamics, 1-D system level simulations, and Digital Twin. The Digital Twin models of battery packs can be developed in order to perform what-if scenarios, to conduct in-depth root cause analyses, to further optimize the cooling system, to make life-time predictions and to optimize operating parameters for thermal management.

Dr. Azita Soleymani is currently holding the director position at ECS Inc. She graduated from Lappeenranta Univ. of Tech., with PhD degree in advanced simulation and modeling of transport phenomena. After graduation she worked as a manager in a Danaher company and Byton Inc.

Short Course 5   Afternoon   1:30 p.m. – 5:30 p.m.
Air Movers and Aeroacoustics for Electronics Cooling
Mark MacDonald, Intel Corporation
This course will survey performance characteristics of various relevant fan types, including axial fans, blowers, crossflow or tangential blowers, volumetric resistance blowers, and other emerging technologies including electronhydrodynamic blowers, synthetic jets, piezo flappers, and micropumps. Emphasis will be placed on understanding the physical mechanisms of operation, best practices for characterization, implementation considerations, and applicable scaling laws (including acoustic scaling laws). The course will also cover aeroacoustics and psychoacoustics (sound quality and ergonomics) for consumer electronics in detail.

Mark MacDonald holds a Ph.D. from Cornell University. Formerly an Adjunct Professor at Portland State University, he is the holder of 45 U.S. patents, 17 of them specific to Air Movers. Dr. MacDonald is a winner of the Martin Hirschorn Prize from the International Acoustics Congress for work on notebook blower acoustics.

SEMI-THERM 36

Short Courses Monday, March 16, 2020

Short Course 6   Afternoon   1:30 p.m. – 5:30 p.m.
Micro-Two-Phase Electronics Cooling…Getting it on its Way
John R. Thome, EPFL

Two-phase flow and flow boiling heat transfer can reliably cool heat fluxes in excess of 500 W/cm2 with heat transfer coefficients nearing 100 kW/m2K with respect to the cold plate's base area. Yet, industry is hesitant to accept this technology on a large scale. Most of the reservations about this approach are easily mitigated with proper design/planning, and the benefits are substantial. In general, a micro-thermosyphon that works passively with gravity-driven flow is used with heat dissipation to a compact air coil. Due to the new "form factor" and huge surface area of the coil compared to an air-cooled heat sink, energy consumption by the fans is greatly reduced. Furthermore, a thermosyphon (no electrical driver or flow controllers) provides high reliability that is commonplace with packages which use two-phase thermal management. This lecture will recount the history and background of two-phase cooling, noting lessons learned along the way. Several case studies will be presented where a design flaw was mitigated and the resulting improvements in performance will be highlighted. At the end of this course, you will be able to successfully design a two-phase cold plate cooled system which improves the reliability, cost of operation, and longevity of your devices.

John R. Thome is Professor-Emeritus of Heat and Mass Transfer at the Ecole Polytechnique Fédérale de Lausanne (EPFL), Switzerland since 1998. He obtained his PhD at Oxford University in 1978. Having retired in July 2018 at the EPFL, he co-owns the consulting/ thermal engineering software company, JJ Cooling Innovation Sàrl in Lausanne. He is also a Visiting professor at Brunel University in London and an Honorary professor at the University of Edinburgh…to keep his "feet" in research while still supervising MS student theses at the EPFL. He recently received the 2019 IEEE Richard Chu ITHERM Award for Excellence in Thermal and Thermo-Mechanic Management of Electronics and the 2019 ASME Allan Krause Thermal Management Medal at InterPack. He is the author of five books on two-phase heat transfer and flow and has over 245 journal papers on macroscale and mircoscale two-phase flow, flow visualization, boiling/condensation heat transfer, flow pattern-based models, and micro-two-phase cooling systems for electronics cooling. He has done numerous sponsored projects with IBM, ABB, Nokia Bell Labs, Carl Zeiss, CERN, etc. He is editor-in-chief of the 16-volume series Encyclopedia of Two-Phase Heat Transfer and Flow (2016-2018). He founded the Virtual International Research Institute of Two-Phase Flow and Heat Transfer in 2014, now with 25 participating universities to promote research collaboration, sharing of experimental and numerical data, and education.

SEMI-THERM 36

www.semi-therm.org

Short Courses Monday, March 16, 2020

Short Course 7   **Afternoon**   1:30 p.m. – 5:30 p.m.

Let's Work Together: How Co-Design Leads to Better Solutions in Thermal Management
Lauren Boteler, Army Research

Laboratory Optimization studies are generally done intradisciplinary rather than interdisciplinary, and this leads to conflict as different fields have different values when it comes to what they want in a packaged solution. Heat sinks in energy dense power electronics are an excellent example of where better communication and co-design models can yield significant improvements to fielded performance with just a small amount of preparation during the design phase. Parameterization and Figure of Merit (FOM) definitions that encapsulate electrical/ thermal/mechanical properties pare down the solution space to a set that represents what all fields want rather than cyclically proposing "optimal" solutions that one or more fields can't possibly accommodate. This course will examine how fielded solutions were truly optimized using novel co-design tools and optimization techniques which span multiple disciplines. The case studies examined will show marked improvement beyond what single-track minded approaches yield, and lessons learned from this course will translate directly to better solutions in your workplace.

Dr. Lauren Boteler leads the thermal and packaging research programs as part of the Advanced Power Electronics group at the U.S. Army Research Laboratory (ARL). She received her Ph.D. degree in mechanical engineering from the University of Maryland. Her work at ARL, beginning in 2005, has focused on electronics packaging and thermal management solutions for a wide range of Army applications. She designs thermal and packaging solutions including 3D chip stacking, power electronics, laser diodes, RF HEMT devices, top side cooling, phase change materials, and additive manufacturing. More recently, she has initiated a research program in Advanced Power Electronics Packaging and Thermal Management which focuses on four main challenges of power electronics packaging: transient thermal mitigation, additive manufacturing, coengineering/codesign, and high-voltage packaging. She was also awarded the 2018 ASME EPPD Woman Engineer of the Year award for her contributions to the electronics packaging community.

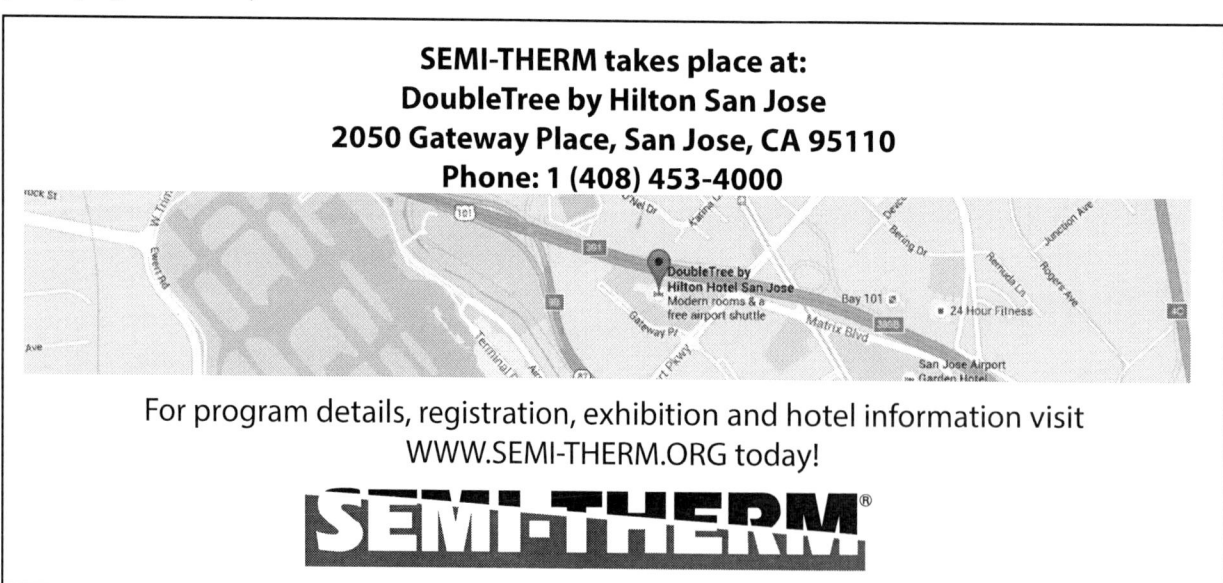

**SEMI-THERM takes place at:
DoubleTree by Hilton San Jose
2050 Gateway Place, San Jose, CA 95110
Phone: 1 (408) 453-4000**

For program details, registration, exhibition and hotel information visit
WWW.SEMI-THERM.ORG today!

SEMI-THERM®

SEMI-THERM 36

Keynote
Tuesday, March 17, 9:10 a.m. - 10:10 a.m.

Innovations in Thermal Management of Electronic Devices

Presenter: Andy Delano
Microsoft

Andy plans to highlight innovations from across the industry and over the last ~ 5-7 years and also talk about what he has found to be effective techniques for innovating over the course of his career.

Andy Delano leads the Microsoft Surface team's thermal architectural and technology efforts. Prior to joining Microsoft in 2012, Andy managed the thermal R&D team within Honeywell's electronic materials division developing and launching highly successful products for the electronics packaging industry. Andy started his career in 1998 as a thermal engineer at Hewlett-Packard designing server and workstation thermal systems. While at HP, Andy was also an adjunct professor at CU and taught heat transfer, thermodynamics, and thermal systems design between 1999 and 2005.

Prior to his career, Andy obtained his Ph.D. in mechanical engineering from Georgia tech in 1998, and his thesis was on a single pressure absorption refrigerator originally patented by Albert Einstein. During the first part of his graduate studies, Andy also worked on the design and production of the 1996 Olympic Torch and spent 6 weeks traveling with the torch relay.

SEMI-THERM 36

Luncheon Speaker
Tuesday March 17

Another day at the office: combining AI, CFD, and Belgian beer

Presenter: Lieven Vervecken
Diabatix

The majority of people do not really know what it is and the word highly overstates the technology, but Artificial Intelligence (AI) has made its entry and it is here to stay. Logically, it was only a matter of time before AI found its way into the field of Computational Fluid Dynamics (CFD). The possibilities with this combination seem endless, but are they really? Which challenges are we facing and how can we overcome them? In this talk I share some of our experiences when working with some of the largest companies in the world with one of the newest technologies in the world.

Lieven Vervecken is co-founder and CEO of Diabatix nv where he is responsible for the general management and the overall vision and strategy of the company. Diabatix is a Belgian technology scale-up specialized in generative design for cooling components that helps multinationals all over the world to push the boundaries in thermal design. Before devoting his work full-time to Diabatix, Lieven completed two master's degrees in engineering and a PhD in the field of Computational Fluid Dynamics from the University of Leuven. Lieven is an experienced speaker at national and international conferences, and former lecturer at the University of Leuven. He is passionate about the limitless possibilities of combining engineering with artificial intelligence technology and takes every opportunity to expand his knowledge in this field.

SEMI-THERM 36

Luncheon Speaker
Wednesday March 18

Bletchley Park: Enigma, Ultra, and the Making of Colossus

Presenter: Dave Saums

The development of what has become known as "signals intelligence" traces back to crude beginnings during World War I, in the United Kingdom. As the dark clouds of political and military moves began to turn into signs of impending winter storms in the late 1930s, efforts were made in the UK, France, and in Poland to begin to more seriously decipher diplomatic and military codes being used by the German government. Similar code-breaking activities were taking shape in the United States in very crude form, and in Germany and Japan. In Poland, a small team of so-called codebreakers had focused on the Enigma code being used by the German government for both diplomatic and military purposes and that team made a striking decision that had enormous implications for the outcome of World War II. The British government, having set up a rudimentary codebreaking office twenty years earlier, moved very slowly to develop a more focused effort to break these diplomatic codes. As September 1939 turned the world again to war, the need for tools and methods to crack both diplomatic and military codes became absolutely critical. A small staff was assigned and this small team moved into an old Victorian family estate in Bletchley Park, less than an hour from London by train. The "Special Relationship" that still exists today between the United Kingdom and the United States grew directly from these seeds of human activity and cooperation. For those who have seen the 2015 movie, The Imitation Game, with Benedict Cumberbatch and Kiera Knightley, this story will begin to sound familiar. The movie gave a very realistic interpretation of those actual events and focused on the story of Alan Turing, an Englishman who is often credited with being one of the fathers of modern computer systems. In reality, the development of mechanical, automated tools to sort huge incoming streams of coded German military and diplomatic message led to what are known as "Turing machines" and, as the volume of data became simply overwhelming, an engineer working for the British Post Office designed and built the first machine dubbed "Colossus" – what is now credited in the computing world as the first digital computing machine, preceding the "Eniac" in the United States by two years.

This presentation will outline the breaking of the German Enigma code (which became a series of different codes, used by different armed forces services), which produced what as titled as top-secret "Ultra" information about German military plans, locations of ships and submarines and battle groups, and how these first mechanized codebreaking machines were devised. There is much more to this story:

- The extraordinary contributions of thousands of young British women, aged eighteen to mid-twenties, for the entire codebreaking effort at Bletchley Park and across the globe;
- The mechanics of the German Enigma machines and the spread of highly complex diplomatic and military codes;

SEMI-THERM 36

- The establishment of the British codebreaking operation at Bletchley Park, where more than ten thousand British, Canadian, and American citizens and military personnel worked in this desperate race to decode thousands of incoming messages per day;
- How the breaking of Enigma was used in the battle against and sinking of the German battleship Bismarck in one of the great sea battles of modern warfare;
- The value of codebreaking to saving the United Kingdom from starvation in the Battle of the Atlantic, with the ultimate defeat of the German U-boat submarine menace;
- The use of Enigma in the greatest tank battle in history, the Battle of Kursk, where thousands of German and Soviet tanks fought in gruesome conditions on the Russian plains;
- The development of codebreaking machinery that led to the "Turing machines" (known as "Bombe" machines) and then to "Colossus" – and the construction of hundreds of these machines in the UK by the British Telegraph Company and in the United States by NCR;
- The fact that of the more than ten thousand people employed at Bletchley Park and connected outstations across the globe, none spoke of the details of what had happened in codebreaking until the mid-1970s, preserving the secret.

This presentation will focus on the technologies employed and short descriptions of hardware developed, as precursors to the modern age of digital computing – but will also illustrate the human contributions to preventing the destruction of the modern democratic world in the 1940s. The connections to technology in today's world rose from what would otherwise have been the ashes of defeat.

Bletchley Park today is an astounding museum of both technical detail and human achievement – the opening chapters in what has become the computing world that we live in today.

Dave Saums has thirty-nine years of technical marketing, product development, and business development experience with advanced thermal materials, thermal components, and twophase liquid cooling systems. Dave has operated a consulting firm focused on thermal materials and components for fourteen years, in addition to twenty-five years' experience with thermal component and materials manufacturers.

SEMI-THERM 36

Embedded Tutorial
Thursday March 19 , 9:10 a.m. - 10:10 a.m.

Additive Manufacturing

Presenter: Ram Ranjan
UTRC

Additive Manufacturing (AM) is an emerging field that enables cost efficient manufacturing of complex design features and reduce the number of parts in component assemblies. For thermal management applications, AM can make structures that optimally reduce pressure drop and reform thermal boundary layers in the coolant flow field, reducing the package's overall thermal resistance to incredibly low levels. Emerging design methods such as topology optimization enable physics-led optimization of thermal components. These structures that were once "academic" are now producible. However, there are rules and guidelines that must be followed to ensure that a part is consistently producible. In this tutorial, topology optimization methods will be introduced as a tool for novel conceptual design of components such as heat sinks and heat exchangers. "Rules of thumb" will be presented which will arm attendees with the information they need to make AM work for them in the most efficient way, i.e. building parts right on the first try with no failures, lower surface roughness, reduced overhang requirements, etc. Attendees will learn about the new possibilities of AM and how best to use this technique to exceed performance over conventional manufacturing methods.

Dr. Ram Ranjan is a principal engineer in the Collins Aerospace program office at UTRC. At UTRC, he currently leads multiple programs on advanced design methods, additive manufacturing, and electronics thermal management related to hybrid electric propulsion. He has more than ten years of experience in the field of electronics thermal management and development of CFD tools for various engineering applications. Dr. Ranjan received his Ph.D. in Mechanical Engineering from Pvturdue University in 2011 and his M.S. and B.S. in Mechanical Engineering from the Indian Institute of Technology Kanpur in 2007.
He has published his research findings in over 25 publications and has five U.S. patents.

SEMI-THERM 36

Evening Tutorial
Tuesday, March 17 7:30 p.m. - 9:00 p.m.

Realistic Thermal Model for Human Skin in Contact with a Wearable Electronic Device

Presenter: Bruce Guenin, Ph.D.

Makers of electronic devices try to provide as much performance and functionality in them as possible, consistent with certain limits for internal chip temperatures. For wearables, the external temperatures of these devices are also critical for user comfort and safety. For accuracy in a thermal model for the wearable device, it is necessary to accurately account for the transfer of heat into human skin. The commonly used ad hoc assumption of an isothermal boundary condition representing the region of contact between a wearable device and human skin is no longer adequate.

In the medical and biological fields, modeling the transfer of heat into or out of living tissue is a mature area of study. The dominant methodology in this regard is referred to as the Pennes biothermal model, named after its creator. It is a conduction model supplemented by a mechanism for cooling the tissue by blood flow, which Pennes called "perfusion." The application of the Pennes model requires that certain specified material properties be measured for each of the different tissue types involved in the heat flow, namely: thickness, thermal conductivity, specific heat, and perfusion rate. In the case of human skin all three layers (epidermis, dermis, and hypodermis) are separately represented.

Despite its wide use in the life sciences, the Pennes biothermal model is virtually unknown in the electronics cooling sector. It's the intent of this presentation to provide sufficient background information and details in its implementation that the attendees will be able to apply it immediately to their work.

Dr. Bruce Guenin has spent many years in the electronics and computer industries, which has given him a broad perspective on macro trends in these fields. His previous affiliations include Oracle, Sun Microsystems, and Amkor. He is a past chairman of the JEDEC JC-15 Thermal Standards Committee and the Semi-Therm Conference. He has been an editor of Electronics Cooling since 1997. His contributions to the thermal sciences have been recognized by receiving the Harvey Rosten Award in 2004 and the Thermi Award in 2010. He received the B.S. degree in Physics from Loyola University, New Orleans, and the Ph.D. in Physics from the University of Virginia. He has authored and co-authored over 90 papers and articles in the areas of thermal and stress characterization of microelectronic packages, electrical connectors, solid state physics, and fluid dynamics and has been awarded 18 patents in these areas.

SEMI-THERM 36

www.semi-therm.org

Panel Discussion
Thursday March 19, 2020 2:00p.m – 4:00p.m.

Liquid Cooling

For the last 10 years, Liquid Cooling has been a "technology that will be widely adopted in the next 2 to 3 years." Are we actually at that place now? What are the barriers to adoption that may keep liquids out of electronics chassis for even longer? What are the "dream features" of a cooling technology that would truly remove thermal constraints in your application area (without breaking any laws of physics)? What are the applications where there is no choice but to use liquid for electronics cooling? This panel will provide a unique perspective on these and other questions, with representatives of the following end users and implementers:

- **ES2 NSF Industry-University consortium**
- **Collins Aerospace**
- **Oak Ridge National Laboratory**
- **Service Logic**
- **Microsoft**
- **Intel**

Panel Members:
Cathy Biber, Intel
Greg Crumpton, Service Logic Corporation
David Grant, Oak Ridge National Laboratory (ORNL)
Alfonso Ortega, Villanova University
Debabrata Pal, Collins Aerospace
Brandon Rubenstein, Microsoft
Moderator: Tim Shedd, ZutaCore

Cathy Biber is a thermal engineer currently architecting systems in Intel's Data Platforms Group, working primarily on server products. She has experience across a wide range of electronics cooling applications.

Greg Crumpton was named Vice President of Critical Environments and Facilities in 2016. In his current role, Greg drives Service Logic's vertical market penetration in the mission critical segment, and oversees EH&S across Service Logic and its operating units. Greg joined Service Logic in 2014 with the sale of AirTight Mechanical, the company he founded and has led since 1999. As founder and president, he built a remarkable company that had a proven track record and expertise in serving the mission critical market throughout the Carolinas and is a foremost expert in mission critical applications and facilities. Prior to founding AirTight, Greg worked as a Division General Manager of Project Management and in sales capacities for McKenney's, Inc.

Greg serves on the advisory boards of Ebullient, LLC a designer and manufacturer of liquid cooling products for data center applications and Atom Power Inc. a designer and manufacturer of advanced electronic circuit breakers.

SEMI-THERM 36

David Grant graduated from the University of Tennessee in 2003 with a B.S. in Mechanical Engineering. He has been at the Oak Ridge National Laboratory (ORNL) since 2009 where he has been involved with the design, construction, and operation of the mechanical systems supporting ORNL's 80,000SF+ of data centers which house Summit, the world's fastest high performance computer, among others. Current work is focused on facility upgrades to enable a future exascale system. He is currently a co-chair of the Energy Efficient HPC Working Group Infrastructure sub-team and is a corresponding member of the ASHRAE TC9.9. David is a registered Professional Engineer with the State of Tennessee and is a Certified Energy Manager (CEM - from the Association of Energy Engineers (AEE)) and a Data Center Energy Practitioner – Specialist (DCEP - from the Department of Energy (DOE)).He has 13 issued patents.

Dr. Alfonso Ortega is the James R. Birle Professor of Energy Technology at Villanova University. He is the Director of the Laboratory for Advanced Thermal and Fluid Systems and the Founding Director of the Villanova site of the NSF Center for Energy Smart Electronic Systems (ES2) founded in 2011. He is currently Associate Director of the NSF ES2 Center.

He received his B.S. from The University of Texas-El Paso, and his M.S. and Ph.D. from Stanford University, all in Mechanical Engineering. He was on the faculty of the Department of Aerospace and Mechanical Engineering at The University of Arizona in Tucson for 18 years. For two years, he served as the Program Director for Thermal Transport and Thermal Processing in the Chemical and Transport Systems Division of The National Science Foundation, where he managed the NSF's primary program funding heat transfer and thermal technology research in U.S. universities.

Dr. Ortega is a teacher of thermal sciences and experimental methods. He is an internationally recognized expert in the areas of thermal management in electronic systems. He has supervised over 40 M.S. and Ph.D. candidates to degree completion, 5 postdoctoral researchers, and more than 70 undergraduate research students. He is the author of over 300 journal and symposia papers, book chapters, and monographs and is a frequent short course lecturer on thermal management and experimental measurements.

He is a Fellow of the ASME and received the 2003 SEMITHERM Thermie Award and the 2017 ITHERM Achievement Award in recognition of his contributions to the field of electronics thermal measurements.

Brandon Rubenstein is the Director of Hardware Development Engineering for Microsoft Azure's Cloud Server Infrastructure group. His team is responsible for providing the IT and supporting infrastructure at Hyperscale on which Microsoft's cloud based platforms operate. Brandon was the director, architect and lead designer for the mechanical and thermal systems for Microsoft's current cloud server product line, which is also known within the Open Compute Project as Project Olympus and Opencloud Server. Before this, he designed the thermal solution for the first generation of Microsoft Surface tablet products. Before joining Microsoft, Brandon was the lead thermal engineer for Hewlett Packard's Enterprise Server Group, developing mechanical and thermal solutions for four generations of Hewlett-Packard's Superdome Enterprise server products over 10 years as well as architecting the HP Apollo liquid cooled "thermal busbar" solution.

Brandon holds over 30 patents and has authored and co-authored several technical papers regarding thermal optimization through modelling. Brandon graduated from Purdue University with a BSME and the University of Wisconsin with an MSME.

www.semi-therm.org

SEMI-THERM 36

Debabrata Pal is a Technical Fellow working at Collins Aerospace, a United Technologies Corporation company. He has B.S, M.S and Ph.D. all in mechanical engineering. He currently leads thermal design of aircraft electrical and electronic systems. Debabrata actively mentors co-op students and engineers. He prepares and teaches classes on thermal management in the Collins Aerospace Technical University. . He has published book chapters, journal papers, conference papers and various patents.

Dr. Timothy A. Shedd is currently Director of Product Management for ZutaCore, Inc. Most recently, he was the Director of the Graduate Program, Supervisor of Entrepreneurship Programs and an Associate Professor of Mechanical Engineering at Florida Polytechnic University. Prior, he was an Assistant, then Associate, Professor of Mechanical Engineering at the University of Wisconsin from 2001 to 2016. In 2012, while still a faculty member, Shedd founded Ebullient, Inc., to commercialize a two-phase cooling system for data centers. He holds a B.S. in Electrical Engineering from Purdue University and M.S. and Ph.D. degrees in Mechanical Engineering from the University of Illinois at Urbana-Champaign. He has received an NSF CAREER award, an ASHRAE New Investigator Award, and a number of teaching and research awards during his academic career. Most recently (2016), Dr. Shedd has been named a Fellow of ASHRAE (the American Society of Heating, Refrigeration and Air-conditioning Engineers).

SEMI-THERM®

SEMI-THERM® 36
The 36th Annual Thermal Measurement, Modeling and Management Symposium
March 16th - 20th, 2020
www.semi-therm.org

DoubleTree by Hilton San Jose
2050 Gateway Place, San Jose, CA 95110
Phone: 1 (408) 453-4000

SEMI-THERM 36

How-To Presentation
Tuesday, March 17 5:00 p.m. - 6:00p.m.

A Tile: A Look at Acoustic Fundamentals and Designs as Applied to Air-Cooled Electronics

Presenter: Herman Chu

As air-cooling design continues to increase in airflow requirement without much relieve in the overall equipment form factor, acoustic design considerations need to be actively engaged at the start of the product development cycle in order to clearly define expectations and deliver the best achievable sound quality.

In this how-to session, the speaker will present acoustic design fundamentals, review logarithmic arithmetic used in calculating sound levels, and review pertinent industry standards in performing acoustic testing for product evaluation.

Herman Chu is classically trained in thermal fluid systems and has over 30 years of industry experience spanning from military aerospace applications to electronic cooling of consumer products, computers and computer servers, mainframes and NEBS compliant networking equipment. His career has taken him to deploy all different kinds of cooling technologies from air cooling to various forms of liquid cooling.

Basic Pumped Refrigerant Cycle Calculations for Cooling IT Loads

Presenter: Joe Marsala
Durbin Group LLC

There is a growing interest in using pumped refrigerant to cool IT loads across various hardware platforms. This how-to session will examine some of the basic first order engineering considerations necessary when evaluating pumped refrigerant as an option. The speaker will present how pumped refrigerant thermodynamic cycles are represented on pressure-enthalpy diagrams, how to calculate refrigerant circulation rates and discuss choice of refrigerant. The four basic components of a pumped refrigerant cycle: refrigerant pump, cold plate, condenser and reservoir will be presented.

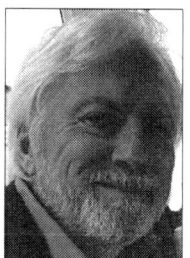

Joe Marsala is the CTO of Durbin Group LLC. He has over 30 years of experience in thermal research, engineering, and business and product management with large and small firms including Rockwell Allen-Bradley, Thermo Electron, Wakefield Engineering and the Gas Research Institute. He is the holder of over 25 issued and pending US and foreign patents and has numerous technical publications. His foundational patents in two phase cooling have been cited in more than 200 subsequent patents. Joe earned his B.S.E. degree in Chemical Engineering from the University of Michigan. He is a licensed Professional Engineer.

SEMI-THERM 36

Mechanical & Aerospace Engineering
The University of Texas at Arlington

UNIVERSITY OF
TEXAS
ARLINGTON

Proudly sponsors the 2020

THERMI Award

Each year, SEMI-THERM honors a person as a Significant Contributor to the field of semiconductor thermal management. The THERMI award is intended to recognize a recipient's history of contributions to crucial thermal issues affecting the performance of semiconductor devices and systems. The 2020 THERMI award is proudly presented to:

Dr. Ross Wilcoxon
Collins Aerospace

Ross Wilcoxon is an Associate Director, Mechanical Engineer in the Collins Aerospace Advanced Technology group in Cedar Rapids, Iowa. He conducts research and supports the development of prototype and production avionics systems for communication, processing, displays and radars. His work is generally related to component reliability, electronics packaging and thermal management with specific areas of research including the development and implementation of glass-based composite coatings, liquid metal cooling, integration of commercial heat pipes into avionics, and determining the reliability of commercial microelectronic components. Dr. Wilcoxon has been a Principal Investigator for research funded by the Office of Naval Research and the Defense Advanced Research Projects Agency. He has 30 US Patents, primarily in microelectronics packaging and thermal management.

Over the past 18 years, Dr. Wilcoxon has served in multiple roles on the SEMI-THERM Program Committee, including Vice-Program/Program/General Chair, Chair of the Technical Committee, head of the Best Paper selection team, and editor for Peer Reviewed papers. He has been an invited speaker at SEMI-THERM, ITherm, IMAPS Thermal ATW and THERMES and has more than forty publications in journals, technical magazines and conferences. Dr. Wilcoxon is also an editor for Electronics Cooling Magazine and has served on engineering advisory boards for South Dakota State University and the University of Iowa. He received a BS in Mechanical Engineering and MS in Engineering from South Dakota State University and a PhD from the University of Minnesota. Prior to joining Rockwell Collins (now Collins Aerospace) in 1998, he was an assistant professor at South Dakota State University.

Thursday March 19, 2020 12:30 p.m.

The 2019 Harvey Rosten Award

Sponsored by Mentor, a Siemens business

For Outstanding Work in the Field of Thermal Analysis of Electronic Equipment:

A Generic Processor Temperature Estimation Method

Baver Ozceylan*[1,] Boudewijn R. Haverkort[2,] Maurits de Graaf[3,] Marco E. T. Gerards[1]

[1]University of Twente, Enschede, the Netherlands, [2]Tilburg University, Tilburg, the Netherlands, [3]Thales Nederland B.V., Huizen, the Netherlands
* Corresponding Author

Baver Ozceylan is a graduate of Middle East Technical University (METU), Turkey (2014, BS and 2017, MS). From 2014 to 2018, he was with the Department of Electrical and Electronics Engineering at METU. Since 2018 he has been a PhD candidate at the Design and Analysis of Communication Systems Group of the University of Twente, The Netherlands. His research interests include mathematical modeling and analyzing, wireless commination systems, energy-efficient and energy-aware algorithms and scheduling, thermal modeling and temperature-aware scheduling.

Boudewijn R. Haverkort (Master and Phd, University of Twente, 1986 and 1991, respectively) is full professor and Dean of the Tilburg School of Humanities and Digital Sciences at Tilburg University, Netherlands. Since 2019. Before moving to Tilburg University, he was a full professor at the University of Twente since 2003, and from 1995 to 2002 he was a professor at RWTH Aachen, Germany. From 2009 to 2013 he was scientific director of the public-private Embedded Systems Institute, an applied research institute focusing on high-tech systems design. His field of interest is very wide, encompassing internet technology, cyber-physical systems, smart energy systems, energy management in data centers, computer performance and reliability evaluation, stochastic model checking, as well as data science. He is a Fellow of the IEEE since 2007, and has published around 200 papers about his scientific work in the above fields, and has chaired a large number of international conferences. Since 2016 he is chairman of the Dutch national research program on big data and applications.

Maurits de Graaf is an experienced Innovation Program and Projectmanager with a thorough scientific background. He has guided many research projects from first concepts to final implementation. He received his PhD in 1994 at the University of Amsterdam for the thesis 'Graphs and Curves on Surfaces'. After a period with the telecommunications research institute KPN Research, he started working with Thales Netherlands B.V. in 1999, mainly in the innovation department. Since 2010 he combines this with a part-time position at the University of Twente as associate professor at the department Mathematics of Operations Research (MOR). He co-authored over 30 publications.

Marco E. T. Gerards received the M.Sc. degrees in computer science and in applied mathematics from the University of Twente, Enschede, the Netherlands, in 2008 and 2011 respectively. He finished his Ph.D. thesis titled "Algorithmic power management: energy minimisation under real-time constraints" in 2014. Then he worked as a postdoc, until 2016 when he became an assistant professor. His research interests are energy management for smart grids and sustainable computing.

The Harvey Rosten Award

The Award is for outstanding work, recently published or in the public domain, which advances the analysis or modeling of thermal or thermomechanical effects in electronic equipment or components, including experiments aimed specifically at the validation of numerical models. The award is in the form of a plaque and a $1000 cash prize. The Award was established by the family and friends of Harvey Rosten, to commemorate his achievements in the field of thermal analysis of electronics equipment, and the thermal modeling of electronics parts and packages. The Award is made annually to encourage innovation and excellence in these and closely related fields.

The recipient is selected by the Selection Committee, made up of eminent practitioners in the electronics-thermal field. The criteria for selection are that the work: represents an advance in thermal analysis or thermal modeling of electronics equipment or components, including experiments aimed specifically at validating numerical models; demonstrates clear application to practical electronics design; demonstrates insight into the physical processes affecting the thermal behavior of electronics components, parts and systems; is innovative in embodying this understanding in either thermal analysis or thermal modeling; takes a pragmatic approach.

SEMI-THERM 36

celsia°

Making Hot Technology Cooler

We are proud to sponsor:

The SEMI-THERM Educational Foundation
Thermal Hall Of Fame

Lifetime Achievement Award

Presented To

Dr. Dereje Agonafer

In Recognition of Significant Contributions
to the Field of Electronics Thermal Management

Dr. Dereje Agonafer is a Presidential Distinguished Professor in MAE at University of Texas at Arlington (UTA) where he heads two centers: Site Director of NSF I/UCRC in Energy Efficient Systems and Director of Electronic Packaging. After receiving his PhD at Howard University, he worked for 15 years at IBM. In 1991, his work was recognized by being awarded the "IBM Outstanding Technical Achievement Award in Appreciation for Computer Aided Thermal Modeling." Since joining UTA in 1999, he has graduated 225 graduate students (a record for the University) including 22 PhDs and currently advising 15 PhDs and 18 MS students. His new initiative is to start a new center called RAMPES (Center for Reliability Assessment in Micro and Power Electronic Systems) for which he has received significant funding including $1.3M for new equipment, 3000 sq ft of new lab space, Assistant and Associate Professor openings to work with him, and research engineer among others. For his contributions, he has received numerous awards including the 2008 Thermi Award, the 2009 InterPACK Excellence Award, the 2014 ITHERM Achievement Award, and the 2019 ASME Heat Transfer Memorial Award. Professor Agonafer was a Martin Luther King Visiting Professor at MIT during the 2007 academic year. He is a fellow of the National Academy of Inventors, the American Association for the Advancement of Science and the American Society of Mechanical Engineers. In 2019, he was elected to the National Academy of Engineering. According to Dean Crouch, "the first current faculty member elected to the Academy." Professor Agonafer is married to his wife Carolyn and they have two children; a son, Dr. Damena Agonafer who is Professor of Mechanical Engineering & Materials Science at Washington University in St. Louis, and a daughter, Dr. Senayet Agonafer, a Radiologist, who works at Lennox Hill Radiology in New York City.

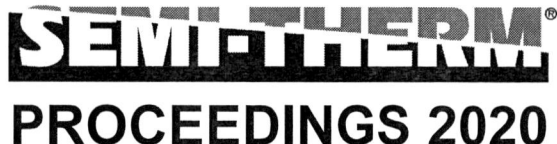

PROCEEDINGS 2020

The PROCEEDINGS from past SEMI-THERM Symposia are available from the IEEE. For publication ordering information, please contact:

Curran Associates
57 Morehouse Lane
Red Hook, NY 12571 USA
Phone: +1-845-758-0400
Web: **www.proceedings.com**

The papers from **1988** through **2019** are available on the IEEE's **IEL/XPLORE** on-line system. Any researcher can use full-text search across the three million papers in the XPLORE database and access the abstracts of previous **SEMI-THERM** papers. Subscribers may download the PDFs of any **SEMI-THERM** papers. Non-subscribers may purchase single copies at a reasonable fee. To access this resource, please visit:

ieeexplore.ieee.org

TABLE OF CONTENTS

SEMICONDUCTOR THERMAL MEASUREMENT, MODELING AND MANAGEMENT SYMPOSIUM

Welcome to SEMI-THERM 36 iii

Committees v

Short Courses vi

Keynote Speaker: Andy Delano, Microsoft xi

Luncheon Speakers: Lieven Vervecken, Diabatix, and Dave Saums, DS&A LLC xii

Embedded Tutorial: Additive Manufacturing, Ram Ranjan, UTRC xv

Evening Tutorial: Realistic Thermal Model for Human Skin, Bruce Guenin xvi

Panel Discussion: Liquid Cooling xvii

How-To Presentations: Herman Chu, Joe Marsala xx

Thermi Award: Dr. Ross Wilcoxon, Collins Aerospace xxi

Rosten Award: Baveer Ozceylan, Boudewijn Haverkort, Maurits de Graff, Marco Gerards xxii

Thermal Hall of Fame Award: Dr. Dereje Agonafer, University of Texas at Arlington xxiii

Session 1: Liquid Cooling
Chair: Timothy Shedd, ZutaCore

Direct Micro-Pin Jet Impingement Cooling for High Heat Flux Applications 1
Vahideh Radmard, Yaser Hadad, Arad Azizi, Srikanth Rangarajan, C. Hiep Hoang, Charles Arvin, Kamal Sikka, Scott N. Schiffres, Bahgat Sammakia, Binghamton University

Shape Optimization of a Pin Fin Heat Sink 10
Thomas Menrath, A. Rosskopf, F. B. Simon, M. Groccia, S. Schuster, Fraunhofer IISB

Wicking Performance Enhancement by Laser Induced Roughness .... *
Sougata Hazra, Farid Soroush, Tanya Liu, Mehdi Asheghi, Kenneth E. Goodson, Stanford University

Session 2: Parallel Session: Data Center
Chair: Kourosh Nemati, Future Facilities Ltd.

Determination of Cost Savings Using Variable Speed Fans for Cooling Servers 17
Minh Tran[1], Nicole Okamoto[2], Hussammedine Kabbani[2], Saeid Bashash[2],
[1]Velodyne Lidar, [2]San Jose State University

On Economic Cooling of Contained Server Racks using an Indirect Adiabatic Air Handler 24
Riccardo Lucchese[1], Michele Lionello[2], Mirco Rampazzo[2], Andreas Johansson[1], Wolfgang Birk[1],
[1]Lulea University of Technology, [2]Università degli Studi di Padova

An Experimental Apparatus for Two-phase Cooling for High Heat Flux Application
using an Impinging Cold Plate and Dielectric Coolant 32
Cong Hiep Hoang, Sadegh Khalili, Bharath Ramakrisnan, Srikanth Rangarajan, Yaser Hadad, Vahideh Radmard,
Kamal Sikka, Scott Schiffres, Bahgat Sammakia, Binghamton University

CFD Investigation of Dispersion of Airborne Particulate Contaminants in a
Raised Floor Data Center 39
Satyam Saini, Pardeep Shahi, Pratik Bansode, Ashwin Siddarth, Dereje Agonafer,
University of Texas at Arlington

Experimental Validation of a Numerically-Optimized Array of Heat Sinks .... *
Georgios Karamanis and Marc Hodes, Transport Phenomena Technologies, LLC

* -- This presentation has no formal paper.

General Guidelines for Commercialization A Small-Scale In-Row Cooled Data Center: A Case Study 48

Yaman Manaserh[1], Mohammad. I. Tradat[1], Ghazal Mohsenian[1], Bahgat G. Sammakia[1], Mark J. Seymour[2], [1]Binghamton University, [2]Future Facilities

Session 3: Consumer Electronics I
Chair: Alex Ockfen, Facebook

Self-Heating Investigation in SOI MOSFET Structures with High Thermal Conductivity Buried Insulator Layers 56

Konstantin Petrosyants and Dmitry Popov, Moscow Institute of Electronics and Mathematics

Testing and Analysis of Improved Thermal Solutions for a Home Wireless Router .... *

Raul Vargas, Justin Dixon, Md Malekkul Islam, Electronic Cooling Solutions

Thermal Acceptability Limits for Wearable Electronic Devices .... *

Mark Andrew Hepokoski[1], Allen Curran[1], Timothy Viola[1], and Alex Ockfen[2], [1]ThermoAnalytics, Inc., [2]Facebook Reality Labs

Transient Thermal Model for Wearable Device in Contact with Human Skin 61

Bruce Guenin, Consultant

DNN-based Fast Static On-chip Thermal Solver 65

Jimin Wen, Stephen Pan, Norman Chang, Wen-Tze Chuang, Wenbo Xia, Deqi Zhu, Akhilesh Kumar, En-Cih Yang, Karthik Srinivasan, Ying-Shun Li, ANSYS, Inc.

Micro-encapsulated Phase Change Materials as Heat Transfer Media in Electronics .... *

John D. Rasberry, Keysight Technologies

Session 4: Air Movers
Chair: Robin Bornoff, Mentor

Advantages of Mixed Flow Air Mover Technology for Upcoming IT Infrastructure Cooling Demand .... *

Wolfgang Laufer, ebm-papst

Inclination Angle Effects on Dual Cool Jet Heat Transfer 76

Sophia Brodish, Matthew Harrison, Ali Haider, Ted Brekken, Peter DeBock, Joshua Gess, Oregon State University

Application of Fan Blade Serration to Reduce Fan Noise .... *

Prathamesh M Ghankutkar, Bart Manufacturing, Inc.

Session 5: Consumer Electronics II
Chair: Alex Ockfen, Facebook

Increased System Performance and Reduced Surface Touch (Skin) Temperature in Mobile Electronics utilizing Composites of Graphite with Ultra-High Spreading Capacity and Insulation with Ultra-Low Thermal Conductivity 82

Mitchell Warren[1], Julian Norley[2], John Allen[1], Jonathan Taylor[2], Lindsey Keen[1], [1]WL Gore & Associates, [2]NeoGraf Solutions, LLC

An Analysis of Temperature Variation Effect on Response and Performance of Capacitive Microaccelerometer Inertial Sensors 91

Jacek Nazdrowicz and Andrzej Napieralski, Lodz University of Technology

Measurement of Performance Characterization of Ultra-Thin Vapor Chamber 97

Wei-Keng Lin, Wen-Hua Zhang, Chien Huang, Ching-Huang Tsai, Kenny Hsaio, T-Global Technology Co., Ltd

* -- This presentation has no formal paper.

Session 6: Two Phase
Chair: Devin Pellicone, ACT

Numerical Investigation of Coolants for Chip-embedded Two-Phase Cooling 105
Pritish R. Parida and Timothy Chainer, IBM Research

Numerical Investigation of Two-Phase Immersion Cooling using FC-72 Dielectric Fluid .... *
Amirreza Niazmand, University of Texas at Arlington

Empirical Study of Oscillating Heat Pipe Heat Spreaders for High Heat Flux Applications .... *
Joe Boswell, ThermAvant Technologies

Experimental Characterization of Refrigerant based Two-Phase Cold Plates to 1000 W: Thermal Metrology and Metrics .... *
Felipe Valenzuela, Villanova University

Design and Optimization of Hollow Micropillar Structures for Enhanced Evaporative Cooling of High-Powered Electronics .... *
Mun Mun Nahar[1], Haotian Wu[1], Zhikai Yang[1], Alexander Austin[1], Jorge Padilla[2], Madhusudan Iyengar[2], Damena Agonafer[1], [1]Washington University in Saint Louis, [2]Google Inc

Actively Cooled Two-phase Cold Plate for High Heat Flux Electronics *
Michael Ellis, Advanced Cooling Technologies, Inc.

Session 7: Automotive/Aerospace/Outdoor/TIM I
Chair: Dave Saums, DS&A LLC

Validated Model Calibration for Simulation Aided Thermal Design 114
Raul Catalin Cioban[1,2] Sz. Szőke[1], Z. Kórádi[1], D. Zaharie-B.[1], C. Leordean[1], [1]Robert Bosch, [1]Babes-Bolyai University

Innovations in Liquid Metal Thermal Interface Materials .... *
Tim Jensen, Indium Corporation

Thermal and EMI Performance of Natural Graphite Sheet Heat Sinks with Embedded Heat Pipes .... *
Xavier Faure[1], Martin Cermak[1], Ali Saket[2], Martin Ordonez[2], and Majid Bahrami[1], John Kenna[3], [1]Simon Fraser University, [2]University of British Columbia, [3]Terrela Energy Systems Ltd.

Developing a Proper Mission Profile to Extend Thermal Margin .... *
Brian Philofsky, Xilinx

Effects of Solder Voiding on the Reliability and Thermal Characteristics of Quad Flatpack No-lead (QFN) Components 124
Ross Wilcoxon, Dave Hillman and Tim Pearson, Collins Aerospace

Calibration of a Detailed FDA Thermal Model to Test Data .... *
Patrick Loney, Northrop Grumman

Session 8: CFD and Measurement Techniques
Chair: Pritish Parida, IBM

Monte Carlo Prediction of PPM Failure Rate using a Parametric Reduced Order Model .... *
Robin Bornoff[1] and Wendy Luiten[2], [1]Mentor, [2]WLC

Thermal Characterization of a Virtual Reality Headset during Transient and Resting Operation 131
Rachel C McAfee, Cole Haxton, Matthew Harrison, Joshua Gess, Oregon State University

Cross Correlation Method for Images Alignment: Application to 4 Buckets Calculation in Thermoreflectance 137
Metayrek Youssef[1], Kociniewski Thierry[2], Khatir Zoubir[1], [1]IFSTTAR, [2]University of Versailles St Quentin

* -- This presentation has no formal paper.

Session 9: Automotive/Aerospace/Outdoor/TIM II
Chair: Adriana Rangel, Cisco

Smart Pole Active Electronics Thermal Solution .... *
Walter Mark Hendrix, SRC Design Solutions, LLC

Experimental Measurement and Finite Element Analysis of Thermal Conductivity of Alumina/Silicone Polymer Composites 143
Masakazu Hattori[1], Kazuaki Sanada[2], Yasushi Kajita[3], [1]Fuji Polymer Industries Co., Ltd., [2]Toyama Prefectural University, [3]Nagoya Municipal Industrial Research Institute

CVD Polycrystalline Diamond for Laser Diode Applications .... *
Firooz Faili, Element Six Technologies

Design of Thermal Metamaterials for Next-Generation Packaging Solutions .... *
Yoonjin Won, University of California Irvine

SUBMIT A PAPER FOR SEMI-THERM 37!

As you further develop a technique or application, consider documenting it for the thermal community. **SEMI-THERM 37** will begin accepting abstracts during the summer (deadline is September 15, 2020). We welcome your submissions! Visit us at **www.semi-therm.org**.

SEMI-THERM 37 is March 16-20, 2021 – *be there!*

---

* -- This presentation has no formal paper.

Direct Micro-Pin Jet Impingement Cooling for High Heat Flux Applications

Vahideh Radmard, Yaser Hadad, Arad Azizi, Srikanth Rangarajan, C.
Hiep Hoang, Charles Arvin, Kamal Sikka, Scott N. Schiffres, Bahgat Sammakia
Department of Mechanical Engineering, Binghamton University – SUNY
Binghamton/ US
Vradmar1@binghamton.edu

Abstract

This study explores the heat transfer from directly manufactured micro-pins undergoing in jet impingement. Micro-pins that are directly attached to the chip eliminate the need for thermal interfaces, and the lids that exist in conventional electronic packages, while also occupying a smaller areal footprint. In this study, the effects of different outlet port positions and pin cross-sections and profiles are examined through simulations. Results from the study show that the outlet port position has a significant effect on hydraulic and thermal resistances. Also, by modifying the height and profile of the pins, a 47% reduction in pressure drop was achieved without sacrificing the thermal performance.

Keywords

High Heat Flux Chips, Jet Impingement Cooling, Direct Cooling, Micro-Pin Fin Heat Sink.

Nomenclature

f	Fluid
s	Solid
Re	Reynolds number
ρ	Density (kg/m^3)
U	Velocity (m/s)
Ψ	Temperature non-uniformity of the chip
μ	Absolute viscosity ($kg/m.s$)
K	Thermal conductivity (W/m. K)
$\dot{q}$	Chip power (W)
R	Thermal resistance (K/W)
R_i	Internal resistance (K/W)
R_e	External resistance (K/W)
R_{j-c}	Junction to chip resistance (K/W)
T	Temperature (K)
p	Pressure ($k.Pa$)
$\dot{W}$	Pump power (W)
c_p	Specific heat capacity ($J/kg.K$)
COP	Coefficient of performance

Introduction

The thermal design powers of graphical processing units (GPUs) and central processing units (CPUs) have been increasing by ~7% per year since 2006 till 2019 [1]. Since temperature has a direct effect on the computational performance and reliability of the chip, this makes maintaining chip temperature challenging. Additionally, die sizes are approaching the reticle limit and the power of single chips is increasing. Therefore, cooling high heat flux chips has become more important than ever. Now, the challenge for thermal engineers is to decrease the thermal resistance of thermal interface materials (TIMs). In order to tackle this problem, various concepts are being introduced from both academia and the industry.

One approach of the industry is to use high thermal conductivity TIMs such as silver pastes and indium [2]. However, previous simulations indicated that even high conductivity TIMs might not be able to keep up with high heat fluxes as thermal resistances become extremely large [3]–[6]. Another approach is to eliminate TIMs altogether. An example of this approach is a bare die pool boiling configuration with dielectrics [7]–[10]. However, this solution is currently not broadly implemented. Recently, it has been demonstrated that by selective laser melting (SLM), heat removal devices such as heat sinks and cold plates can be directly fabricated onto the chip without conventional TIMs [3], [4]. In this process, a Sn-Ag-Ti interlayer alloy is used to bond silicon die to metal structures. SLM provides localized heating that initiates the reaction of titanium in the interlayer alloy with silicon, which results in the strong silicide bond. Afterwards, high conductivity alloys, such as copper or aluminum, can be printed on top of the interlayer alloy using a powder bed fusion technique to form various type of heat removal devices. Another example is silicon micro-channels fabricated by complex lithography-based techniques, though these processes do not scale well, as this processing is generally viewed as costly and disruptive by industry [11]. This paper looks at the potential cooling performance of a 1000-watt single chip with heat sink directly printed onto silicon by SLM via numerical simulations.

There is a considerable amount of fundamental studies on hydrodynamics and heat transfer of single and two-phase jet impingement configurations for cooling applications [12]–[14]. Fin geometries, including circular, square, hydrofoil, and elliptical, were fabricated by Micro-Electro-Mechanical Systems (MEMS) microfabrication techniques and experimentally studied in a single-phase jet impingement configuration [15]. Results indicated that out of all the shapes, circular pin fins and square pin fins were able to achieve the highest heat transfer coefficients, with the circle slightly outperforming square [16].

Numerical simulations have been broadly employed in order to further understand the heat transfer and hydrodynamics involved in jet impingement configurations. Studies have been conducted within various accuracies from Reynolds-averaged Navier–Stokes (RANS) models [17], [18] to large eddy simulations (LES) [19]–[21] and direct numerical simulations

(DNS) [22]–[24]. Most of these numerical analyses were conducted under the framework of parametric studies, often including the effect of distance of nozzle from substrate to diameter of nozzle (H/D) ratio, Reynolds (Re) number, and flow rate ($\dot{Q}$) on transport properties (mass, momentum, energy). Enhanced surfaces, including bumps, dimples, micro-fins and micro-pillars, have been broadly investigated because they increase surface area and also modify thermal and hydrodynamic boundary layers [19]–[23]. To the best of the authors' knowledge, the optimal placement of outlet ports and pin profiles have not been studied yet. This study aims to investigate the effectiveness of geometrical parameters like outlet port position and various pin profiles on the thermal and hydraulic performance of the cold plate. Assessing the role of these factors will help designers to develop configurations suited to their needs. The main interest of this approach is eliminating TIMs and improving resistance with a direct micro-pin jet impingement design.

The following sections propose a thermal management device design for high heat-flux chips based on direct attachment of the heat sink to the chip, followed by a numerical analysis of the cooling performance.

1. Geometry

Figure 1 shows the schematic geometry of the heat-sink with exaggerated pin sizes. The sides and upper part are made of plastic and include inlet and outlet ports. Jet parameters are fixed with a square inlet of 3 mm x 3 mm.

The lower part shows the pin fins that are directly attached to the 2 cm x 2 cm chip. A total power of 500W is specified for the chip. The cross-section of the micro-pins used in the current study are square for the sake of computational expediency. To reduce the thermal resistance and pressure drop, the pins are locally shortened in a linear manner such that they are shortest at the center (1.25 mm) and highest at the edges (5 mm). Figure 2 compares the resistances to heat transfer from the chip to the incoming fluid for a direct pin cold plate and a regular cold pla

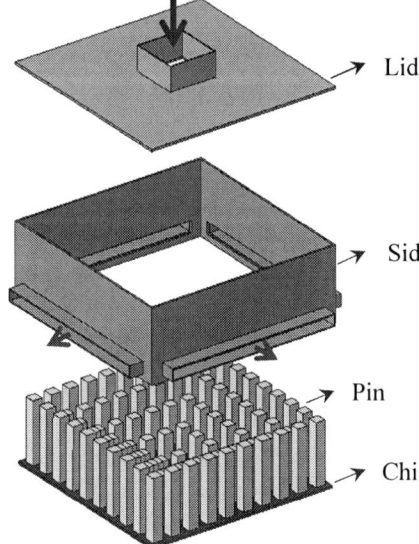

Figure 1: Schematic geometry of the heat-sink with exaggerated pin sizes. (Produced by

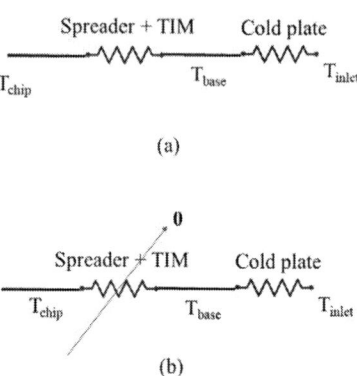

Figure 2: The resistances for the heat flow from chip to incoming fluid for (a) regular cold plate arrangement (b) printed pins on the chip.

2. Numerical computation
2.1. Basic assumption

The following assumptions are made based on the computational domain:

- An incompressible, single phase, and steady fluid flow has been assumed for this study.
- The effect of gravity is considered negligible.
- The flow is turbulent.
- The thermo-physical properties of the solid and coolant are assumed constant and obtained from thermo-dynamic tables at the defined inlet temperature.
- Radiation heat transfer is assumed to be zero.

2.2. Governing equations

The heat transfer consists of conduction in the pins and forced convection from the surface area of the fins and substrate.

The three-dimensional governing equations for an incompressible, steady state condition can be addressed in the following form:

Incompressible continuity equation of liquid phase:

$$\nabla . \vec{u} = 0 \qquad (1)$$

Momentum equation of liquid phase:

$$\rho_f \, (\vec{u}.\nabla) \, \vec{u} = -\nabla P + \mu_f \, \nabla^2 \vec{u} \qquad (2)$$

Energy equation of liquid phase:

$$\rho_f \, c_{p,f} \, (\vec{u}.\nabla) \, T_f \, \vec{u} = K_f \, \nabla^2 T_f \qquad (3)$$

Energy equation for solid phase:

$$\nabla^2 T_s = 0 \qquad (4)$$

Turbulent kinetic energy equation:

$$\rho \vec{u}. \quad \nabla k = \quad \nabla.\left(\left(\mu + \rho \frac{C_\mu\, k^2}{\sigma_k\, \varepsilon}\right)\nabla k\right) + \rho C_\mu \frac{k^2}{\varepsilon}\,(\nabla \vec{u} +$$
$$(\nabla \vec{u})^T)^2 - \rho \varepsilon \qquad (5)$$

The rate of dissipation of turbulent kinetic energy equation:

$$\rho \vec{u}. \nabla \varepsilon = \nabla.\left(\left(\mu + \rho \frac{C_\mu\, k^2}{\sigma_\varepsilon\, \varepsilon}\right)\nabla \varepsilon\right) + \rho\, C_{\varepsilon 1}\, C_\mu\, k\, (\nabla \vec{u} +$$
$$(\nabla \vec{u})^T)^2 - \rho C_{\varepsilon 2} \frac{k^2}{\varepsilon} \qquad (6)$$

2.3. Numerical domain

Symmetry is apparent in both the geometry of the heat sink and the pattern of the flow/thermal boundary condition. Using the advantage of symmetry, the computational domain is reduced to one-quarter of the unit cell of assemblage in order to simplify the analysis (Figure 3). A Cartesian grid with hexahedral cells is used for both the solid and liquid portions. A grid independence study has been done based on the pressure drop and thermal resistance. The result of the grid study (Figure 4) demonstrates that increasing the number of cells beyond 11 million leads to a 2.3% and 0.9% variation in the equations of pressure drop and thermal resistance. All results reported subsequently have used approximately 11 million level of grid refinement.

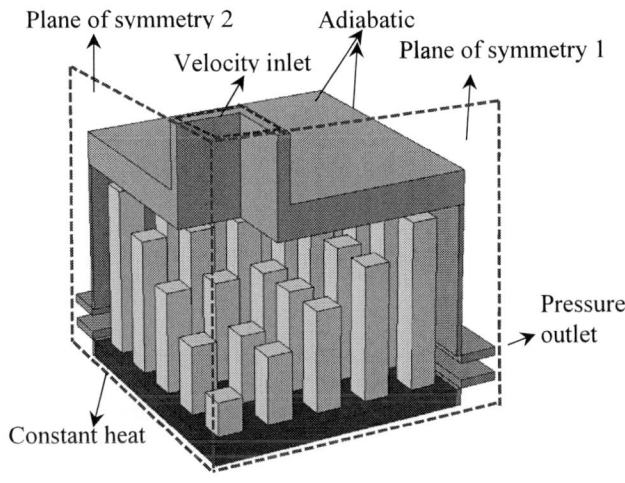

Figure 3: Schematic views of computational domain with boundary conditions.

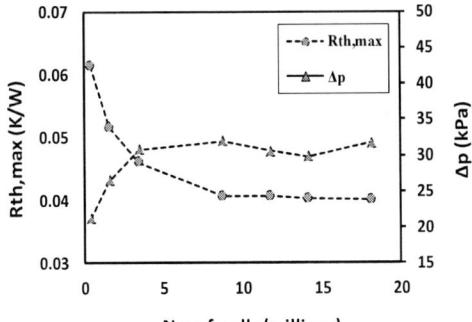

Figure 4: Grid independence study with respect to pressure and thermal resistance

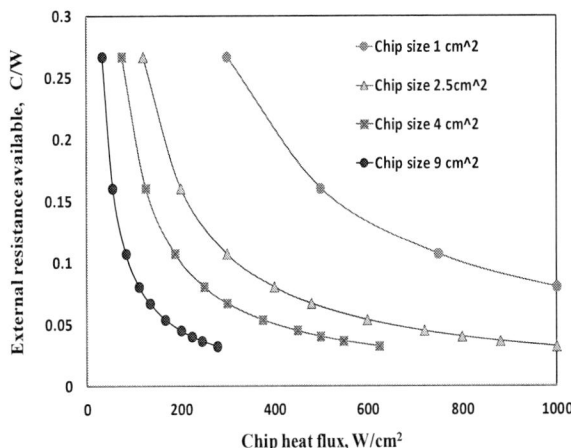

Figure 5: Cooling limits for the printed fins technology for various chip sizes for a thermal budget of 80° C.

2.4. Material properties

The coolant used in this simulation is dielectric water with an inlet temperature of 300K. Material properties were obtained from thermodynamics tables based on the defined inlet temperature (Table 1). Silver was chosen as the material for the printed micro-pin fins, since it has high conductivity. Given that the printing volume is small, there will be a small amount of silver needed to print the pins onto the silicon device of 2cm x2cm. With that in mind, the cost is not unreasonable. Our trends and general geometry are also applicable to copper and aluminum.

	$c_p\left(\frac{J}{kg.K}\right)$	$k_f\left(\frac{W}{m.K}\right)$	$\rho\left(\frac{kg}{m^3}\right)$	$\mu\left(\frac{kg}{m.s}\right)$
Coolant	4179	0.613	997	0.00085
Silver	230	406	10500	-

Table 1: Thermo-physical properties of the coolant and solid.

2.5. Numerical method

The three-dimensional governing equations of heat transfer and fluid flow characteristics were solved through the finite volume method. The second-order upwind and structured grid system were applied to discretize the governing equations. The standard model was used to simulate the flow characteristics for the turbulent flow condition. The commercial CFD program 6Sigma ET has been utilized to construct the numerical simulations.

Results and discussion

The intent of this study is to investigate the effect of the pin fin cross-sectional area, pin profiles and the outlet port position on the hydraulic and thermal resistance of the heat sink. In electronic cooling for a heat sink:

The total resistance = R_i + R_e (7)

The external resistance is the resistance from the spreader surface to the flow. The internal resistance constitutes the sum of resistances:

$$R_i = R_{J-C} + R_{TIM} + R_{spreading} \qquad (8)$$

The advantage of fabricating cooling directly onto a chip is the lack of internal resistance from TIMs and lids.

Performance is characterized with three parameters: thermal resistance, temperature non-uniformity, and thermodynamic pumping power. From Figure 5, the maximum available thermal resistance for various chip heat fluxes and chip sizes are plotted.

1. Thermal resistance is considered as an important variable to evaluate the thermal performance of the heat-sink. It can be obtained by measuring the temperature difference at the surface of the chip.

$$R_{th} = \frac{T_{max}^{chip} - T_{in}}{\dot{q}} \qquad (9)$$

2. Temperature non-uniformity of the chip is another thermal performance criterion that can depict the temperature gradient on the chip surface.

$$\Psi = \frac{T_{max}^{chip} - T_{in}}{T_{min}^{chip} - T_{in}} \qquad (10)$$

3. Thermodynamic pumping power is one of the primary criterions considered for this study. It is the product of the heat sink pressure drop and volumetric flow rate. This indicates that increasing the pressure drop will increase the operational cost due to the higher amount of energy consumed by the pump when it compensates. Larger pumps are needed to provide a higher pressure drop, which leads to higher initial and current costs.

$$\dot{w}_p = \Delta p . Q \qquad (11)$$

The influence of mass flow rate on the thermal resistance and pump power was simulated on the cold plate with side oulets and uniform full pin height (5 mm) on the chip. Figure 6 shows that thermal resistance decreases with an increasing flow rate. This is attributed to the fact that by increasing the mass flow rate, velocity is also increased, which leads to an improvement in advective heat transfer. However, one of the disadvantages of increasing flow rate is that pumping power increases with an increase in flow rate. Figure 7 illustrates that the temperature profile of the chip becomes more uniform with an increasing flow rate. An inlet flow rate of 8.33e-5m^3/s (5 lit/min) was selected for analyzing the performance.

A numerical model was developed to find the optimal outlet port position. We selected thermal resistance, temperature non-uniformity, and thermodynamic pumping power as the criterion to evaluate the performance of the heat sink.

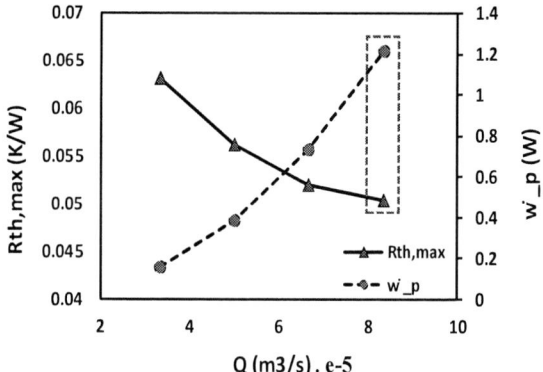

Figure 6: The effect of flow rate on the thermal resistance and pump

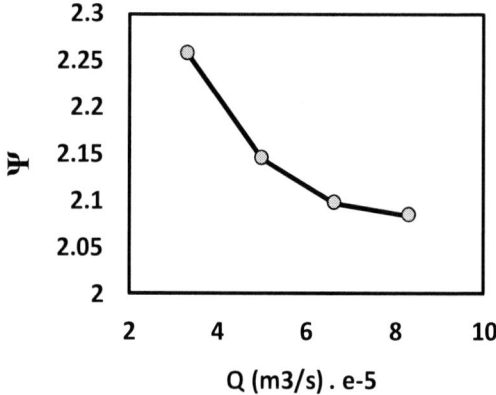

Figure 7: The effect of flow rate on the temperature non-uniformity of chip.

Figure 8 shows the schematic geometry of two configurations with different outlet port positions. The outlet size of case (a) and (b) was 2 mm × and 0.8 mm × 19.5 mm, respectively. Two pin profiles were simulated for each case (Figure 9). In one profile, the pins have a uniform full height of 5 mm. In the other profile, the pins are locally shortened to 1.25 mm in the center of heatsink and linearly increase to 5 mm at the edges. The purpose of locally shortening the pins is to reduce the pressure drop while maintaining an acceptable thermal performance. Generally, the total pressure drop of the heat sink depends on the following factors: the friction factors, the heat sink geometry, and velocity. Considering a full height pin profile, transferring the outlet ports from the top to the sides of the heat sink leads to a 61% and 19% reduction in the pressure drop and thermal resistance, respectively. Added to that, locally shortening pins in the case with side outlets, resulted in a 47% reduction in pressure drop without a significant drop in thermal behavior (Table 2).

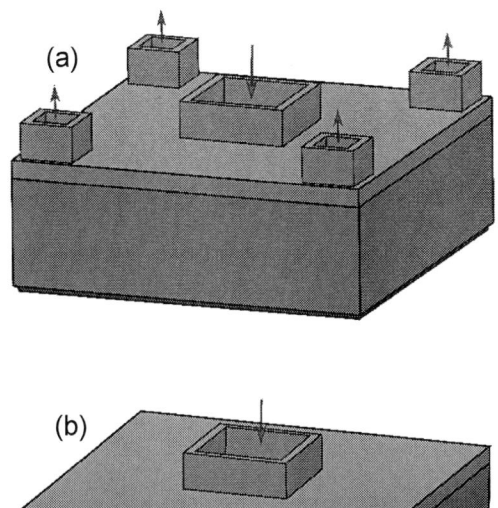

Figure 8: Schematic geometry of the heat-sink for (a): a top outlet port, and (b) a side outlet port

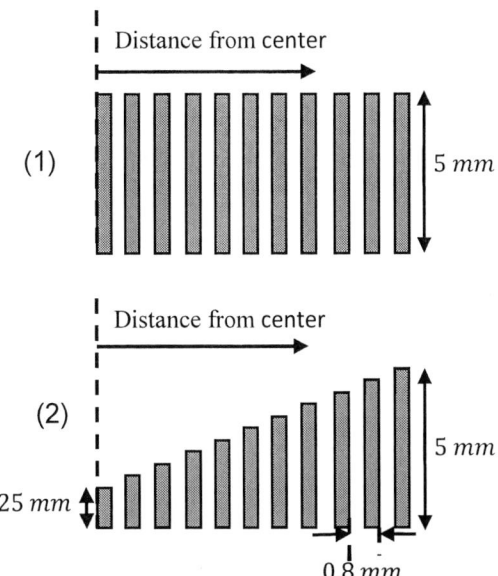

Figure 9: Schematic geometry of (1) a full height pin profile and (2) a locally shortened profile

	T_{chip}^{Max} (K)	T_{chip}^{ave} (K)	T_{chip}^{min} (K)	ΔP (kpa)	R_{th}^{Max} (K/W)	Ψ
Case. a (full pins)	328.5	317.7	311.4	78.3	0.057	2.5
Case. a (shortened pins)	330.5	318.3	311.6	60.1	0.061	2.6
Case. b (full pins)	322.9	317.6	311	30.7	0.046	2.1
Case. b (shortened pins)	325.2	317.9	311.1	16.3	0.050	2.3

Table 2: Results obtained from the numerical simulation for the both cases (a) and (b).

The pin effectiveness for a chip surface with constant heat flux has been calculated based on following equation:

$$\varepsilon_p = \frac{\Delta T_{pins}}{\Delta T_{without\ pins}} = \frac{T_{max}^{chip,with\ pins} - T_{in}}{T_{min}^{chip,without\ pins} - T_{in}} \leq 1 \quad (12)$$

The pin effectiveness for the case design of side outlets and pins with uniform pin height of 5 mm, is 0.56.
In order to discern the effect of cross-flow on the hydraulic and thermal performance of the heat sink, configurations with varied pin cross-sectional area were simulated.

In order to discern the effect of cross-flow on the hydraulic and thermal performance of the heat sink, configurations with varied pin cross-sectional area were simulated. These configurations with side outlet port positions had a uniform pin height of 5 mm and a fixed pin pitch of 0.8 mm as shown in Figure 9. We found that the thermal resistance decreases with larger pins (Figure 12). By increasing the pin cross-section size by 125% (the pin-pitch is fixed) led to improvements in thermal resistance of 37.8% accompanied 179% increases in pump power (Figure 13). One explanation for this is that the conduction resistance of a pin is inversely proportional to its thickness, thus thick pins have a lower conduction resistance than thin pins. Another explanation is that larger pins with the same size pin pitch have a greater surface area of contact between solid and liquid.

Additionally, by increasing the size of the pin cross-section, the flow gap between the pins decreases, which results in a higher channel velocity. All these factors could account for the increased thermal performance. The temperature profile of chips with different pin cross-sectional areas are compared in Figure 14. Thicker pins with smaller flow gaps allow for a more uniform temperature profile. However, by increasing the size of the pin cross-section, the flow passes through a narrower area, which results in higher friction and increases the pressure drop. The subsequent increase in pump power will consume more energy, which in turn increases the operational costs. In the case of constant pressure drop design constraints, designs with smaller flow gaps may perform poorer.

The coefficient of performance (COP) can be defined as the ratio of heat transfer rate to work required. COP can be used as metric to evaluate a cooling system:

$$COP = \frac{\dot{q}}{\dot{w}_p} = \frac{f(R_{th})}{f(\triangle p)} \quad (13)$$

A higher COP equates to lower operating costs and an efficient design. The COP analyses shows that by increasing the pin cross section area the COP is decreasing, however the temperature uniformity on the surface of chip and thermal resistance is improving (Figure 15). Therefore, it is important to consider the pump power along with the thermal performance for the design of a heat-sink.

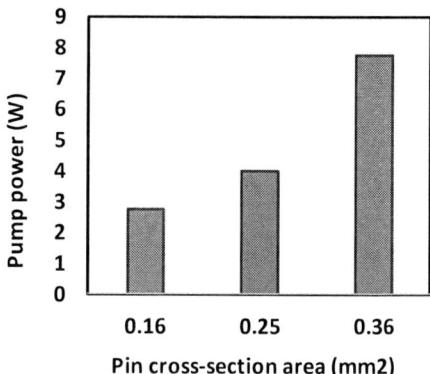

Figure 13: The effect of pin cross-section area (fixed pitch) on the pump power.

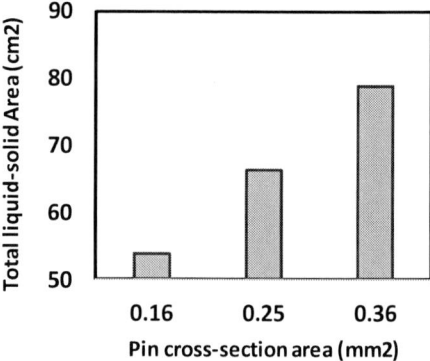

Figure 11: The effect of pin cross-section area (fixed pitch) on the total liquid-solid area

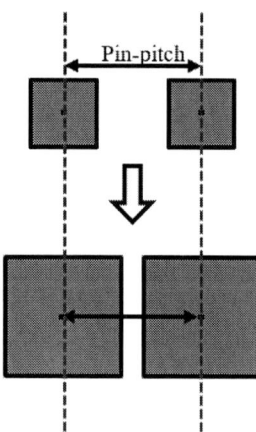

Figure 10: Schematic figure showing increasing cross-section area while the pin pitch is fixed.

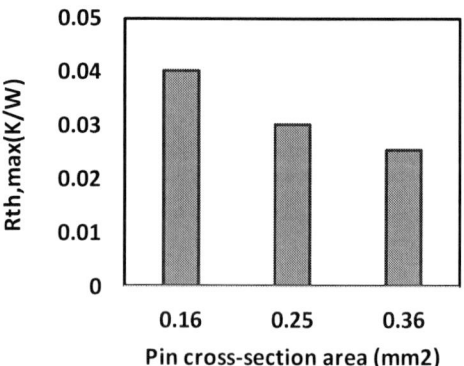

Figure 12: The effect of pin cross-section area (fixed pitch) on the thermal resistance.

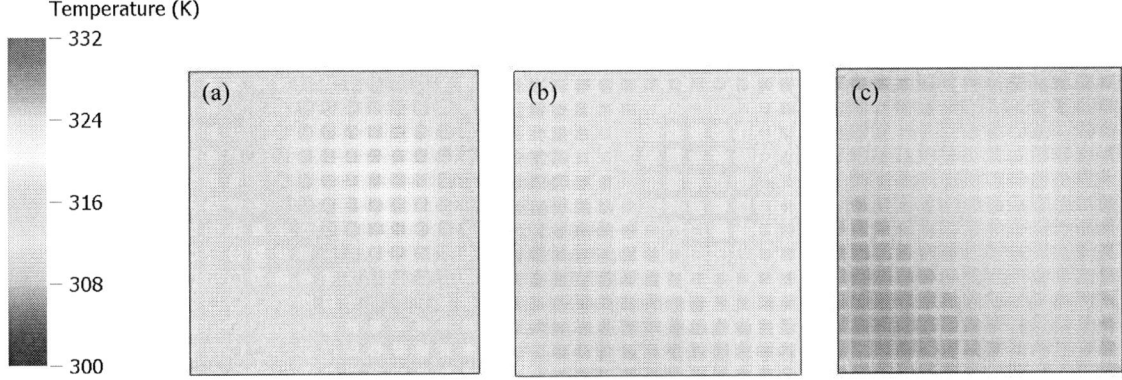

Figure 14: The temperature profile of chip for the cases with different pin cross-sectional area (a) 0.4 mm x 0.4 mm, (b) 0.5 mm x 0.5 mm and (c) 0.6 mm x 0.6 mm, while the pin pitch is kept fixed.

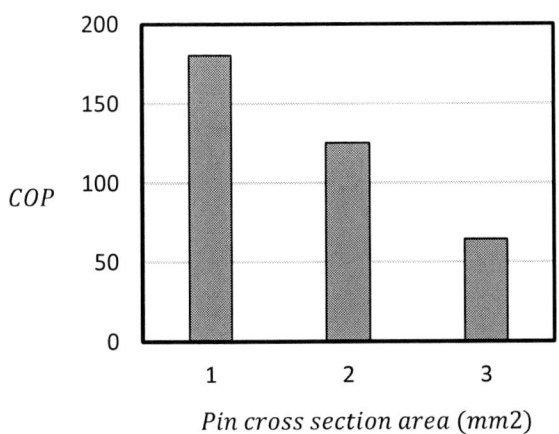

Figure 15: The effect of pin cross-section area (fixed pitch) on COP of the system.

Conclusion

In this study, a micro-pin fin heat sink was analyzed for design parameters such as flow rate, the crossflow, and outlet port position. The configuration with full height pin profile and side outlets (case (b) with profile of (1)) provided a 19% reduction in thermal resistance and a 61% reduction in pressure drop. By locally shortening pins inside the design with side outlets (case b), a 47% decrease in pressure drop was achieved without a significant change in the thermal performance. Other critical factors that were investigated in this study were the size of the crossflow and pin cross-section area. It was illustrated that thicker pins with smaller flow gaps leads to an improvement in the temperature uniformity on the surface of chip and thermal resistance of system but it can decrease the COP values.

Acknowledgments

This work is supported by SRC (Task 2878.006). Many thanks to Kammal Sikka, Charles Arvin, and Vibhash Jha for their useful suggestions. S.N.S. and A.A. acknowledge the NSF Award IIP-1738793 and the Integrated Electronics Engineering Center (IEEC) at the State University of New York at Binghamton. The IEEC is a New York State Center for Advanced Technology with funding from New York State through the Empire State Development Corporation.

References

[1] L. Su, "Delivering the future of high-performance computing (AMD), HotChips Conference, Stanford University CA, US," 19-Aug-2019.

[2] D. Saums, B. Jarrett, A. C. Mackie, and J. Ross, "Thermal Management Materials Choices," Indium Corporation, Indium Corporation Tech Paper.

[3] A. Azizi, M. A. Daeumer, J. C. Simmons, B. G. Sammakia, B. T. Murray, and S. N. Schiffres, "Additive Laser Metal Deposition Onto Silicon for Enhanced Microelectronics Cooling," in 2019 IEEE 69th Electronic Components and Technology Conference (ECTC), 2019, pp. 1970–1976, doi: 10.1109/ECTC.2019.00302.

[4] A. Azizi, M. A. Daeumer, and S. N. Schiffres, "Additive laser metal deposition onto silicon," Additive Manufacturing, vol. 25, pp. 390–398, Jan. 2019, doi: 10.1016/j.addma.2018.09.027.

[5] X. C. Tong, Advanced Materials for Thermal Management of Electronic Packaging, vol. 30. New York, NY: Springer New York, 2011.

[6] K. M. Razeeb, E. Dalton, G. L. W. Cross, and A. J. Robinson, "Present and future thermal interface materials for electronic devices," International Materials Reviews, vol. 63, no. 1, pp. 1–21, Jan. 2018, doi: 10.1080/09506608.2017.1296605.

[7] W. B. Anderson, "Module for protecting and cooling computer chip die mounted on a thin film substrate and a chassis for conduction cooling of such modules," US5274530A, 28-Dec-1993.

[8] M. J. Lowry, "Immersion cooling apparatus for a power semiconductor device," US8094454B2, 10-Jan-2012.

[9] I. Mudawar, "Direct-immersion cooling for high power electronic chips," in [1992 Proceedings] Intersociety Conference on Thermal Phenomena in Electronic Systems, Austin, TX, USA, 1992, pp. 74–84, doi: 10.1109/ITHERM.1992.187743.

[10] R. D. Nelson, S. Sommerfeldt, and A. BarCohen, "Thermal performance of an integral immersion cooled multichip module package," in [1993 Proceedings] Ninth Annual IEEE Semiconductor Thermal Measurement and Management Symposium, 1993, pp. 8–18, doi: 10.1109/STHERM.1993.225336.

[11] B. Dang et al., "Integration and Packaging of Embedded Radial Micro-Channels for 3D Chip Cooling," in 2016 IEEE 66th Electronic Components and Technology Conference (ECTC), 2016, pp. 1271–1277, doi: 10.1109/ECTC.2016.60.

[12] K. Esmailpour, A. Azizi, and S. M. Hosseinalipour, "Numerical study of Jet Impingement Subcooled Boiling on the Superheated Surfaces," Scientia Iranica, vol. 0, no. 0, pp. 0–0, Jul. 2018, doi: 10.24200/sci.2018.20693.

[13] S. Ghadi, K. Esmailpour, S. M. Hosseinalipour, and A. Mujumdar, "Experimental study of formation and development of coherent vortical structures in pulsed turbulent impinging jet," Experimental Thermal and Fluid Science, vol. 74, pp. 382–389, Jun. 2016, doi: 10.1016/j.expthermflusci.2015.12.007.

[14] M. J. Rau and S. V. Garimella, "Confined Jet Impingement With Boiling on a Variety of Enhanced Surfaces," Journal of Heat Transfer, vol. 136, no. 10, p. 101503, Jul. 2014, doi: 10.1115/1.4027942.

[15] S. Ndao, Y. Peles, and M. K. Jensen, "Effects of pin fin shape and configuration on the single-phase heat transfer characteristics of jet impingement on micro pin fins," International Journal of Heat and Mass Transfer, vol. 70, pp. 856–863, Mar. 2014, doi: 10.1016/j.ijheatmasstransfer.2013.11.062.

[16] P. Naphon and S. Wongwises, "Investigation on the jet liquid impingement heat transfer for the central processing unit of personal computers," International Communications in Heat and Mass Transfer, vol. 37, no. 7, pp. 822–826, Aug. 2010, doi: 10.1016/j.icheatmasstransfer.2010.05.004.

[17] M. K. Isman, E. Pulat, A. B. Etemoglu, and M. Can, "Numerical Investigation of Turbulent Impinging Jet Cooling of a Constant Heat Flux Surface," Numerical Heat Transfer, Part A: Applications, vol. 53, no. 10, pp. 1109–1132, Jan. 2008, doi: 10.1080/10407780701790078.

[18] N. Zuckerman and N. Lior, "Impingement Heat Transfer: Correlations and Numerical Modeling," Journal of Heat Transfer, vol. 127, no. 5, pp. 544–552, May 2005, doi: 10.1115/1.1861921.

[19] S. Gao and P. R. Voke, "Large-eddy simulation of turbulent heat transport in enclosed impinging jets," International Journal of Heat and Fluid Flow, vol. 16, no. 5, pp. 349–356, Oct. 1995, doi: 10.1016/0142-727X(95)00050-Z.

[20] T. Cziesla, G. Biswas, H. Chattopadhyay, and N. K. Mitra, "Large-eddy simulation of flow and heat transfer in an impinging slot jet," International Journal of Heat and Fluid Flow, vol. 22, no. 5, pp. 500–508, Oct. 2001, doi: 10.1016/S0142-727X(01)00105-9.

[21] F. Beaubert and S. Viazzo, "Large eddy simulations of plane turbulent impinging jets at moderate Reynolds numbers," International Journal of Heat and Fluid Flow, vol. 24, no. 4, pp. 512–519, Aug. 2003, doi: 10.1016/S0142-727X(03)00045-6.

[22] Y. M. Chung and K. H. Luo, "Unsteady Heat Transfer Analysis of an Impinging Jet," Journal of Heat Transfer, vol. 124, no. 6, pp. 1039–1048, Dec. 2002, doi: 10.1115/1.1469522.

[23] H. Hattori and Y. Nagano, "Direct numerical simulation of turbulent heat transfer in plane impinging jet," International Journal of Heat and Fluid Flow, vol. 25, no. 5, pp. 749–758, Oct. 2004, doi: 10.1016/j.ijheatfluidflow.2004.05.004.

[24] T. Dairay, V. Fortuné, E. Lamballais, and L.-E. Brizzi, "Direct numerical simulation of a turbulent jet impinging on a heated wall," J. Fluid Mech., vol. 764, pp. 362–394, Feb. 2015, doi: 10.1017/jfm.2014.715.

[25] J. Ortega-Casanova and F. J. Granados-Ortiz, "Numerical simulation of the heat transfer from a heated plate with surface variations to an impinging jet," International Journal of Heat and Mass Transfer, vol. 76, pp. 128–143, Sep. 2014, doi: 10.1016/j.ijheatmasstransfer.2014.04.022.

[26] R. Brakmann, L. Chen, B. Weigand, and M. Crawford, "Experimental and Numerical Heat Transfer Investigation of an Impinging Jet Array on a Target Plate Roughened by Cubic Micro Pin Fins1," Journal of

Turbomachinery, vol. 138, no. 11, p. 111010, Nov. 2016, doi: 10.1115/1.4033670.

[27] P. Naphon, S. Klangchart, and S. Wongwises, "Numerical investigation on the heat transfer and flow in the mini-fin heat sink for CPU," International Communications in Heat and Mass Transfer, vol. 36, no. 8, pp. 834–840, Oct. 2009, doi: 10.1016/j.icheatmasstransfer.2009.06.010.

[28] P. Naphon and S. Klangchart, "Effects of outlet port positions on the jet impingement heat transfer characteristics in the mini-fin heat sink," International Communications in Heat and Mass Transfer, vol. 38, no. 10, pp. 1400–1405, Dec. 2011, doi: 10.1016/j.icheatmasstransfer.2011.08.017.

Shape Optimization of a Pin Fin Heat Sink

T. Menrath, A. Rosskopf, F. B. Simon, M. Groccia, S. Schuster

Fraunhofer IISB, Schottkystraße 10, 91058 Erlangen, Germany

+49 9131 761 235, thomas.menrath@iisb.fraunhofer.de

Abstract

The increasing power density of electronic components, the demand for higher overall system efficiencies and new manufacturing methods lead to a continuous expansion of thermal management boundaries. The aim of the presented work is to gather the current boundaries for optimizing the shape of a pin fin heat sink in terms of thermal resistance, pressure drop and coefficient of performance (COP, fraction of thermal power by pumping power). These new shapes can be fabricated by additive manufacturing (AM), which opens up significantly more design freedom in comparison to common methods like milling or extrusion. The optimization was performed by a Genetic Algorithm (GA) based on either a computational fluid dynamics (CFD) simulation or a metamodel approximating these results. The goal was to investigate the Pareto frontier of the physical limits. A reduction in computational effort was achieved by subdividing a reference, state of the art cooling system, into a smaller, representative subsystem. Depending on the pin shape and with the reference boundary conditions, an increase of 55% of COP was verified by measurement.

Keywords

Shape Optimization, Additive Manufacturing, CFD Simulation

1. Introduction

In modern power converters and inverters, wide bandgap devices (like silicon carbide and gallium nitride) enable an increase of electrical efficiency compared to standard silicon based systems. Therefore, the thermal management has to be extended not only to address performance, power density and reliability demands, but also focus on low pumping power to meet high overall system efficiency.

From another perspective, there is typically a given total pressure drop for each component within a cooling circuit. The task for the heat sources within each component is to allocate a certain amount of pressure drop to it, depending on the target component temperature. Typically, there are additional limitations, like minimum spacing, minimum structure thickness and heat spreading capabilities.

In [1] an analytic approach has shown that there is an optimum to the COP for fin height, fin thickness and fin spacing. Increasing the COP of a heat sink any further, with boundaries like high power density, minimum spacing and manufacturing capabilities, would need the pin shape to be altered. Previous works such as [2] have shown that the shape of pin fins is crucial to the heat sinks COP.

The capabilities of additive manufacturing (AM) create a variety of new possibilities to manufacture optimized, complex structures. The shape of pin fins is no longer limited to common cylindrical pin fins, constant cross sections along the height, like in [2], or tapered shapes. Due to an increasing variety of materials available for these AM processes, especially with materials of high thermal conductivity like copper, new components for thermal applications are possible.

Compared to the topology optimization presented in [3] and [4], which only results in a single specific design, the aim of this work was to achieve multiple pin fin shapes with different properties each. The gathered results should be applicable to other tasks with conformable boundary conditions, but different optimization goals.

2. Numerical Method

Initially, a steady state conjugate convective heat transfer simulation of a simplified power converter, cooled by an emulsion of water and glycol, was performed.

2.1. Reference simulation setup

The reference pin fin heat sink configuration is shown in Figure 1. The setup consists of three individual half bridges (one power switch high side with a diode and one low side with a diode), on a DCB substrate, mounted on top of a single heat sink. The heat sink has standard cylindrical pin fins shown in Figure 1. This configuration and pin definition was used as a reference for pressure drop Δp, thermal resistance R_{th} and COP for the following pin fin modifications. The three parallel half bridge modules in Figure 1 are displaying a color map of the resulting temperature distribution. All necessary physical properties for this steady state CFD simulation are shown in Table 1 for rigid materials and Table 2 for the coolant.

Rigid Material	Thermal conductivity (W/mK)
Copper (sintered)	340
AlSi10 (sintered)	130
AlMg5	150
AlN	180
Silicon	94
Thermal grease	1
Solder	50

Table 1: Physical properties of ridig materials

Emulsion property at	65 °C	35 °C
Thermal conductivity (W/mK)	0.387	0.373
Specific heat capacity (J/kgK)	3406	3287
Density (kg/m³)	1062	1080
Dynamic viscosity (kg/ms)	0.0014	0.0029

Table 2: Physical properties of the water glycol emulsion at 65 °C and 35 °C

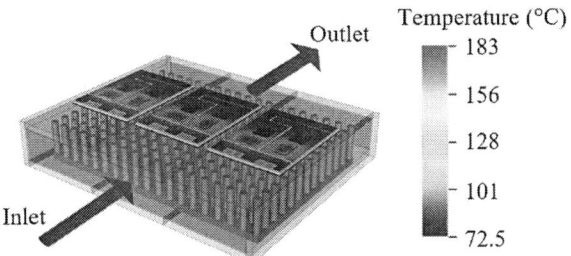

Figure 1: Reference heat sink displaying a color map of the resulting temperature distribution of each half bridge

A single run of the reference system simulation took 3.8 hours (8 cores in parallel, Xeon X5560 2.8 Ghz, 168 Gb RAM), with the commercial Computational Fluid Dynamics (CFD) Tool 6SigmaET by Future Facilities, which is specialized to electronics cooling simulation. The total solver run time emphasizes the need for parallelization and further system complexity reduction, as multiple pin shapes will be simulated. For that reason, the heat sink was subdivided into equal "unit cells". Each one of these cells consists of a centered pin fin with a top and a bottom plate. The fluid inlet and outlet are located on opposing sides of the cell, while the two remaining sides are of symmetrical boundary conditions. The entire heat sink only consists of one type of these unit cells, with a certain pin shape.

2.2. Unit cell simulation setup

Addressing both inline and staggered pin orders, a compromise for the unit cell boundary definition had to be found to run only a single CFD simulation.

The highest pressure drop gradient for inline pin order is located at the first pin row as shown in Figure 2. These pins are exposed to a free stream. For a staggered pin order, the front face of each pin is facing the highest fluid velocity, even higher than the free stream velocity, which leads to even higher pressure drop gradients. These pins are additionally exposed to the lowest fluid temperature at that same pin row, which leads to higher heat sink efficiency, compared to inline pins. A compromising boundary definition is required so that the pin is exposed to a free stream inlet condition, with no leading pin in front.

The total of these simplifications and generalizations lead to the loss of a direct transition from the single unit cell simulation results to a full heat sink. It was suggested that the relation between the different shapes in terms of pressure drop, thermal resistance and therefore COP will remain comparable, when transitioning from a unit cell to a full heat sink. The benefits of this are less computational effort, while having both inline and staggered sufficiently addressed in only one simulation run.

The pin fin is defined by ten parallel, equidistant ellipse cross sections, (Figure 3) and results in 20 degrees of freedom (two radii per ellipse). Both radii of each ellipse are limited by the pin fin spacing and minimum structure thickness as shown in Table 1.

Ellipses were chosen to define the cross section due to the fact, that only two parameters are needed for definition, while still having better thermal and flow performance than cylindrical pins. This was experimentally shown by [5] and [6].

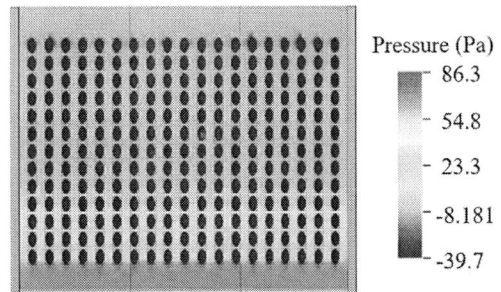

Figure 2: Reference system simulation result of the pressure distribution

Even though there is no direct transition from a unit cell to a full heat sink, its boundaries were defined in a way that they could later be combined. Therefore, the base plate temperature was defined to be 90 °C, with a temperature difference ΔT of 25 K to the coolant. The simulation result was consequently the thermal power $\dot{Q}$ each pin can exchange to the coolant fluid and the resulting pressure drop Δp of that pin shape. With these values the thermal resistance $R_{th} = \Delta T/\dot{Q}$ and $COP = \dot{Q}/P_{pump} = 1/(R_{th} \cdot \Delta p \cdot \dot{V})$ were calculated. As already stated by [Guan], COP needs an additional constraint, otherwise the reduction of the volume flow rate, would lead to higher COP. Thermal power or a target component temperature is a practical constraint to address this. The boundary parameters were defined upon consultation with the research project partners, the manufacturing process capabilities, or calculated based on the reference system, like the flow rate $\dot{V}$.

Parameter	Value
Base plate temperature	90 °C
Inlet flow rate	0.32 l/min
Inlet temperature	65 °C
Pin height	10 mm
Pin width perpendicular	0.6 mm – 2.0 mm
Pin width	0.6 mm – 3.4 mm
Pin pacing perpendicular	2 mm
Pin spacing	0.6 mm
Base plate section	4 mm x 4 mm

Table 3: Boundary conditions and geometrical limits of the unit cell simulation

2.3. Optimization results

The two objectives of optimization study were minimizing pressure drop and thermal resistance. The shape of two randomly chosen individual pin fin unit cells is shown in Figure 3 along with the resulting temperature distribution of the conjugate heat transfer simulation. These two were part of the initial group of geometries and are not optimized. The computational effort for each unit cell performed on the previously mentioned machine was less than 3 min. This is a reduction by a factor of about 70 compared to the reference simulation.

The CFD simulations were automated by a self-developed framework, which enables for parallelization, license management and performance tweaking. This framework was built around 6SigmaET, which was opened up for this automated access by the developers at Future Facilities.

The strong non-linearity of both objectives and the high amount of parameters is predestined for genetic optimization approaches. Due to the fact, that the strength pareto evolutionary algorithm 2 (SPEA-II) has already demonstrated its potential for a large variety of applications [7], this implementation is used in all following optimization runs.

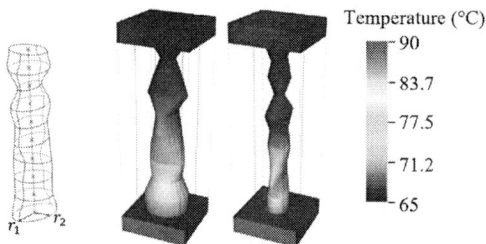

Figure 3: Pin shape definition on the left and two random samples of altered pin shapes with the temperature distribution on the right (not optimized)

In the first two runs (Figure 4, GA Run 1 and 2) the optimization is directly performed based on the CFD analysis of the unit cells. Within the optimization, the SPEA2 algorithm iteratively improves the geometric shape of the pin fin by manipulating the radii of the ellipses. Consequently, several thousand of CFD simulations have to be performed to get the resulting Pareto frontier.

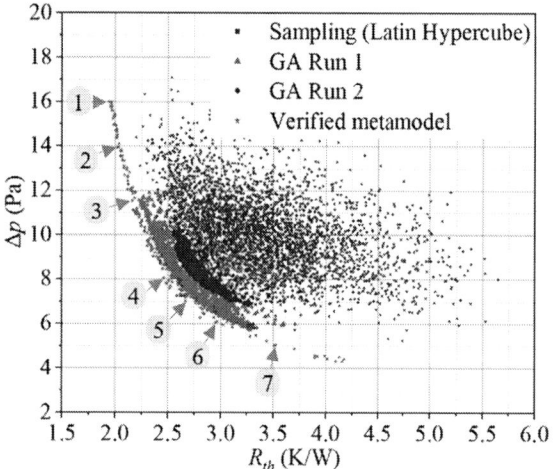

Figure 4: Pressure drop and thermal resistance results for each individual of the optimization study

In a separate, third optimization run, initially a representative sampling (based on the Latin Hypercube algorithm) of the parameter space was calculated and corresponding one thousand geometric designs were evaluated with the help of CFD simulations. The results of both objectives is visualized in Figure 4 and used to generate a metamodel. The resulting model represents and approximates the input to output relation and therefore approximates the CFD simulation of the unit cell. Due to the fact, that the evaluation of the metamodel consumes only a split second of calculation time, the SPEA-II algorithm can perform a high amount of optimization loops. The resulting geometric candidates where

verified by separate CFD simulations and yield the Pareto frontier provided in Figure 4.

2.4. Discussion

The 2 mm cylindrical pin was used as a reference to evaluate the new pin shapes in terms of pressure drop, thermal resistance and COP. Table 4 shows the results and calculated values.

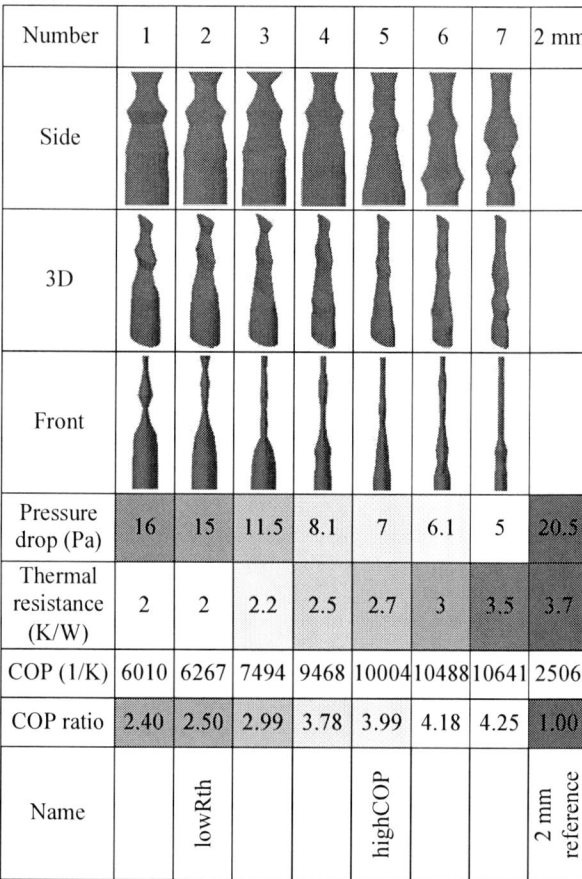

Number	1	2	3	4	5	6	7	2 mm
Side								
3D								
Front								
Pressure drop (Pa)	16	15	11.5	8.1	7	6.1	5	20.5
Thermal resistance (K/W)	2	2	2.2	2.5	2.7	3	3.5	3.7
COP (1/K)	6010	6267	7494	9468	10004	10488	10641	2506
COP ratio	2.40	2.50	2.99	3.78	3.99	4.18	4.25	1.00
Name		lowRth			highCOP			2 mm reference

Table 4: Samples of pin geometries along Pareto frontier from side, front and 3D view.

The shapes along the Pareto frontier evolve, starting at low pressure with very thin shapes like number 7. The bottom ellipse is then increased to its maximum with a short tapered section on top. After that form 5 to 1 this maximum size bottom section increases in height. The shape of the pins is not the optimum, as a GA is only capable of narrowing towards the optimum. Therefore artefacts may persist, if their influence to thermal resistance or pressure drop is rather small. Nevertheless, these shapes have to be further investigated by local optimization. The contrary behavior of pressure drop and thermal resistance and therefore thermal performance is also visible.

2.5. Reference System simulation with pin shapes

Two pin shapes were chosen to verify that the relation of pressure drop and thermal resistance of these shapes compared to the 2 mm cylindrical pin can also be obtained at a full heat sink level.

The first shape is of very low thermal resistance but higher pressure drop (lowRth, number 2). The second shape (highCOP, number 5) is a compromise of both. Additionally a geometrical smooth version (smooth) was designed, to have an additional option that can be manufactured not only by AM, if needed. This pin was also designed following the evolution of the pin shapes, towards lower thermal resistance. Therefore, the thick bottom section was extended and then tapered down to minimum width compared to lowRth. Consequently this pin is supposed to reduce thermal resistance further, with an increase of pressure drop. The full comparison of all pin shapes and important orders is given in Table 5.

	lowRth	highCOP	smooth	2 mm	2 mm staggered	smooth staggered	maximum ellipses
ΔT (K)	102	109	98	103	95	96	96
Δp (Pa)	51	35	71	131	223	115	129
COP (1/K)	2299	3176	1710	891	566	1085	974
ΔT ratio	1.0	1.1	1.0	1.00	0.9	0.9	0.9
Δp ration	0.4	0.3	0.5	1.00	1.7	0.9	1.0
COP ratio	2.6	3.6	1.9	1.00	0.6	1.2	1.1

Table 5: Comparison of the four inline pin shapes and staggered ordered full heat sinks

The comparison in Table 5 especially with regard to COP lacks the fact, that component temperatures are not equal. To account this, flow rate would have to be adjusted. For the highCOP pin shape, increasing flow rate would lead to higher pressure drop and both would reduce COP. Nevertheless smooth staggered and 2 mm inline are compared, the benefits are obvious. The smooth staggered heat sink has 7 % lower component temperature, 12 % less pressure drop and resulting in 21 % higher COP. The lowRth pins has even more benefits. At the same component temperature, it has 60 % less pressure drop and a 159 % higher COP.

3. Experimental methods

Reducing errors for any values of interest is important when designing a measurement setup. In this case, the temperature and pressure drop distribution over the defined flow rate range from 2 l/min-10 l/min had to be measured.

Thermal interconnection is depending on surface finish, thermal interface material and its thickness after mounting, along with other boundaries that have additional dependencies. These could be for example, how many thermal cycles at which

temperature the interface has experienced. On this account, four sensors were places in grooves directly underneath a wide spread heat source, conducting the entire backside of the pin area. These grooves were completely filled with a thermal grease to thermally connect the type K thermocouples. They were not glued to the heat sinks, to reuse the same sensors for each heat sink at the same position, which will lead to a more accurate temperature distribution. Additionally the sensors could be reapplied multiple times on the same heat sink, to reduce attachment errors.

A full size heater was chosen to further reduce error due to the thermal interface, which has an increasing influence when the heat source decreases in size.

Measurements were conducted to verify the optimization results. The four pin designs used for the reference system simulation were chosen to manufacture full heat sinks of. The measurement setup was adopted from previous projects and therefore the base plate size had to be.

The manufacturing of copper heat sinks was still in development within the research of this paper. Hence, all four designs were manufactured by Selective Laser Melting (SLM) from aluminum AlSi10 base material and additionally the cylindrical pin heat sink was milled from an aluminum alloy (AlMg5). The milled version serves as reference to account the difference in surface finish of SLM parts and plain surface suggested by simulation. This will lead to different pressure drop and thermal performance, as the surface roughness of the SLM heat sinks is a lot higher. Physical properties are given in Table 1.

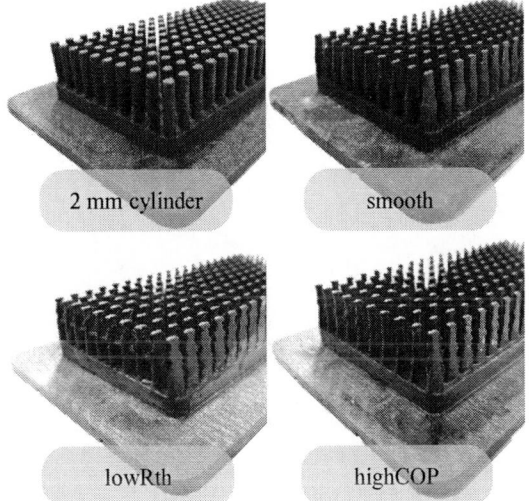

Figure 5: SLM manufactured heat sinks made up of the four pin shapes

Accurate absolute temperature measuring is difficult, shown uncertainties and because of the difference in surface finish, it will not result in any advantage for this measurement series. Hence, the focus was to verify the temperature distribution over a given flow rate range and different pin shapes. The same four sensors were used for each heat sink. In addition, other system errors will also be reduced by focusing on temperature difference, like an unequal heat distribution of the heaters.

Figure 6 shows a cross section of the heat sink setup, the device under test (DUT) and the four thermocouple grooves with one sensor displayed. Six cartridge heaters embedded inside of an aluminum block were in parallel, producing 44 W each and a total of 264 W thermal power.

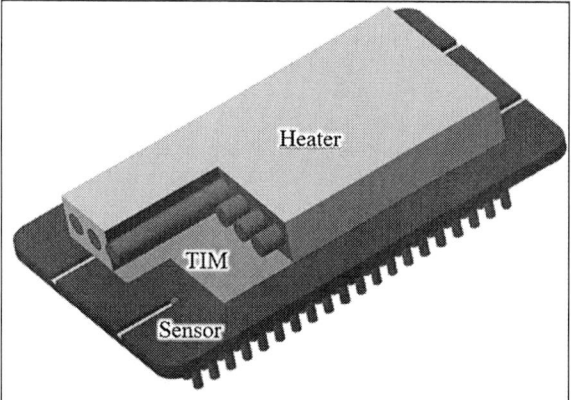

Figure 6: DUT assembly for heat sink measurement, with six cartridge heaters inside of an aluminum block, which is mounted on top of the heat sink with a thermal interface material (TIM). A virtual sensor is showing the thermocouple position inside of the groove, underneath the heater block.

Additionally Figure 7 shows the entire measurement setup. The DUT is right in the middle, baseplate facing upwards and the heater block was clamped on top with thermal grease in between. The two ports perpendicular to the flow direction left and right of the DUT are mounted across the heat sink for pressure drop measurement.

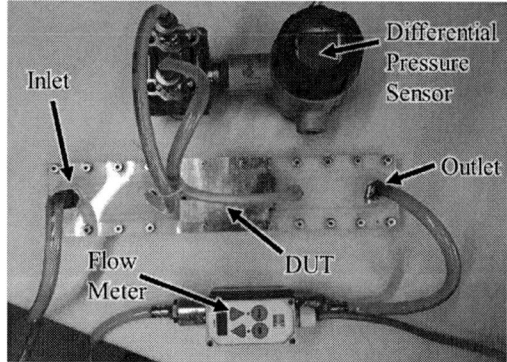

Figure 7: Measurement setup

Close to the inlet port, plate fins were integrated to equalize the fluid flow across the entire channel. The differential pressure sensor is a Honeywell ST3000, which was calibrated prior and a few times in between the measurements. The flow sensor is a paddle-wheel Kobold DF series.

To compensate for small changes of inlet temperature and flow rate, 30 seconds and a total of 3900 sets of data were recorded at each measurement point. This also reduces errors due to digital analog conversion by oversampling.

Each heat sink was measured several times, ramping up and down flow rate and reassembling the thermocouples, heater block and heat sinks applying new thermal grease.

3.1. Measurement setup simulation

Simulations were done for the entire measurement setup and flow rates ranging from 2 l/min to 10 l/min by 2 l/min steps. In accordance to the measurement setup, four virtual sensors were installed where the thermocouples were embedded and also thermally connected by thermal grease.

The inlet temperature was reduced to 35 °C to reduce temperature difference between the DUT and the Environment. Hence, this will reduce the thermal power conducted to anything in contact with the heater block, which would otherwise lead to additional error for thermal measurement. A summary of the results is given in Figure 8.

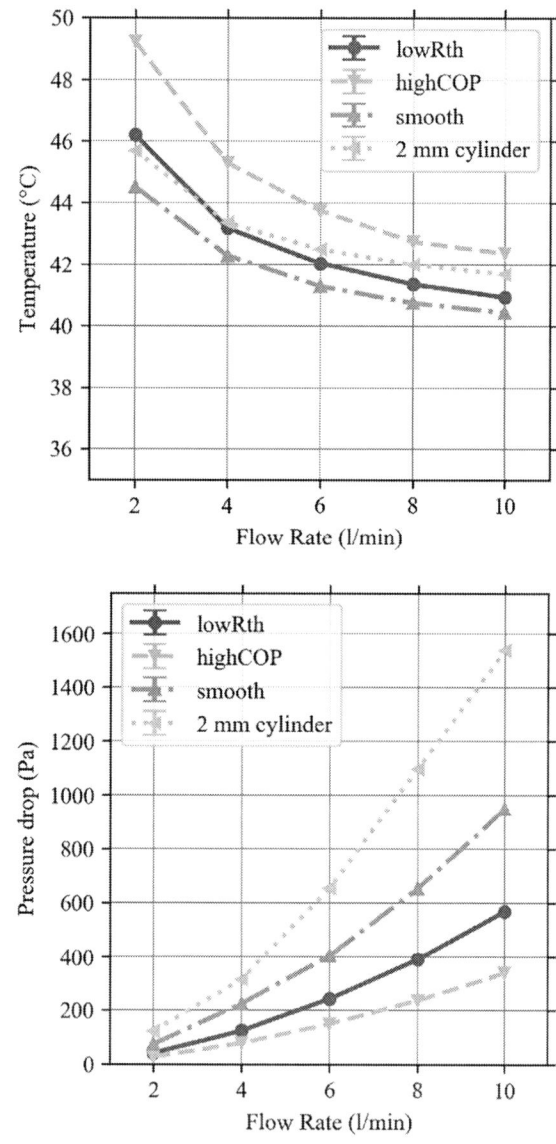

Figure 8: Temperature (top) and pressure drop (bottom) distribution over the flow rate range from 2 l/min to 10 l/min for all four heat sinks

In Table 2 the fluid properties at 35 °C are shown. Table 6 shows the summary of simulation values at 3.2 l/min, which is

14

the flow rate to achieve the equivalent fluid velocity of the reference system.

3.2. Measurement results

The smooth pin measurement series is shown in Figure 9. The flow rate deviation has influence on both the thermal error and the pressure drop error. Combined the temperature deviation was about 1 K at low flow rates.

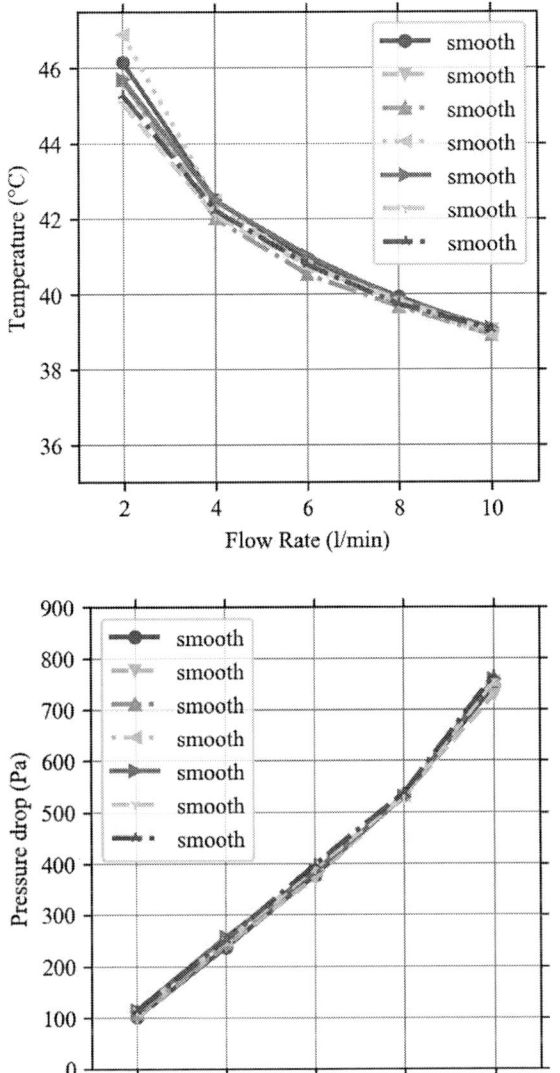

Figure 9: Seven measurement series of the smooth pin shape. Three times reassembled each ramping up and down flow rate

All following measurement results are calculated root mean square (RMS) values, which were achieved by ramping up and down the flow rate and reassembling the whole setup several times.

A summary of the measurement results of all four designs is shown in Figure 10. The cylindrical pins tends to have the lowest possible heat sink temperature with increasing flow rate, but its pressure drop is higher than that of the smooth version.

	2 mm	lowRth	smooth	highCOP
Temperature (°C)	45	45	43	47
Pressure drop (Pa)	210	80	150	50
Temperature rise (K)	10	10	8	12
COP (1/K)	2357	6188	4125	8250
COP ratio	1	2.6	1.8	3.5

Table 6: Simulation results of the measurement setup of all four pin shapes

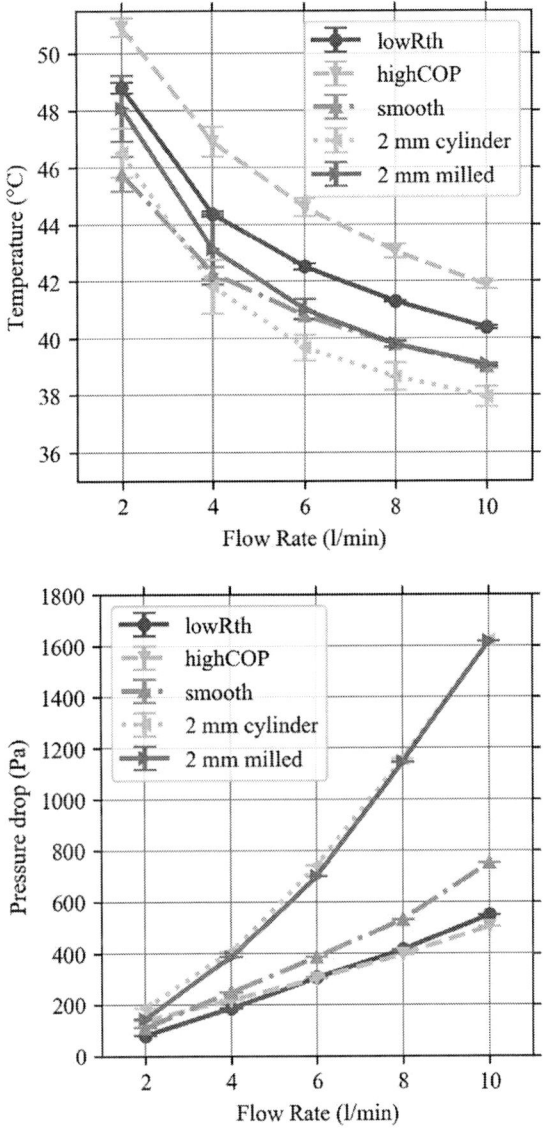

Figure 10: Measurement RMS results for temperature (top) and pressure drop (bottom) of all five heat sinks with error bars.

The pressure drop of the highCOP pin is equal or even higher than that of the lowRth pin. This might be due to manufacturing. The surface roughness of this pin is a lot higher,

15

compared to the other shapes and the geometrical deviation is higher.

4. Comparison of optimization, measurement setup simulation and measurement results

The equivalent flow rate to compare optimization, reference simulation and measurement is 3.2 l/min. As already stated COP comparison needs an additional constraint, like equal component temperature or thermal power. For the measurement setup, optimization and reference heat sink simulation that is only true for the smooth pin shape, without any adjustment. Values are shown in table Table 7. The optimization predicted an increase of COP to 179 %, the reference simulation 192 % and the simulation of the measurement setup 180 %. Finally the measurements approved the increase to 160 % of the 2 mm reference COP.

Adjusting the flow rate to meet the same temperature rise, the lowRth pin performs slightly better than the 2 mm reference. The highCOP pin drops down to about 50 % of the COP of the reference due to the high flow rate resulting high pressure drop.

	2 mm	lowRth	smooth	highCOP
Temperature (°C)	44	47	44	49
Pressure drop (Pa)	280	120	180	180
Temperature rise (K)	14	17	14	19
COP (1/K)	1283	2540	1995	1486
COP ratio	1.00	2.0	1.6	1.2

Table 7: Measurement results of all four pin shapes at reference flow rate of 3.2 l/min

There is a difference in total temperature change between the simulation and the measured 2 mm milled pin. The simulation predicts a lower total difference over the flow rate range. This might be due to the heater cartridges, which have an unequal distribution of thermal power along their length, which was not modeled in the simulation.

5. Conclusions and further work

This study investigated the improvement capabilities of pin fin heat sinks. A reference system was subdivided into single pin unit cells to reduce computational effort of the following optimization, which is based on multiple CFD runs. A compromising definition of the boundary conditions was chosen to meet both inline and staggered pin orders. The shape of this single pin inside of the unit cell was then optimize for two objectives, minimizing thermal resistance and pressure drop.

Two individuals of the Pareto frontier, along with an additional design based on the optimization results and the 2 mm cylindrical reference pin, was then used to create full heat sinks. These were compared under the boundaries of the reference system as well as under a defined range of volume flow rate from 2 l/min– 10 l/min. An increase of COP by 60 %, was measured for the smooth pin shape in comparison to the 2 mm pin, at equal temperature rise and with the reference boundary conditions.

The effect of total thermal power and thermal conductivity of the heat sink to the relation between the different pin shapes within the flow rate range has to be further investigated.

Acknowledgments

The author is grateful for the support by his project partners, within the research project SiCool. A special thanks to Alpha-Numerics and Future Facilities, that supported this investigation with a special access to and additional licenses of their commercial CFD software called 6SigmaET. Thanks to all colleagues at Fraunhofer IISB, which have supported this work with great effort.

References

1. Guan, D., "Thermal Analysis of Power Electronic Modules", Dissertation, 2011
2. Reddy, S. R., G. S. Dulikravich, "Multi-Objective Optimization of Micro Pin-Fin Arrays for Cooling of High Heat Flux Eelectronics", International Mechanical Engineering Congress & Exposition IMECE15, 2015
3. Dede , E. M., S.N. Joshi, F. Zhou, "Topology Optimization, Additive Layer Manufacturing, and Experimental Testing of an Air-Cooled Heat Sink", Journal of Mechanical Design, 2015
4. Oevelen, T. V., M. Baelmans, "Application of Topology Optimization in a Conjugate Heat Transfer Problem", OPT-I International Conference on Engineering and Applied Sciences Optimization, 2014
5. Ota, T., S. Aiba, T. Tsuruta, M. Kaga, , "Forced Convection HeatTransfer From an Elliptical Cylinder," Bull. JSME,26212, pp. 262–267.5, 1983
6. Ota, T., H. Nishiyama, Y. Taoka, "Heat Transfer and Flow Around an Elliptical Cylinder," Int. J. Heat Mass Transfer,2710, pp. 1771–1779., 1984
7. Rosskopf, A., S. Volmering, S. Ditze, C. Joffe, E. Baer, "Autonomous circuit design of a resonant converter (LLC) for on-board chargers using genetic Algorithms", IEEE Transportation Electrification Conference and Expo (ITEC), 2018

Determination of Cost Savings Using Variable Speed Fans for Cooling Servers

Minh Tran[1], Nicole Okamoto[2], Hussammedine Kabbani[2], Saeid Bashash[2]

[1] Velodyne Lidar
5521 Hellyer Ave
San Jose, CA 95138
minhnhutquangtran@gmail.com

[2]San Jose State University
Department of Mechanical Engineering
One Washington Square
San Jose, CA 95192-0087
Nicole.Okamoto@sjsu.edu

Abstract

This paper addresses how different control schemes for fans used to cool servers can affect energy consumption. An analytical model that includes physical parameters of a heat source/heat sink/fan was developed. As a case study, an air duct with a DC fan, heat sink and heat source was built. Experiments were performed to generate physical parameters used as inputs to the model. Three fan control methods -- variable speed, on/off and low/high speed fan controls -- were modelled using MATLAB and Simulink to compare fan speed behavior, target chip desired temperature and saving in fan electrical energy. Potential savings in air conditioning power were also estimated. Simulation results show that variable speed fan control can save more energy than other fan control types, especially on/off control. The model developed can also be used by practitioners to explore the trade-offs between fan and air conditioning power.

Keywords

variable speed fans, thermal management, data centers

Nomenclature

c - viscous damping of the fan
c_p - specific heat of the chip
E - thermal energy
i - electrical current
J - mass moment of inertia
K_b - back EMF constant
K_t - motor torque constant
L - electrical inductance
m_{ch} - mass of the chip
Q - heat transfer rate (W)
Q_{tot} - heat transfer (J)
R - electrical or thermal resistance
T - temperature
V - voltage
V.S. - variable speed
α - constant coefficient used for drag force
$\dot{\theta}$ - fan motor angular speed
$\ddot{\theta}$ - fan motor angular acceleration
τ - torque

1. Introduction

Data center power consumption comes largely from two sources, the room infrastructure and the IT equipment. Servers are the main components of the data center IT. A widely accepted method for evaluating power usage effectiveness of a data center is via a PUE value. Power Usage Effectiveness (PUE) is defined as a ratio of total energy used by the data center compared to the energy used by the IT equipment:

$$PUE = \frac{total\ data\ center\ energy\ usage}{IT\ equipment\ energy\ usage}$$

A typical enterprise data center built with an outdated computer room building standard has a PUE value between 2 and 3 [1]. This means that the power supply for the infrastructure already takes up 50% to even 70% of the total power supply for the data center. This is very inefficient by modern standards of data center design. Typical PUE values for huge data centers built today by companies like Google, Facebook and Yahoo are about 1.1 [2].

The PUE in a data center can be reduced by using variable speed fans to replace on/off or low/high speed fan controls in server systems. At first glance this method seems to only minimize fan power, which is a small portion of power usage. However, variable speed fans may allow the room air temperature to be increased, which would reduce the energy needed for cooling the air in the server room. According to research, if we can increase the temperature of supplied cooling air by 1°F (0.56°C), then infrastructure energy use may be reduced by 4% [3]. In fact, in 2011, American Society of Heating, Ventilation and Air Conditioning approved new building standards that help reduce power consumption by encouraging higher ambient temperature in many types of data centers [4].

Another reason to focus on improving power efficiency in servers by using variable speed fans is because of the "Energy Proportional Server Efficiency" concept [2]. This concept states that computing efficiency not only depends on a server's computational results but also on the total amount of energy being consumed by the server to do the computing. If we can make a server's power consumption proportional for all workload conditions, then energy efficiency can be improved. This idea can be viewed with the two graphs in Figure 1. The top graph describes power consumption of inefficient servers that consume a large amount of power, about 70% of peak power, while running at idle mode. When servers with variable speed fans are used, the power consumption could look more like the bottom graph.

Research shows that most servers in 2007 consumed almost the same amount of power from 0 to 100% utilization [2]. Some servers operate at idle mode or low utilization for long periods of time. It is desirable to enhance the efficiency of these servers by implementing variable speed fans so that power consumption is more proportional to the server's use.

Vogel, et al. [5] were able to improve aerodynamic performance by at least 20% by developing air flow mockups that included experimental flow measurements. The idea of optimizing fan performance clearly is not new. Therefore, the objective of this work is to develop a new model to predict potential energy savings by utilizing variable speed fans and to examine trade-offs between fan and air conditioning power. The results are compared to low/high and on/off fan speed control.

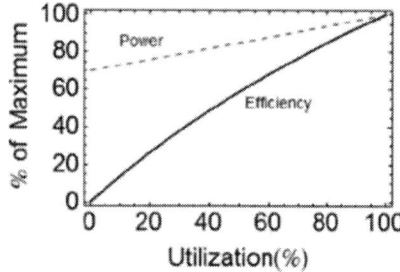

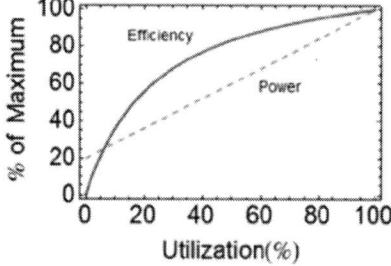

Figure 1: The relationship between power consumption and efficiency of servers for traditional and proportional control. [2]

2. Methodology

An analytical model of a heat source and heat sink under variable heat loads was developed. Analysis for a specific system was performed using MATLAB and Simulink. Experiments were conducted to determine important parameters in the model, and a variety of example heat load signatures over time were examined.

2.1 Heat Sink Modelling

A simple heat source (chip)/heat sink geometry was modelled as shown in Figure 2. A DC fan is assumed to cool the package.

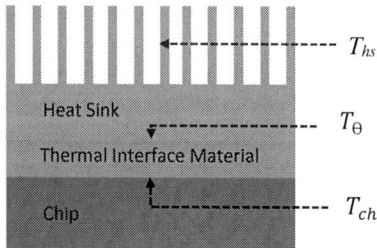

Figure 2: Chip package with heat sink model

Conservation of energy was used, setting the amount of energy increase in the chip equal to the heat generated in the chip minus the heat transferred away from the chip.

$$\frac{d(E_{built\ up\ in\ chip})}{dt} = \frac{d(E_{generated\ by\ chip})}{dt} - \frac{d(E_{transfered\ away\ from\ chip})}{dt}$$

$$\frac{d(m_{ch}*c_P*(T_{ch}-T_{ch0}))}{dt} = Q_{gen} - Q_{out}$$

Assuming the thermal paste has a negligible thermal mass, the equation above can be simplified to:

$$m_{ch}c_p\dot{T}_{ch} = Q_{gen} - (T_{ch}-T_{hs})/R_{tot}$$

$$\dot{T}_{ch} = [Q_{gen} - (T_{ch}-T_{hs})/R_{tot}]/m_{ch}c_p \quad \textbf{(E.1)}$$

The model represented by Equation E.1 was numerically simulated using Simulink. Experiments were performed for an example system to calculate certain parameters.

2.2 Fan Modelling

A standard DC motor circuit was analyzed to develop the differential equation showing the relationship between voltage applied and fan speed. Figure 3 shows the electrical diagram that represents a regular DC motor circuit of the fan:

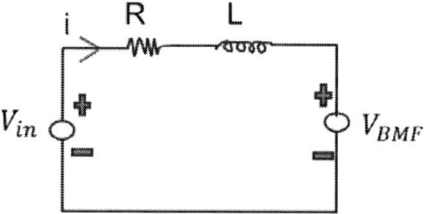

Figure 3: Motor electrical circuit diagram

Kirchhof's law states that the sum of voltages in a series loop must be zero:

$$V_{in} - V_R - V_L - V_{BMF} = 0$$

$$V_{in} - i*R - L\frac{di}{dt} - K_b\dot{\Theta} = 0$$

Here $\dot{\Theta}$ is the fan angular speed and K_b the back EMF constant of the fan motor. For a small DC motor, its inductance value L is negligible; then the equation above can be reduced to:

$$V_{in} - i*R - K_b\dot{\Theta} = 0 \quad \textbf{(E.2)}$$

Next, mechanical modeling of the fan is considered. A differential equation derived from Newton's second law was developed to relate parameters such as fan speed, motor torque and torques that are against fan rotation, as shown in Figure 4.

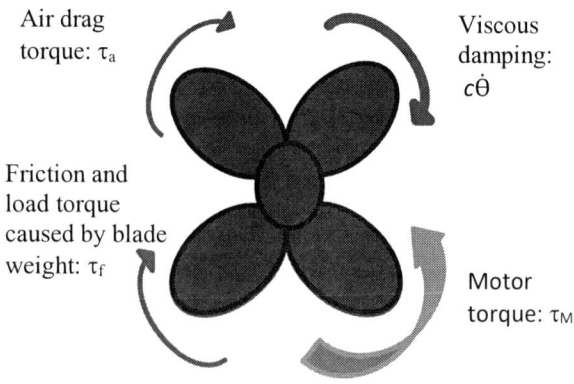

Figure 4: Fan mechanical modeling diagram

Applying Newton's second law to get a mechanical differential equation we get the following:

$$\sum Moments = J\ddot{\theta}$$

$$\tau_M - \tau_a - \tau_f - c\dot{\theta} = J\ddot{\theta}$$

$$K_t i - \alpha\dot{\theta}^2 - \tau_f - c\dot{\theta} = J\ddot{\theta} \qquad \textbf{(E.3)}$$

Fan speed can be correlated to the thermal resistance either from the manufacturer of the heat sink or by experimental testing. At zero fan speed, there is only natural convection. As a result, thermal resistance of the heat sink and paste is at a maximum, called R_0. As fan speed increases, this total thermal resistance decreases exponentially to a minimum, called R_{min}. A general formula for total thermal resistance, R_{total}, can be derived:

$$R_{total} = (R_0 - R_{min})e^{-a\dot{\theta}} + R_{min} \qquad \textbf{(E.4)}$$

In this equation, R_0, R_{min} and a are all constants which can be found experimentally or from curve fitting data provided by the heat sink manufacturer.

2.3 Numerical Modelling

To get the relationship between fan voltage input and temperature of the chip, three differential equations, E.1, E.2, E.3 and the function E.4 were solved simultaneously by Simulink. Two Simulink diagrams were developed for three different fan speed control types--on/off, low/high and variable speed fan control. Figure 5 shows the diagram used for on/off and low/high fan speed control case.

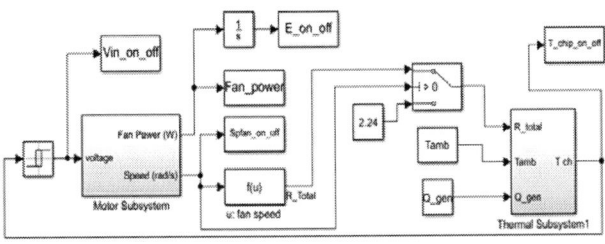

Figure 5: Simulink block diagram for on/off fan control

For variable speed fan control, a PID block was used to produce voltage input for the fan. Figure 6 shows the Simulink block diagram for variable speed fan control. Many blocks from two control diagrams are the same since they are representing the same physical system. They both receive temperature feedback from the chip so that voltage input can be adjusted accordingly. The only difference between the two diagrams is that one uses a simple relay block to turn on and off or to switch voltage supply back and forth between a fixed low and a fixed high value, and the other uses a PID block to adjust voltage supply proportionally to temperature of the chip.

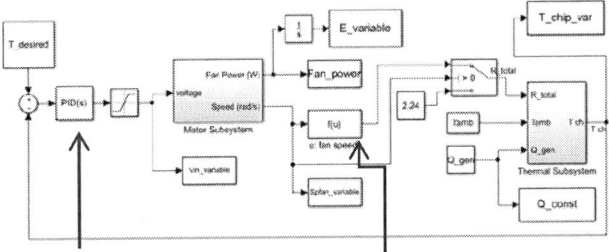

PID control block   Function relating fan speed and thermal resistance

Figure 6: Simulink block diagram for variable speed fan control

2.4 Experimental Apparatus

A simple physical system that includes a heat sink placed in an air duct with little bypass was built. It was used to obtain real system parameters of the DC fan, heat sink, and heat source to include in the numerical models. Figure 7 shows the experimental set up that includes a Plexiglas air duct with cross section of 89 mm x 89 mm, a dc fan at the duct inlet, a temperature sensor and a heat sink. Solidworks Flow Simulation was used to design an optimum contraction angle at the flow inlet that provides uniform flow at the inlet to the heat sink. Two power resisters placed under the heat sink were able to generate a variable heat load from 0 to 100 W. The heat sink shown is made by aluminum with the dimension of 77(L) x 77(W) x 70(H) mm. Its advertised maximum thermal cooling performance to stay below typical maximum case temperatures without using forced convection is 40W.

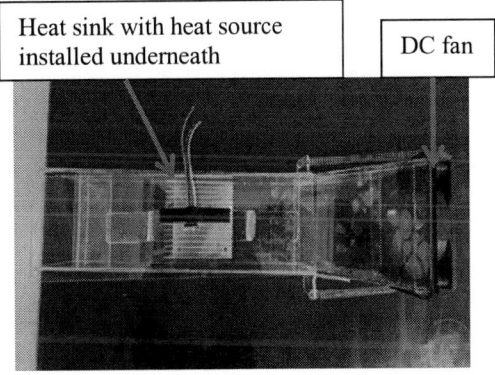

Figure 7: Physical system prototype

a) **Obtaining DC Fan Parameters:**

The fan subsystem's characteristic parameters are:

J - mass moment of inertia of the fan

K_t - torque constant of fan motor

K_b - back EMF constant - assumed to equal to K_t

α - constant coefficient used for air drag force

c - damping constant of motor

τ_f - friction torque which depends on motor's structure

If the fan is running at steady state, meaning angular acceleration is zero, Equation 3 is reduced to:

$$K_t i - \alpha \dot{\theta}^2 - \tau_f - c\dot{\theta} = J\ddot{\theta} = 0$$

Or: $(\alpha \dot{\theta}^2 + c\dot{\theta} + \tau_f) / K_t = i$

Electrical current and fan speed are related to each other by a quadratic function with these three coefficients: α/K_t, c/K_t and τ_f/K_t. The approach used for finding these parameters is to find the K_t first and then all other coefficients by using a quadratic curve fit of experimental data. To find K_t, the fan was kept at zero speed by connecting a fan blade to a spring scale while current was still supplied for the fan. This condition allowed for the extraction of stall torque and the stall current. As a consequence, K_t can be obtained through this equation:

$$\tau_{stall} = K_t i_{stall}$$

Eight tests for the fan used in the experimental apparatus showed an average K_t of 0.0288 Nm/A, with a standard deviation of 0.0045 Nm/A. This K_t value is reasonable for a small DC motor fan. From the K_t value and quadratic curve fitting based on fan speed and corresponded measured current data, other parameters were found to be $\alpha = 1.154 \times 10^{-7}$ kgm^2/rad^2, c=2.597x10^{-6} kgm^2/rad s, and τ_{dy_f}=0.000747 Nm. The resulting equation for current as function of fan speed had an R^2 value of 0.9925.

Another fan parameter, the motor coil electrical resistance, can be determined when fan speed is kept at zero. The average coil resistance was measured to be 13.05 Ohm with a standard deviation of 0.17 Ohms.

Finally, the last parameter of the fan subsystem that needs to be defined is moment of inertia, J. Since this parameter is linked to acceleration, data samples for fan speed over time were captured during both the transient time duration and steady state time duration. To acquire those data, a camera was used to record the fan speed from zero to its steady state speed. Then this video file was viewed with video software which has the capability of time tracking for any specific image frames. A polynomial equation relating fan speed to time was generated with an R^2 value of 0.9999.

This polynomial equation was used to calculated the time constant for this particular DC fan. The moment of inertia of the fan J, was then found to be 5.874x10^{-5} kgm^2, based on the calculated time constant.

b) Obtaining Thermal System Parameters

Experiments were also performed to determine the thermal resistance of the heat sink and thermal paste as a function of fan speed by measuring the heat input to the power resister and the temperature of the power resistor case and ambient air, as shown in Equation E.5.

$$Q = \Delta T/R_{total} = (T_{case} - T_{ambient})/R_{total} \quad \textbf{(E.5)}$$

Under natural convection, for this particular experimental setup, $R_{total}=R_0=2.24$ °C/W. When the fan was running, the thermal resistance followed Equation E.6 with velocity of the fan in radians/s.

$$R_{total} = 0.952 e^{(-0.0123\theta)} + 0.4 \quad \textbf{(E.6)}$$

Fan angular speed was measured using a digital photo tachometer. According to Equation E.1, there are only two more characteristic parameters belonging to the thermal subsystem that need to be defined. These are:

m_{ch} - mass of chip
c_p - specific heat of the chip or power resistor, the heat source

For the first parameter, mass of chip (or mass of the power resistor in this case) was measured to be 46 g. To find c_p, heat specific capacity of this particular power resistor, experiments were run where heat was added to a well-insulated power resistor, and the surface temperature of the resistor was measured with a thermocouple. The value of c_p could then be found using Equation E.7.

$$Q_{tot} = mc_p \Delta T \quad \textbf{(E.7)}$$

After an initial ramp-up period of 28 s, the measured value of c_p became a constant at 740 J/kgK (standard deviation of 5 J/kgK) over the next 24s.

3. Simulation results

The parameters determined from the experiments were input into the MATLAB and Simulink models for all simulations. An ambient temperature of 25°C was assumed along with a maximum desired chip temperature of 60°C. Fan voltage minimums were 0 V in on/off control case and 5.4V in low/high case as well as in the variable speed case. At maximum speed in three cases, fan voltage requirements were all equal to 12 V, which is the maximum fan voltage supply. Two heat load signatures were simulated. In all equations, time is in seconds.

1) Constant heat generation of 60, 70, or 80W (heat generation functions 1, 2, and 3 in Table 1)

2) Two different sine wave generation (heat generation functions 4 and 5 in Table 1):

a. f = 60 + 20*sin(πt/150) (W)

b. f = 60 + 20*sin(πt /300) (W)

These are merely example heat generation signatures that were input. A function describing heat generation measured in a real system could be input to the model to determine performance of a real system. Power savings were estimated from two sources: fan power and air conditioning. The air conditioning energy saving calculation comes from the research of Mark Monroe of Sun Microsystems who estimated that "for every degree Fahrenheit increased in air temperature, 4% energy cost can be saved" [3]. While on/off control is rarely used these days, it is included as a comparison to illustrate how savings can be predicted using this methodology. Results are shown in Table 1. Simulations were run with three different temperature bands for on/off and low/high speed control, as defined in Table 2.

Heat Generation Function	Setting 1				Setting 2				Setting 3			
	On/Off		Low/High		On/Off		Low/High		On/Off		Low/High	
	Fan	A/C	Fan	A/C	Fan	A/C	Fan	A/C	Fan	A/C	Fan	A/C
1	57.0%	0.0%	0.0%	0.0%	71.0%	14.0%	-35.0%	14.0%	58.0%	29.0%	-92.0%	29.0%
2	60.0%	0.0%	40.0%	0.0%	36.0%	14.0%	4.0%	14.0%	-8.0%	29.0%	-58.0%	29.0%
3	0.4%	0.0%	0.7%	0.0%			0.0%	14.0%				
4	57.0%	0.0%	27.0%	0.0%	40.0%	14.0%	-6.0%	14.0%	27.0%	29.0%	-25.0%	29.0%
5	57.0%	0.0%	27.0%	0.0%	40.0%	14.0%	0.0%	14.0%	27.0%	29.0%	-20.0%	29.0%

Table 1: Summary of energy savings using variable speed fans under three different settings

	Setting 1	Setting 2	Setting 3
$T_{setpoint}$ for V.S	59.5	60	60
T_{on}	60	62	64
T_{off}	59	58	56
T_{amb} for on/off and low/high	25	25	25
T_{amb} for V.S	25	27	29

Table 2: Temperature bands evaluated (°C)

From simulation results at the three different settings for chip and ambient temperature listed above, some conclusions can be drawn. First, in general, variable speed control is the most effective energy saving method among three control types, while on/off control is the least effective method. This conclusion is well known and is the reason why on/off control is rarely used. At the maximum constant heat generation situation (function 3 in Table 1), fan energy savings decrease since even in variable speed control, fans need to operate at constant maximum speed at all time to keep the chip temperature at the required level. Second, there is a considerable trade-off in energy savings between fan power and A/C power in low/high and variable speed control methods under Setting 3 when temperature bands were largest. Variable speed fans can keep the chip very close to the target temperature of 60°C, while there are larger temperature bands for on/off and low/high control. As a result, the data center ambient air temperature for variable speed fans can increase in comparison, reducing power requirements for air conditioning. An alternate analysis could have compared savings with the same ambient temperatures for all three types of fan controls. In that case, there would have been no air conditioning power savings, but fan power savings using variable speed fans would have been significant since those fans could have been run at lower speeds.

Under Setting 3 (largest temperature band for low/high control and highest ambient temperature for V.S.), while A/C energy saving is high for V.S. control compared to low/high, V.S control required a lot more fan power to operate. Moreover, in Setting 3, the chip temperature couldn't be kept at the required safe margin stably under maximum heat generation condition. Setting 2 is more reasonable and a

more balanced condition for V.S fan control to work most effectively among the three temperature settings. In Setting 2, V.S fan control offers no savings in fan energy, but there is about 14% of A/C energy that can be saved.

Figure 8 illustrates the differences between low/high and variable fan speed control for the constant 60 W case (heat generation case 3) and Setting 3 – a large temperature band and increased ambient temperature for V.S.

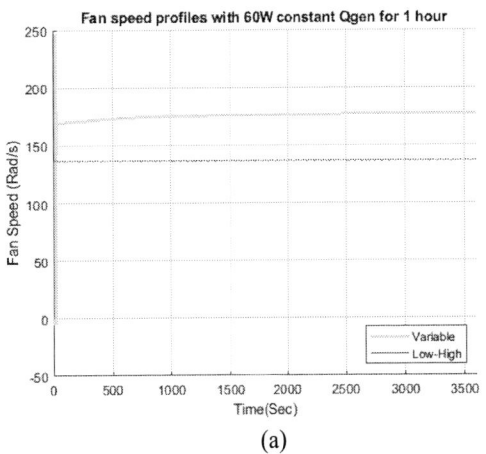

(a)

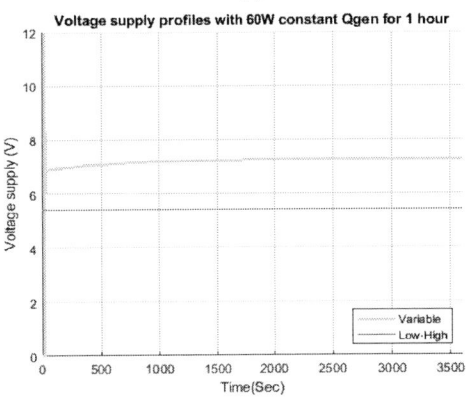

(b)

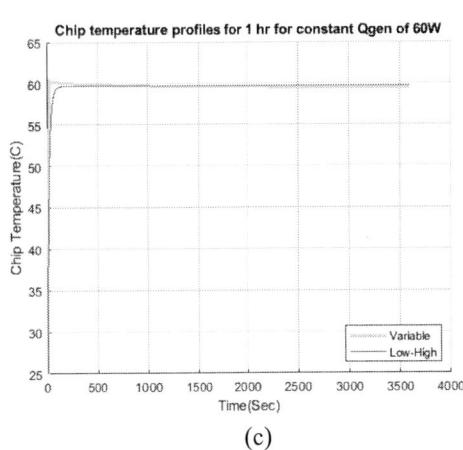

(c)

Figure 8: (a) Fan speed, (b) fan voltage, and (c) chip temperature profiles under Setting 3 for variable and low/high temperature control

Here we see that both control schemes work well for this low heat load and setting. The variable speed case consumes more fan power, but the higher ambient temperature in this setting means lower A/C power consumption. Clearly, a trade-off between the two exists.

Results for the sine wave generation of heat flux are also shown in Table 1, in the last two rows. These cases are for an average heat generation of 60W +/- 20W. While heat generation will never exhibit an exact sine wave function, these results illustrate what can happen with regular heat fluctuations. In this case, fans for all type of controls were on at full speed at the high point of the sinusoidal wave. The period of fluctuation did not affect energy savings since the percentage of time at each heat flux level remained the same, regardless of period. Figure 9 shows fan voltage supply and fan speed over 600 s for the last row and Setting 1 shown in Table 1 for low/high and V.S. control, to illustrate the type of results that the model can predict.

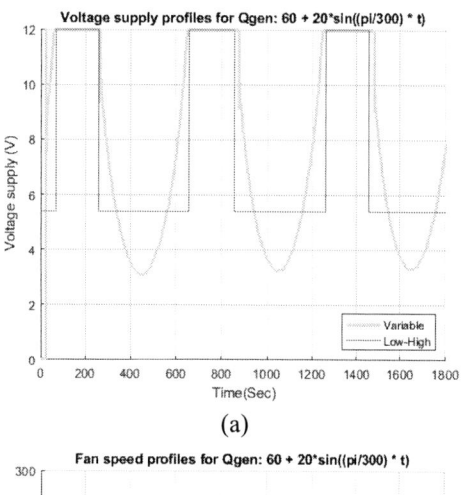

(a)

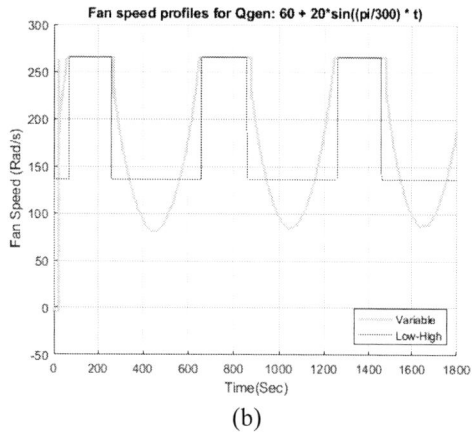

(b)

Figure 9: (a) Fan voltage and (b) speed for sinusoidal heat flux with a 600 s period and 56-64°C low/high temperature band.

Exponential power functions were examined to illustrate the effects of startup. In the simulation, power started at zero and increased exponentially to 80 W. During startup, fans do not turn on until a set minimum chip temperature is reached, 60°C. In the simulations performed, power reached 80 W before the fans turned on. Thus, fan energy savings looked

very similar to the constant power generation case. Figure 10 shows the fan power consumption in case of an exponential power generation with a time constant of 40 s. Fan power savings for the three control schemes are illustrated in Table 3. Conditions for this table were set to be 25°C for ambient temperature in all cases, 59.5°C for desired temperature in variable speed case, and 59°C-60°C temperature band for on/off and low/high speed cases.

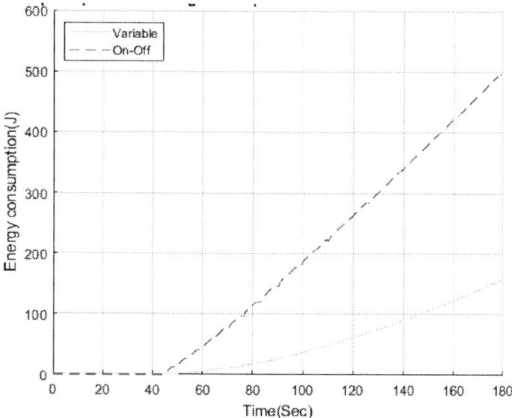

Figure 10: Energy consumption for a time constant of 40 s.

When 0 volts was used as a minimum voltage supply it was hard to control the chip temperature without considerable overshooting for all the heat generation conditions tested in this project. By keeping the fan's minimum speed running at the minimum required voltage for turning on the fan, the chip temperature can be controlled much more easily without much overshooting.

Q_{gen}	Energy Savings		Time Duration
	on/off	low/high	
Expo 10 s	17%	12%	over 3 minutes
Expo 40 s	54%	42%	

Table 3: Fan energy savings using variable speed fans during start-up

4. Conclusions

The simulations performed here show that variable speed fans can result in significant savings in fan energy over on/off and low/high speed fan controls. As expected, on/off fan control was the least efficient method among the three control types. The smaller temperature fluctuations that occur with variable speed fans may also allow data center air temperatures to be increased. Potential savings in air conditioning power may dwarf savings in fan energy usage. Case studies were performed for a specific fan/heat sink package with example heat load signatures. The model presented could be used by practitioners who wish to analyze trade-offs between different control systems and ambient temperatures. Practitioners would need to input their own fan and heat sink parameters as well as their typical heat load signatures.

While not analyzed in this paper, an additional benefit in using variable speed fans is the reduction in thermal stress due to smaller temperature fluctuations. That in turn may extend the lifetime of some electronic components.

Potential future work could examine the total cost of ownership for different heat sink solutions. Solutions that are initially more expensive (more sophisticated heat sinks, liquid cooling, etc.) will typically require less fan/pump energy and may allow for higher ambient temperatures, lowering A/C costs.

References

1. Victor Avelar, Dan Azevedo, Alan French, eds., "PUE: A Comprehensive Examination of the Metric," White Paper #49, The Green Grid, 2013.
2. Gough, C., Steiner, I., and Saunders, W., "Why Data Center Efficiency Matters," Energy Efficient Servers: Blueprints for Data Center Optimization, Apress, pp. 1-20, 2015.
3. Miller, Rich. "Data Center Cooling Set Points Debated". DataCenter Knowledge. Informa USA, INC. 24 Sep.2007. Web. Retrieved 2 April 2018.
4. ASHRAE Technical Committee 9.9. "2011 Thermal Guidelines for Data Processing Environments– Expanded Data Center Classes and Usage Guidance". American Society of Heating Refrigerating and Air Conditioning Engineers' Whitepaper. 2011.
5. Vogel, M., Chen, T., Doan, S., Harrison, H., and Nair, Rajeesh, "New Approach to System Server Air Flow/Thermal Design Development, Validation, and Advancement in Green Fan Performance," 26th Annual IEEE Semiconductor Thermal Measurement and Management Symposium (SEMI-THERM), 2010.

On Economic Cooling of Contained Server Racks using an Indirect Adiabatic Air Handler

Riccardo Lucchese[1], Michele Lionello[2], Mirco Rampazzo[2], Andreas Johansson[1], Wolfgang Birk[1]

[1]Luleå University of Technology, Dep. of Computer Science, Electrical and Space Engineering
97187 Luleå, Sweden
<riccardo.lucchese|andreas.johansson|wolfgang.birk@ltu.se>

[2] Università degli Studi di Padova, Dep. of Information Engineering
via Gradenigo 6, 35313, Padova, Italy
<lionello|rampazzom@dei.unipd.it>

Abstract

We study the economic operation of a free-cooling setup in which an Indirect Adiabatic Air Handler (IAAH) recovers heat from an array of server racks placed in a contained aisle. For this setting we propose two different control policies: in the first approach, the airflow supply rate to the racks is maintained constant while only the process-side operations of the IAAH are optimized. In the second approach, also the room-side rate is updated adaptively. Building on calibrated models of the IAAH and the servers, we design experimental trials considering different outdoor temperatures and humidity conditions as well as varying computational workloads. The in silico analysis contributes actionable insights on the optimal thermal and cost operations of the system.

Keywords

Data center thermal models, Indirect adiabatic cooling, Server thermal models, Energy efficient cooling.

1. Introduction

The current scale at which information flows require storage, sharing and analysis has put the efficient operation of data centers under the spotlight [1]. In this respect, a popular performance index is the Power Usage Effectiveness (PUE), namely the ratio of the total electrical power absorbed by a data center over the power consumption due to computing only [2]. With industry average PUEs above 1.4 [3], the impact of data center cooling on the European electrical grid is estimated above 1.5% of the distributed power [4]. The adoption of both *free-cooling* technologies and novel provisioning policies has the potential to benefit substantially this state of practice [5], [6], [7]. Indeed, control oriented models and Computer-Aided Control System Design (CACSD) tools are enabling the critical analysis of relevant thermal phenomena [8], [9] and the design of novel controllers that outperform static policies tuned against nominal conditions [9], [10], [11]. Model-based analyses are moreover instrumental to informing which technological components should be prioritized, aiding the development of thermal guidelines such as [12].

Here, we build on a calibrated virtual process and study the *economic* controls over varying outdoor conditions and

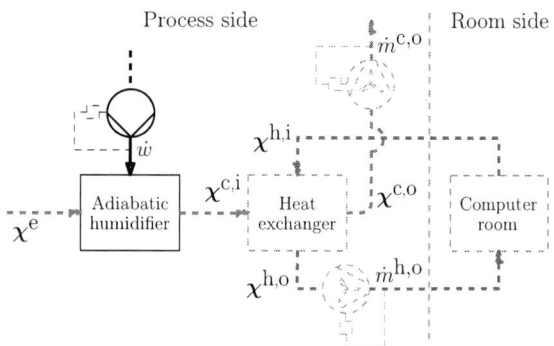

Figure 1: Simplified block diagram of the indirect free-cooling setup. An *external* airflow (in blue) is humidified before reaching a crossflow heat exchanger where heat is recovered from the *internal* airflow (red).

compute workloads. We consider a free-cooling application in which an Indirect Adiabatic Air Handler (IAAH) is paired to an array of 12 contained server racks (see Figure 1). In this setup, the heat load is recovered by circulating a *process*-side airflow (also called *external*) and a *room*-side airflow (*internal*) through a crossflow heat exchanger (HX). Past the heat recovery step, the external air is exhausted immediately and the internal air is drawn back into the room. The peculiarity is an adiabatic humidifier which precools the external air, allowing the unit to be effective even with hot-side temperatures below the outdoor ones. In practice, the internal air rate $\dot{m}^{h,o}$ is constant and oversized, limiting the overall energetic efficiency. Here, we specifically address the case of manipulable $\dot{m}^{h,o}$ to meet current trends seeking more flexibility.

Statement of contributions. We address an emerging technological trend among data center hyper-scalers. Our contribution is summarized as follows:

- We propose a novel modeling of the complete cooling setup which pairs an IAAH unit with a contained array of server racks. We build on accurate and experimentally validated, nonlinear, thermal models that have been made available recently [13], [14].
- We propose and discuss in detail two different actu-

ation policies. In the first case, only the process side of the IAAH system is optimized, trading a simpler controller for lower performance. In the second case, a more holistic approach allows also the room-side actuator to be controlled adaptively.

- We report in silico experimental results, obtained from benchmarking the proposed control strategies, that i) demonstrate the cost saving potential of more holistic policies, and ii) suggest relevant directions toward achieving cooling efficiency.

Organization of this manuscript. Section 2 places this work in the existing literature. Section 3 describes the mathematical model of the cooling setup. Section 4 details two optimizing flow provisioning strategies. Section 5 discusses the numerical results. Finally, Section 6 collects concluding remarks and future directions.

2. Literature review

Traditionally, cooling efficiency improvements have been achieved through advances in the hardware. More recently, supervisory controllers are increasingly deployed to attain the performance by acting on software-only components. Within this trend, [15] develops a broad survey of power modeling and prediction techniques. [16] devises a supervisory optimizer aimed at indirect free-cooling units. [14] proposes to use Model-Based Repeated Optimization (MBRO) to minimize the process-side operation cost of a IAAH. [9] devises a comprehensive data center control framework that accounts for both the compute and cooling resources but limits its analysis to static airflow provisioning policies and approximate linear thermal dynamics. Nonlinear thermal models of the servers (where the bulk of the heat load is produced) have been devised in, for example, [11], [13]. In particular, [13] proposes a network oriented formalism that results in accurate temperature predictions. Incidentally, we report also on gradient-free optimizers, such as [17], which are effective but typically exhibit slow convergence. To our best knowledge, there is a lack of attention on the modeling and control of the *complete* cooling system (including both the cooling unit and the servers under *dynamic* provisioning policies,) impeding a fair study of the performance limitations induced by the servers' thermal envelope.

3. Mathematical model of the cooling setup

In this section we outline the mathematical modeling of our data center cooling setup. We address the main components of the plant (including the humidifier, the heat exchanger, and the temperature dynamics of the servers) by building on top of calibrated First-Principles-Data-Driven (FPDD) models that have been experimentally validated in [13] and [14]. In what follows, for the sake of a compact notation, we let

$$\chi^{\star} = \left(\dot{m}^{\star}, x^{\star}, \varrho^{\star} \right) \qquad (1)$$

Table 1: Nomenclature.

Sym.	Description	Unit
Parameters		
n	number of deployed servers	adim.
λ	ratio of internal air losses	adim.
$\overline{x}^{\mathrm{cpu}}$	max. safe CPU temperature	K
$\underline{w}, \overline{w}$	lower and upper bounds on $\dot{w}$	l/s
$\underline{m}^{\mathrm{e}}, \overline{w}^{\mathrm{e}}$	lower and upper bounds on $\dot{m}^{\mathrm{e}}$	kg/s
$\underline{m}^{\mathrm{h,o}}, \overline{w}^{\mathrm{h,o}}$	lower and upper bounds on $\dot{m}^{\mathrm{h,o}}$	kg/s
Signals		
$\dot{w}$	water volumetric rate at the humidifier	l/s
χ^{e}	inflow air properties of external air	†
$\chi^{\mathrm{c,i}}$	inflow air prop. at HX's cold-side	†
$\chi^{\mathrm{h,i}}$	inflow air prop. at HX's hot-side	†
$\chi^{\mathrm{h,o}}$	outflow air prop. at HX's hot-side	†
χ^{loss}	air prop. of the room-side losses	†
$\chi_j^{\mathrm{srv,i}}$	inflow air prop. at j-th server	†
$\chi_j^{\mathrm{srv,o}}$	outflow air prop. at j-th server	†
x^{srv}	global state vector of the servers	⋆
W	global CPU workload matrix	adim.
Material properties		
c_{p}	specific heat at constant pressure of air	J/(kg K)
x^{wb}	wet-bulb temperature of air	K
ϱ^{sat}	specific humidity of saturated air	$\frac{\mathrm{g_{wv}}}{\mathrm{kg_{da}}}$
Cost mappings		
$p^{\mathrm{fan},*}$	external (* = ext), internal (* = int), server (* = srv) fan power absorption	kW
p^{hum}	humidifier power absorption	kW
p^{leak}	CPU leakage power dissipation	kW
J	total cooling power consumption	kW

†: A tuple of a mass rate and a thermo-hygrometric (TH) state has three entries with units kg/s, K, and $\mathrm{g_{wv}}/\mathrm{kg_{da}}$.

⋆: Each server accounts for four predominant thermal inertiae. Therefore, $x^{\mathrm{srv}} \in \mathbb{R}^{4n}$ and each entry has unit $[K]$.

denote the mass rate and thermo-hygrometric (TH) state (temperature and specific humidity) of the airflow indicated by ⋆. The symbols and units in use throughout this manuscript and our implementation are summarized in Table 1[1].

The overall thermal and flow model is written in terms of input/output relations at the humidifier, the heat exchanger, and the servers (see Figures 1 and 2):

$$\begin{cases} \chi^{\mathrm{c,i}} = f^{\mathrm{hum}}(\chi^{\mathrm{e}}, \dot{w}) & \text{(humidifier)} \\ \chi^{\mathrm{h,o}} = f^{\mathrm{hx}}(\chi^{\mathrm{c,i}}, \chi^{\mathrm{h,i}}) & \text{(heat exchanger)} \\ \dot{x}^{\mathrm{srv}} = f^{\mathrm{srv}}(x^{\mathrm{cpu}}, \chi^{\mathrm{h,o}}, W) & \text{(servers' dynamics)} \\ \chi^{\mathrm{h,i}} = f^{\mathrm{rack}}(\chi^{\mathrm{h,o}}, x^{\mathrm{srv}}) & \text{(rack arrays)} \end{cases} \qquad (2)$$

where f^{hum}, f^{hx}, f^{rack} are static mappings, and f^{srv} is the joint temperature dynamics of the servers. We stress that (2) is justified on the basis of time-scale separation between the process and room-side dynamics. In what follows, the

1. For convenience, we will also make use of more familiar units in the text (such as Celsius degrees [°C] and relative humidity (RH) [%])

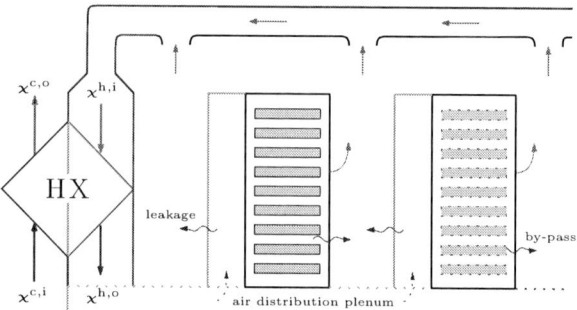

Figure 2: The cooling resources are supplied to an array of 12 server racks using a contained cold-aisle. The supply airflow ($\chi^{h,o}$) returns to the hot side of the HX ($\chi^{h,i}$) by either leaking from the containment or by crossing the server enclosures (absorbing the heat loads).

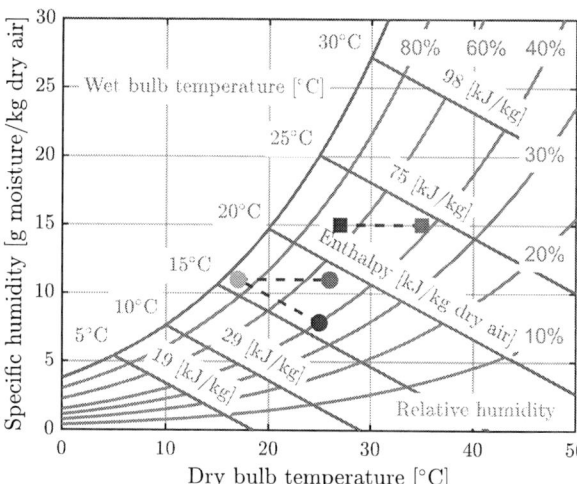

Figure 3: Psychrometric chart showing typical temperature and humidity transformations of the external and internal airflows. The external air at 25°C and 40% RH (blue circle) is adiabatically cooled to the TH state 17°C and 90% RH (green circle). Heat transfer across the heat exchanger increases this temperature to the exhaust value of 26°C (red circle). At the same time, the internal air temperature drops across the heat exchanger from 35°C (red square) to 27°C (blue square).

manipulable variables are associated to the airflow rates[2] $\dot{m}^e, \dot{m}^{h,o}$ (moved by the process and room fans) and the water rate $\dot{w}$ (at the humidifier), all of which can be modulated continuously (see Figure 1).

3.1. Air distribution model

Within the scope of this work, we address a simplified air distribution model based on heat and mass conservation, schematized in Figure 2. In particular, the model includes 12 server racks, mimicking the deployment of 4 Open Compute Triplet cabinets [18]. Each rack shelves 42 identical Windmill V2 servers, totaling 84 Central Processing Units (CPUs) per rack, and $n = 504$ deployed servers. We assume that the whole supply of internal air, drawn from the hot-side of the HX, is directed to the rack array. From here, the cooling air may *leak* directly into the room (without absorbing heat from the electronic equipment) by either escaping through the cold-aisle containment or the shelving racks [19], [20]. We capture the aggregate flow leakages by the ratio of air that does not contribute to cooling:

$$\chi^{loss} \doteq \left(\lambda \dot{m}^{h,o}, x^{h,o}, \varrho^{h,o} \right), \tag{3}$$

where $\lambda = \lambda^{leak} + \lambda^{bypass} > 0$ is a small positive number. The airflow χ^{loss} is then recirculated through the hot-side of the HX, with the effect of decreasing the flow distribution efficiency on the room-side. By mass conservation we have $\dot{m}^{h,i} = \dot{m}^{h,o}$, and by assuming negligible condensation and humidification phenomena we moreover obtain $\varrho^{h,i} = \varrho^{h,o}$. Denote with $\chi_j^{srv,i}, \chi_j^{srv,o}$ the mass rate and TH state at the inlet and outlet of the generic j-th server. The main air supply is shared among the n servers:

$$x_j^{srv,i} = x^{h,o}, \quad \varrho_j^{srv,i} = \varrho^{h,o}, \quad j = 1, \ldots, n. \tag{4}$$

Moreover, taking the supply losses into account, we get

$$\sum_{j=1}^{n} \dot{m}_j^{srv,i} = \dot{m}^{h,o} - \dot{m}^{loss} = (1 - \lambda)\dot{m}^{h,o}, \tag{5}$$

2. Notice that $\dot{m}^e$ quantifies the mass rate of *dry* air drawn from the outdoor environment and is equal to the mass rate of dry air $\dot{m}^{c,o}$.

where the individual cooling flow *requirements*, $\dot{m}_j^{srv,i}$, are set by local controllers at the servers (see Section 3.4). Finally, the room's return temperature is evaluated through the following heat balance

$$x^{h,i} = \frac{c_p^{h,o} \lambda \dot{m}^{h,o} x^{h,o} + \sum_{j=1}^{n} \left(c_p^{srv,o} \dot{m}^{srv,o} x^{srv,o} \right)}{c_p^{h,i} \left(\dot{m}^{h,o} + \sum_{j=1}^{n} \dot{m}^{srv,o} \right)}, \tag{6}$$

where the $c_p^{h,i}$ denotes the specific heat of air at temperature $x^{h,i}$.

Due to space constraints, we do not address a room temperature model with energy storage as in, for instance, [14]. Instead, we focus on the novelty of treating the contained cold-aisle scenario.

3.2. Model of the adiabatic humidifier

Spray nozzles within the humidifier pulverize water at a volumetric rate $\dot{w}$ into the airflow drawn from the outdoor environment, χ^e. This process involves negligible variations in enthalpy and wet bulb temperature. In particular, the outflow temperature $x^{c,i}$ is transferred from x^e toward the wet-bulb temperature $x^{wb}(x^e, \varrho^e)$ along the isenthalpic lines of Figure 3. Like other mass, and heat transfer phenomena, adiabatic humidification can be characterized in terms of ε-Number of Transfer Units (NTU) effectiveness [21]. A wetting and thermal effectiveness can be defined by analogy

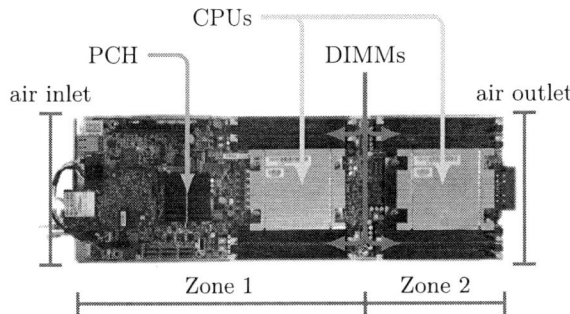

Figure 4: Top view of a Windmill V2 server platform. The cooling airflow moves from the inlet (on the left) to the outlet (on the right) while crossing the two CPUs and the corresponding heat sinks in series.

with a classical heat exchanger operating with a fictitious moist fluid

$$\varepsilon^{\text{hum}} = \frac{x^{c,i} - x^e}{x^{wb}(x^e, \varrho^e) - x^e} = \frac{\varrho^{c,i} - \varrho^e}{\varrho^{\text{sat}}(x^e, \varrho^e) - \varrho^e}, \quad (7)$$

where $\varrho^{\text{sat}}(x^e, \varrho^e)$ is the specific humidity of air at saturated conditions. The value of ε^{hum} follows from the NTU, which is itself a nonlinear correlation of $\chi^e, \dot{w}$ and depends on the physical parameters of the system (see [21] for details). f^{hum} in (2) corresponds to evaluating the NTU, the effectiveness, and then solving (7) algebraically for the outflow properties.

3.3. Model of the heat exchanger

The model for the rate of heat recovery at the HX is based on correlations from [22] and involves the Reynolds and Prandtl numbers of the process and room-side airflows, together with several parameters that require calibration. For the sake of a simpler implementation, the heat rates are evaluated by applying the P-NTU formalism for counterflow heat exchangers. Formally, this involves evaluating a single analytic expression with a structure independent of the heat capacities on each side of the HX, and in which two non-dimensional thermal effectiveness terms account for the inflow conditions at the cold and hot side [23]. The NTU correlations have been estimated for a relevant range of operation using data from an extensive experimental campaign targeting a commercial IAAH. See [14] for details.

3.4. Dynamical model of a single server unit

We adopt an accurate temperature modeling of Windmill V2 servers (see Figure 4) that has been proposed and experimentally validated in [13]. The modeling is built on a network oriented formalism in which a graph of interacting dynamical nodes captures the storage of energy at the self-heating components, the rates of heat and mass exchanges, and the airflow temperatures. Let $\mathcal{N}$ denote the set of network nodes for the generic j-th server. Then each node $\nu \in \mathcal{N}$ is endowed with an *inflow air temperature* x^i_ν, a

temperature state x^p_ν, and an *outflow air temperature* x^o_ν. The thermal energy balance at node ν is moreover affected by the local rate of heat *conduction* q^{cd}_ν, heat *convection* q^{cv}_ν, and heat generation p_ν. A family of low-order polynomial mappings $\varphi_{h\nu}(\dot{m}^{\text{srv,i}}_j) : \mathbb{R}_{\geq 0} \to \mathbb{R}_{\geq 0}, h, \nu \in \mathcal{N}$, is introduced to approximate the *effective* mass rate of cooling air that circulates from node h to node ν. The generic temperature model at node ν relates the above quantities through a set of nonlinear static and dynamic constraints:

$$x^i_\nu(t) = \frac{\sum_{h \in \mathcal{N}} c^o_h \varphi_{h\nu}(\dot{m}^{\text{srv,i}}_j) x^o_h(t)}{c^i_j \sum_{h \in \mathcal{N}} \varphi_{h\nu}(\dot{m}^{\text{srv,i}})}$$

$$d_\nu \dot{x}^p_\nu(t) = q^{\text{cd}}_\nu(t) + q^{\text{cv}}_\nu(t) + p_\nu(t) \quad (8)$$

$$x^o_\nu(t) = \frac{c^i_{p,\nu}}{c^o_{p,\nu}} x^i_\nu(t) - \frac{q^{\text{cv}}_\nu(t)}{c^o_{p,\nu} \sum_{h \in \mathcal{N}} \varphi_{h\nu}(\dot{m}^{\text{srv,i}}_\nu)}$$

where d_ν is the node's heat capacity and $c^i_{p,\nu}$ ($c^o_{p,\nu}$) is the specific heat at constant pressure of the inflow (outflow) air entering (exiting) the ν-th node. The rates of heat transfer due to conduction and convection (q^{cd}_ν and q^{cv}_ν) are formally captured using Fourier's and Newton's laws, accounting for any nonlinearities (such as flow rate dependencies) in the rate coefficients. The heat generation term is formally approximated through an affine mapping of the chip's *workload* (or *utilization*) and temperature as in [11]

$$p_\nu(w_\nu(t), x^p_\nu(t)) = p^{\text{idle}}_\nu + w_\nu(t) p^{\text{dyn}}_\nu + p^{\text{leak}}_\nu(x^p_\nu(t)). \quad (9)$$

In (9), p^{idle} and p^{dyn} are positive characteristic constants of the chip and the server's physical design. The variable $w_\nu \in [0, 1]$ models the computational *workload* (or chip *utilization*) and is defined as the average ratio of the chip's active cycles over the totality of available computing cycles during one second. The leakage dissipation p^{leak}_ν captures the temperature-dependent effect

$$x \mapsto p^{\text{leak}}_\nu(x) = \alpha_\nu \frac{x - \overline{x}^{\text{leak}}}{\overline{x}^{\text{leak}} - \underline{x}^{\text{leak}}}, \quad (10)$$

where $\underline{x}^{\text{leak}} = 303.15\,\text{K}$ and $\overline{x}^{\text{leak}} = 350.15\,\text{K}$ are arbitrary interpolation abscissas and $\alpha_\nu > 0$ is a characteristic constant of the chip[3].

To model the Windmill V2 platform we specialize (8) considering a network with 6 nodes (recall Figure 4): $\{1\}$ the inlet, $\{2, 3\}$ the CPU and heat exchanger in zone 1, $\{4, 5\}$ the CPU and heat exchanger in zone 2, and $\{6\}$ the outlet. The model accounts thus for the four predominant thermal inertiae (nodes 2 to 5). We refer to [13] for the details on the full dynamical model and the adopted data-driven calibration methodology. We stress that the estimated model predicts the temperature of both CPUs and the exhaust air with a root-mean-square error of 0.97 degrees under varying workload conditions, across our validation data sets.

A global temperature state for the servers $\boldsymbol{x}^{\text{srv}} \in \mathbb{R}^{4n}$ is defined by orderly stacking all the state variables x^p_ν, $\nu = 2, 3, 4, 5$, relative to the n server models. Similarly a global

---

3. More in detail, the leakage term (9) measures the variations in leakage power dissipation relative to operating conditions at 30°C.

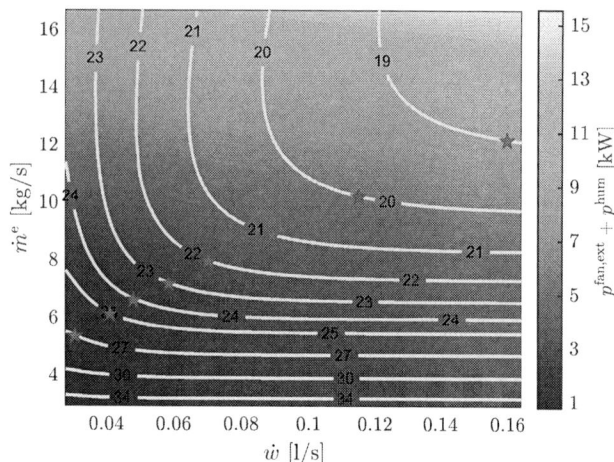

Figure 5: Process-side power consumption of the IAAH. The iso-temperature curves correspond to the supply temperature $x^{\mathrm{h,o}}$: the same heat recovery rate is achieved by different set-points and at different operation costs (the optimal settings are indicated with stars). The simulation was generated with $x^{\mathrm{e}} = 25\,^{\circ}\mathrm{C}$, $\mathrm{RH}^{\mathrm{e}} = 40\,\%$, $\dot{m}^{\mathrm{h,o}} = 13.25\,[\mathrm{kg/s}]$, and a room-side heat load of $75\,\mathrm{kW}$.

workload matrix $W \in [0,1]^{2n}$ is obtained by aggregating the load of each pair of CPUs at each server. The global dynamics (2) are then propagated given W and the air inflow properties and flow rate requirements $\chi_j^{\mathrm{srv,i}}, j = 1, \ldots, n$. As a by-product, we obtain the heat convection rate into the cooling airflow and thus the outflow variables $\chi^{\mathrm{srv,o}}, j = 1, \ldots, n$.

4. Model-based economic cooling control

We devise a 2-level control system. At the higher level, the plant supervisor uses knowledge of the model to optimize the rates at which air and water are provisioned while, at the same time, satisfying the cooling demand. At the lower level, model-free regulators are tasked to operate the system at the optimal set-points. The supervisory control loop operates as follows. The workload and flow requirements are continuously monitored using standard out-of-band management interfaces (such as Intelligent Platform Management Interface (IPMI)). At each supervision period (for example, 1 minute), the optimal set-points are computed by solving a nonlinear control problem exploiting knowledge of the model and of the monitored variables, in a one-step look-ahead fashion.

4.1. Operation cost model

In what follows, the minimization objective $J(\cdot)$ is the economic cost of operating the *complete* cooling setup. That is, the total power consumption spent by the IAAH and any

power dissipation terms induced by the cooling policy[4]

$$J\left(\dot{m}^{\mathrm{e}}, \dot{w}, \dot{m}^{\mathrm{h,o}}\right) \doteq \underbrace{p^{\mathrm{fan,ext}}(\dot{m}^{\mathrm{e}}) + p^{\mathrm{hum}}(\dot{w})}_{\text{process side}}$$
$$+ \underbrace{p^{\mathrm{fan,int}}(\dot{m}^{\mathrm{h,o}}) + p^{\mathrm{racks}}(\dot{m}^{\mathrm{e}}, \dot{w}, \dot{m}^{\mathrm{h,o}})}_{\text{room side}} \quad (11)$$

where the cooling cost at the racks accounts for the power absorption of the local server fans and that due leakage at the CPUs:

$$p^{\mathrm{racks}}(\dot{m}^{\mathrm{e}}, \dot{w}, \dot{m}^{\mathrm{h,o}}) = \sum_{j=1}^{n} p_j^{\mathrm{leak}}(\dot{m}^{\mathrm{e}}, \dot{w}, \dot{m}^{\mathrm{h,o}})$$
$$+ \sum_{j=1}^{n} p_j^{\mathrm{fan,srv}}(\dot{m}^{\mathrm{e}}, \dot{w}, \dot{m}^{\mathrm{h,o}}). \quad (12)$$

The external and internal fan profiles (including the flow rate and power consumption) have been estimated from real data by regressing the polynomial model proposed in [24]. The server fan profiles have been estimated using lower order polynomial regressors. In both cases, the cubic terms dominate the power curves, demonstrating agreement with the fan affinity laws. Notice that the fan power curves are convex, monotonic increasing functions of the desired flow rates. Moreover, the power consumption of the humidifier fits an affine model of the rate $\dot{w}$. While the exact model parameters are an IP of the industrial partner, we demonstrate the cost and cooling performance of the IAAH graphically in Figure 5 for one instance of the boundary conditions.

4.2. The two-variables controller with constant room-side airflow rate

In the simpler two-variables Optimal Flow Provisioning (OFP) strategy, the internal airflow rate is maintained constant. The volumetric flow rate at the humidifier and the mass rate of external air are instead manipulated to minimize J:

$$\mathrm{OFP}_2 \doteq \begin{cases} \widehat{\boldsymbol{u}} \in \underset{\dot{m}^{\mathrm{e}}, \dot{w}}{\operatorname{argmin}}\ J(\dot{m}^{\mathrm{e}}, \dot{w}, \dot{m}^{\mathrm{h,o}}) \\ \text{given:} \\ \quad \dot{m}^{\mathrm{h,o}} \quad \text{(nominal room rate)} \\ \text{subject to:} \\ \quad \Gamma \quad \text{(exogenous inputs)} \\ \quad \Psi \quad \text{(constraints)} \end{cases} \quad (13)$$

In (13), the exogenous inputs include the compute workload and the outdoor air properties over the supervision interval:

$$\Gamma \doteq \left\{W, x^{\mathrm{e}}, \varrho^{\mathrm{e}}\right\}. \quad (14)$$

---

4. To stress the role of J as a control objective, we write the cost terms as functions of the optimization variables.

The shared set of constraints Ψ is instead formally defined as

$$\Psi \doteq \begin{cases} \text{(I/O constraints)} \\ \underline{w} \le \dot{w} \le \overline{w} \\ \underline{m}^{\mathrm{e}} \le \dot{m}^{\mathrm{e}} \le \overline{m}^{\mathrm{e}} \\ \boldsymbol{x}^{\mathrm{srv}} \le \overline{x}^{\mathrm{cpu}} \\ \text{(static and dynamic models)} \\ \text{see (2) and Section 3.1} \end{cases} \quad (15)$$

Notice that on the room-side, a trade-off between the rate and the temperature of the supply air must be taken into account to allow the CPUs to reject the excess heat at package temperatures below the threshold $\overline{x}^{\mathrm{cpu}} = 80°\mathrm{C}$. Notice moreover that the flow requirements $\chi_j^{\mathrm{srv,i}}$ are set by server-local flow provisioning controllers: similarly to [13], we define $\dot{m}_j^{\mathrm{srv,i}}$ as the minimum mass rate (given Γ) that operates the CPUs of the j-th server below the critical temperature $\overline{x}^{\mathrm{cpu}}$. We also observe that the fixed rate $\dot{m}^{\mathrm{h,o}}$ in (13) must be chosen considering the worst-case heat load. When the setup is not operated at peak load, the cooling effect of the *surplus* airflow rate

$$(1-\lambda)\dot{m}^{\mathrm{h,o}} - \sum_{j=1}^{n} \dot{m}_j^{\mathrm{srv,i}} > 0, \quad (16)$$

must be taken into account. Here we model this condition by letting the surplus rate be split uniformly across the n server enclosures, ensuring that the mass balance (5) obtains.

4.3. A holistic three-variables provisioning controller

In the more flexible three-variables control strategy the aim is as before. However, now, also the mass rate of the room air is addressed as a manipulable variable by the optimizer

$$\mathrm{OFP}_3 \doteq \begin{cases} \widehat{\boldsymbol{u}} \in \underset{\dot{m}^{\mathrm{e}}, \dot{w}, \dot{m}^{\mathrm{h,o}}}{\mathrm{argmin}} \ J(\dot{m}^{\mathrm{e}}, \dot{w}, \dot{m}^{\mathrm{h,o}}) \\ \text{subject to:} \\ \underline{m}^{\mathrm{h,o}} \le \dot{m}^{\mathrm{h,o}} \le \overline{m}^{\mathrm{h,o}} \\ \Psi, \Gamma \end{cases} \quad (17)$$

Section 5 demonstrates substantial differences in the control policy when the system is supervised by OFP_3.

4.4. Discussion

We stress that both sets of static and dynamic constraints in (13) and (17) involve significant and non-removable nonlinearities. The latter follow from the underlying semi-empirical correlations and the material properties. We stress moreover that the cost terms in (11) relative to the actuators are in fact convex; however, this property is lost once the leakage dissipation is taken into account. As for this latter term, while leakage is often neglected as part of the cooling cost, it constitutes a significant temperature-dependent contribution as shown later in Section 5. Finally, within our present scope of evaluating the benefits of more holistic

Setting	Value	Unit
n	504 (1008 CPUs)	adim
λ	0.1	adim.
x^{e}	$[20, 30]$	°C
RH^{e}	$[50, 90]$	%
$\varrho^{\mathrm{h,o}}$	$8 \cdot 10^{-3}$	$\frac{\mathrm{g_{wv}}}{\mathrm{kg_{da}}}$
$\dot{w}$	$[0, 0.05]$	l/s
$\dot{m}^{\mathrm{e}}$	$[3, 17]$	m³/s
$\dot{m}^{\mathrm{h,o}}$	$[3, 17]$ $\star$	m³/s
W	{*Low, Medium, High*}	adim

$\star$: The room-side rate is fixed at $12\,[\mathrm{kg/s}]$ in the case OFP_2.

Table 2: Parameters used in the experiments.

control approaches, we assume that the exogenous inputs Γ are measured at the start of the supervision period and remain constant over its length. It is then worth stressing that the proposed methodology does not negotiate the workload, as in, for instance, [25]. Instead, the cooling is optimized for the desired heat load.

5. Numerical experiments

This section reports the performance of controllers OFP_2 and OFP_3 across different temperature and humidity conditions of the external air and computational workloads. Since the literature lacks of standardized data sets for testing efficiency in this kind of scenarios, we propose to evaluate these control strategies over three different workload conditions, *Low*, *Medium*, and *High* corresponding to choosing CPU utilizations $W_k, k = 1, \ldots, 2n$, of 0.25, 0.5, and 0.75 respectively. This methodology is justified by the results in [26], where it is shown that the average heat load is the most affecting variable in terms of the server's flow rate requirements, while the transient behavior of the utilization traces has a smaller effect. For the following analysis, we moreover adopt the IAAH's and server model calibrations obtained in [13], [14], [27]. To solve the nonlinear constrained problems OFP_2 and OFP_3 we use Particle Swarm Optimization (PSO), a stochastic strategy in which a population of candidate solutions moves across the search space to grant global and local search capabilities [28]. The complete experimental summary is available in [29].

5.1. Control performance of OFP_2 and OFP_3

The optimal controls of OFP_2 corresponding to the *High* workload scenario are shown in Figure 6. As intuition suggests, the effort at the process-side fan increases with the temperature of the external air. However, in order to attain the CPU temperature constraint in (15), the humidifier is tasked to sustain high volumetric rates also for high RH^e, despite the less favorable wetting effectiveness. Allowing flexible settings of the internal air rate induces significant differences in the control decision. See, for instance the optimal controls of OFP_3 in Figure 7: as the outdoor humidity increases, the optimizer now prefers to decrease

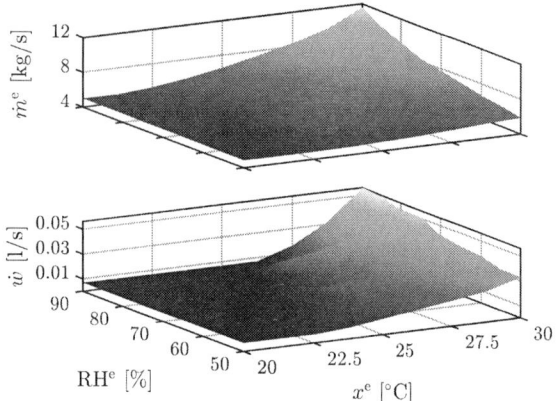

Figure 6: Optimal controls computed by the process-side optimizer OFP$_2$ in (13) under the *High* workload scenario. The internal room supply rate is maintained constant at the nominal value 12kg/s, suggested by the industrial partner. Dark blue and light yellow indicate, respectively, proximity to the lower and upper control bounds.

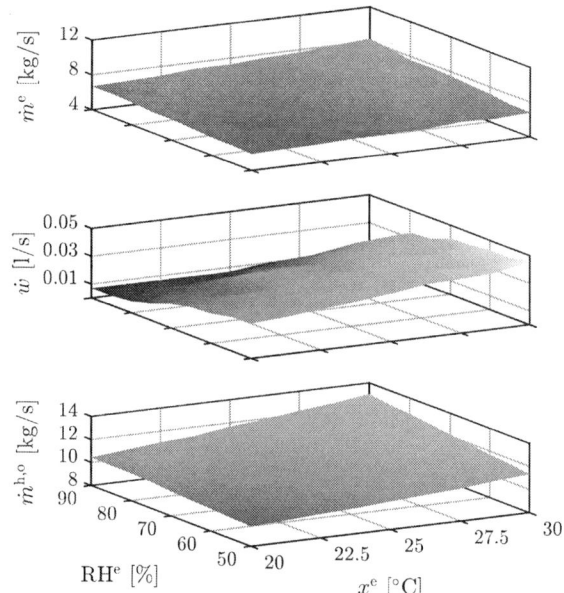

Figure 7: Optimal controls computed by the global optimizer OFP$_3$ in (17) under the *High* workload scenario (same as in Figure 6).

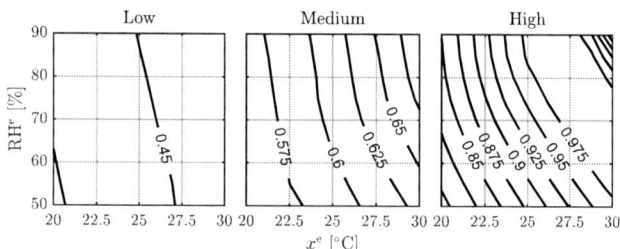

Figure 8: Contour plot of index (18) under the three workload scenarios.

the rate of pulverized water at the humidifier. The corresponding actuation effort is greater in dry air conditions, and smaller as the benefit of adiabatic cooling decreases. More in general, we note that OFP$_3$ operates within reduced ranges for the process-side fan and the humidifier rates while simultaneously being able to exploit these smaller ranges to attain better efficiency. Figure 8 summarizes the energetic benefit of allowing manipulable room-side rates as captured through the relative efficiency index

$$\eta \doteq \frac{\widehat{J}_3(x^e, \mathrm{RH}^e, W)}{\widehat{J}_2(x^e, \mathrm{RH}^e, W)}, \qquad (18)$$

namely, the ratio of the optimal cooling cost attained by OFP$_3$ over the optimal cost of OFP$_2$. In the modeled setting, the three-variables strategy reduces the overall cooling power up to 20% under the High workload scenario and up to 50% under Low workload. This margin remains significant across the experimental plan and reduces in proximity of the harshest heat load, x^e and RH^e conditions, where the constant room supply rate used by OFP$_2$ approaches the optimal value.

5.2. The role of leakage as a cost term

The contribution of the additive leakage dissipation accounts for more than 20% of the overall cooling cost. Depending on the operation point, this cost exceeds the consumption of the humidifying and fan actuation costs (cf. Table 2 in [29]). Nevertheless, the experimental analysis suggests that the optimal cooling controls lead the CPU temperatures near the upper temperature constraint of 80°C, maximizing the leakage dissipation. Indeed, in the considered setup, the pressure loss across the hot side of the HX substantially increases the room-side flow provisioning cost, marginalizing the benefits of accounting for leakage in (11).

5.3. Implications on the design of efficient IAAH cooling setups

This parametric study suggests a significant benefit in adopting flexible internal air rates. Inspection of Figures 6–7 and Tables 1–2 in the suppletive document [29], reveals that the control effort applied to the room side by OFP$_2$ can be spent more efficiently on the process-side across the whole set of experimental scenarios. Moreover, when the IAAH is operated by OFP$_3$, the optimal input ranges on the process side are both shifted to lower amplitudes and their span is reduced compared to OFP$_2$. Downsizing of the actuators on the process side could open to design and running-cost savings by reducing the acquisition costs and by improving the operation efficiency across the smaller ranges of mass and volumetric rates. Finally, within the experimental conditions, we observed that the optimal share of the power budget that is allocated to the room-side fan belongs in the range from 50% to 75%. This suggests that model-free regulators of IAAH implementations, operating in close conditions, could benefit from tuning their control policies to strive for the optimal power allocation.

6. Conclusions

We investigated the economic operation of a *free-cooling* system that pairs an IAAH with an array of server racks installed in a contained aisle. In this setting, we evaluated a holistic control strategy that supervises both the process and room sides of the cooling unit, and we compared its cost performance against that of a simpler but less flexible solution in which only the process side is optimized. An in silico parametric study revealed important insights on the optimal power allocation among the IAAH's actuators on the process and room sides, highlighting the importance of manipulable internal airflow rates. Moreover, the economic significance of the holistic control strategy was demonstrated, reinforcing the role of model-based analysis as a means to inform technology development directions. For instance, knowing the actuation ranges of the optimal controls can inform better sizing decisions, decreasing the acquisition costs while increasing the average operation efficiency. Future directions include the study of *direct* free-cooling scenarios in which the lack of a heat exchanger drastically reduces the pressure loss across the room-side circuit, potentially inducing a different trade off between the room-side control effort and the total CPU leakage dissipation.

References

[1] K. Kant, "Data center evolution: A tutorial on state of the art, issues, and challenges," *Computer Networks*, 2009.

[2] V. Avelar, D. Azevedo, and A. French, "PUE: A Comprehensive Examination of the Metric," 2012.

[3] Uptime Institute, "Data Center Industry Survey," Tech. Rep., 2014.

[4] M. Avgerinou, P. Bertoldi, and L. Castellazzi, "Trends in Data Centre Energy Consumption under the European Code of Conduct for Data Centre Energy Efficiency," *Energies*, 2017.

[5] C. Patel, C. E. Bash, R. Sharma, M. Beitelmal, and R. Friedrich, "Smart Cooling of Data Centers," in *InterPACK*, 2003.

[6] M. K. Patterson, "The effect of data center temperature on energy efficiency," *IEEE Conference on Thermal and Thermomechanical Phenomena in Electronic Systems*, 2008.

[7] Y. Fulpagare and A. Bhargav, "Advances in data center thermal management," *Renewable and Sustainable Energy Reviews*, 2015.

[8] K. Skadron, M. R. Stan, K. Sankaranarayanan, W. Huang, S. Velusamy, and D. Tarjan, "Temperature-Aware Microarchitecture: Modeling and Implementation," *ACM Transactions on Architecture and Code Optimization*, 2004.

[9] L. Parolini, B. Sinopoli, B. H. Krogh, and Z. K. Wang, "A cyber-physical systems approach to data center modeling and control for energy efficiency," *Proceedings of the IEEE*, 2012.

[10] Q. Tang, S. K. S. Gupta, and G. Varsamopoulos, "Thermal-Aware Task Scheduling for Data Centers Through Minimizing Heat Recirculation," in *IEEE Conference on Cluster Computing*, 2007.

[11] Z. Wang, C. Bash, N. Tolia, M. Marwah, X. Zhu, and P. Ranganathan, "Optimal fan speed control for thermal management of servers," in *ASME InterPACK*, 2009.

[12] ASHRAE Technical Committee 9.9, "Thermal Guidelines for Data Processing Environments," *American Society of Heating, Refrigerating, and Air-Conditioning Engineers Inc*, 2011.

[13] R. Lucchese and A. Johansson, "On Energy Efficient Flow Provisioning in Air-Cooled Data Servers," *Control Engineering Practice*, 2019.

[14] A. Beghi, G. Dalla Manna, M. Lionello, M. Rampazzo, and E. Sisti, "Energy-efficient operation of an indirect adiabatic cooling system for data centers," in *IEEE American Control Conference*, 2017.

[15] M. Dayarathna, Y. Wen, and R. Fan, "Data center energy consumption modeling: A survey," *IEEE Communications Surveys & Tutorials*, 2016.

[16] M. Ogawa, H. Fukuda, H. Kodama, H. Endo, T. Sugimoto, T. Kasajima, and M. Kondo, "Development of a cooling control system for data centers utilizing indirect fresh air based on Model Predictive Control," in *IEEE Congress on Ultra Modern Telecommunications and Control Systems*, 2015.

[17] R. Lucchese, M. Lionello, M. Rampazzo, M. Guay, and K. Atta, "Newton-like Phasor ESC," in *IFAC NOLCOS*, 2019.

[18] S. Furuta, "Open Compute Project Server Chassis and Triplet," Tech. Rep., 2011.

[19] M. Tatchell-Evans, N. Kapur, J. Summers, H. Thompson, and D. Oldham, "An experimental and theoretical investigation of the extent of bypass air within data centres employing aisle containment, and its impact on power consumption," *Applied Energy*, 2017.

[20] X. Zhang, C. M. Healey, J. W. VanGilder, and Z. R. Sheffer, "Compact modeling of data center air containment systems," 2013.

[21] N. Fonseca and C. Cuevas, "Experimental and theoretical study of adiabatic humidification in HVAC&R applications," *Ingeniare*, 2010.

[22] G. Polley and M. Abu-khader, "Interpreting and Applying Experimental Data for Plate-Fin Surfaces : Problems with Power Law Correlation," *Heat Transfer Engineering*, 2006.

[23] R. Shah and D. Sekulic, *Fundamentals of Heat Exchanger Design*, 2003.

[24] F. Wang, H. Yoshida, and M. Miyata, "Total Energy Consumption Model of Fan Subsystem Suitable for Continuous Commissioning," *ASHRAE Transactions*, 2004.

[25] L. Parolini, E. Garone, B. Sinopoli, and B. Krogh, "A hierarchical approach to energy management in data centers," in *IEEE Conference on Decision and Control*, 2010.

[26] R. Lucchese and A. Johansson, "On server cooling policies for heat recovery: exhaust air properties of an Open Compute Windmill V2 platform," in *IEEE CCTA*, 2019.

[27] M. Rampazzo, M. Lionello, A. Beghi, E. Sisti, and L. Cecchinato, "A static moving boundary modelling approach for simulation of indirect evaporative free cooling systems," *Applied Energy*, 2019.

[28] J. Kennedy, "Particle Swarm Optimization," 1995.

[29] R. Lucchese, M. Lionello, M. Rampazzo, A. Johansson, and W. Birk, "On economic cooling of contained server racks using an indirect adiabatic air handler: Results from numerical experiments." [Online]. Available: https://staff.www.ltu.se/~ricluc/Documents/lucchese-stsyn20-addendum.pdf

An Experimental Apparatus for Two-phase Cooling of High Heat Flux Application using an Impinging Cold Plate and Dielectric Coolant

Cong Hiep Hoang, Sadegh Khalili, Bharath Ramakrisnan, Srikanth Rangarajan, Yaser Hadad, Vahideh Radmard, Kamal Sikka, Scott Schiffres, Bahgat Sammakia.
Department of Mechanical Engineering, Binghamton University-SUNY
Binghamton, New York, US
hhoang2@binghamton.edu

Abstract

Increasing heat transfer rates over smaller area and volumetric footprints are a challenge to electronics cooling. Conventional air-based cooling solutions require larger space for heatsink fins and impose challenges on chassis design. Those challenges of air cooling encourage a shift to a more efficient liquid cooling solution. The more efficient heat transfer performance in a small volume is attributed to high specific heat and latent heat associated with a liquid coolant. In this research, a bench level two-phase experimental setup was built to study the heat transfer and hydraulic performance of a jet impinging cold plate using a dielectric coolant HFE 7000. A copper block heater arrangement was used to mimic a computer chip with a footprint of $1" \times 1"$ (6.45 cm^2). Heat transfer performance and pressure drop were compared in single- and two-phase operation at three volumetric flow rates 1.0, 1.25, and 1.75 LPM corresponding to mass fluxes 2005, 2506, and 3509 kg/m^{2}s respectively. The tests were carried out for heat fluxes ranging from 5 to 70 W/cm^2. It is found that thermal resistance remains unchanged in the single-phase regime at lower heat fluxes but then decreases significantly when boiling occurs at high heat fluxes. At lower heat fluxes and high heat fluxes, thermal resistance decreases with flow rate. However, at moderate heat fluxes (27-58 w/cm^2), the thermal resistance is independent on flow rate. This behavior is attributed to the dominance of nucleate boiling at moderate heat fluxes. Therefore, thermal performance at moderate heat fluxes depends on cold plate design (nucleate site density) rather than flow rate. Pressure drop decreases slightly (less than 7 %) in single phase regime due to the reduction of viscosity with temperature. When two-phase boiling occurs, pressure drop is observed to rise considerably (over 70 %). A high power of 450 w was dissipated at operable base temperature (31 to 85 $^{\circ}$C) and pressure drop (7 to 24 kpa).

Keywords

Two-phase cooling; microchannel; boiling

Nomenclature

T_{in}	Temperature of coolant entering the cold plate (K)
T_{out}	Temperature of coolant leaving the cold plate (K)
T_b	Average base temperature (K)
Q	Heat picked by the coolant (W)
ONB	Onset of nucleate boiling
SiNWs	Silicon nanowires
P_{in}	Pressure at the inlet (Pa)
P_{out}	Pressure at the outlet (Pa)
$\dot{V}$	Volumetric flow rate (m^3/s)
R_{th}	Specific Thermal Resistance (K.cm^2/w)

1. Introduction

The advancements in design of faster and smaller components in the electronics industry have called for high heat transfer rates over smaller area and volumetric regions. What will likely be required for the next generation of high-density electronic devices in data center is the re-emergence of thermal management using the liquid cooling method [1]. In addition to solutions to high heat flux and smaller scale, significant savings in data center operation can be obtained with liquid cooling. By adopting hybrid cooling where the liquid cooling and air cooling are combined to address the rack level thermal problems, chiller energy savings can be improved by 40 % [2]. There are two ways in which liquid cooling is implemented in cold plates: single phase liquid cooling systems and two-phase cooling systems. Performance of two-phase flow boiling is significantly greater than that of its single-phase in both thermal performance and pumping power [3]. Two-phase cooling systems utilize the latent heat, resulting in larger cooling capacities for two-phase cold plates and more uniform cold plate temperatures.

Jet-impingement and parallel flow in microchannel are two of the most common flow configurations for two-phase cooling in electronic applications. Recently, many researches have focused on a hybrid module that combines these two flow configurations. This combined flow configuration in two-phase cooling can dissipate very high heat fluxes and results in lower pressure drop compared to parallel flow micro-channel heat sinks. Drummond et al. [4] built a setup with a hierarchical impinging manifold microchannel heat sink array for high-heat-flux two-phase cooling of electronics. In order to eliminate thermal resistance due to the contact and conduction resistance, the microchannels were etched directly on the surface of heated substrate. The 5 mm x 5 mm silicon chip area was discretized into a 3x3 array of microchannel heat sinks, which were cooled by hierarchical manifold distributor using the engineered dielectric liquid HFE-7100 as the working fluid. The heat sink with 15μm × 300μm channels was shown to dissipate maximum base heat fluxes up to 910 W/cm^2 with an acceptable pressure drop (below 162 kpa) and chip temperature (lower than 125 $^{\circ}$C). Myung et al. [5] studied single-phase and two-phase cooling performance of a hybrid micro-channel/slot-jet impingement module using HFE-7100 as the working fluid. Experimental results of single-phase cooling were compared with numerical results and results of two-phase cooling were compared with analytical correlations. It was found that increasing micro-channel/fin height for low velocities reduces the temperature gradient along the micro-channel. However,

did not show an obvious delay in the onset of nucleation. Wang also suggested a study of a much higher mass flux to investigate the influence of flow rate on onset of nucleate boiling. This

1. Reservoir
2. Pump
3. Flow Meter
4. Inline Heater
5. Thermocouple
6. Pressure Sensor 1
7. Camera 1
8. Mock Package
9. Data Acquisition System
10. Camera 2
11. Pressure Sensor 2
12. Thermocouple 2
13. Heat Exchanger
14. Julabo LH40 Chiller

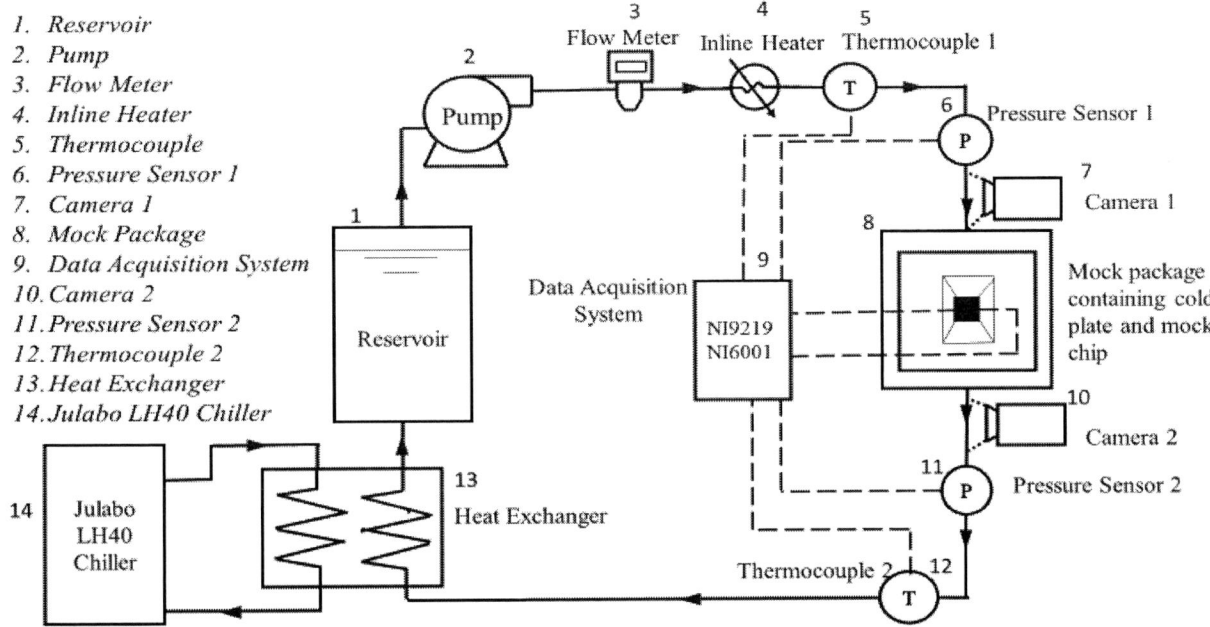

Figure 1. Schematic Diagram of the setup

this influence of fin height was less significant at high Reynolds numbers. The combination of micro-channel and slot-jet can enhance surface temperature uniformity with less than 2 °C degree of temperature gradient. The data for heat transfer coefficient was validated by a correlation of hybrid microchannel/micro-circular jet module.

Although dielectric coolants have a much lower thermal conductivity and much higher price; than water, they are electrically non-conductive and therefore preferred when working with sensitive electronics. In case of a coolant leakage, dielectric fluids can protect circuits from being shorted and from an electrical discharge. The boiling point of HFE 7000 is 34 C degree at 1 atm, which is much lower than typical maximum operating chip temperature. This low boiling point is a reason why this coolant attracts attention from researchers for two-phase electronics cooling applications. Yang et al. [6] investigated flow boiling heat transfer of HFE 7000 in nanowire-coated microchannels. They concluded that, at heat fluxes lower than 120W/cm², the heat transfer coefficient and CHF were enhanced by 344% and 14.9% respectively. The enhancement of heat transfer coefficient was due to the thin-film evaporation in the entire channels when the flow pattern was dominated by capillarity-induced annular flow. Wang et al. [7] carried out experimental studies with HFE 7000 as a working fluid in SiNWs embedded manifold microchannels. The maximum heat flux of the experiment was 141.5 W/cm² with relatively small mass fluxes of 116 kg/m²s and 235 kg/m²s. It was shown that a higher mass flow rate of HFE 7000

good performance in flow instability and high heat flux dissipation resulted in the potential for application in high powered electrical devices.

In the present article, a sophisticated experimental apparatus was designed to explore the heat transfer and hydraulic performance of an impinging microchannel cold plate using HFE 7000. The experiment with coolant HFE 7000 was carried out with a large heated surface of 6.45 cm² and maximum dissipated heat flux was around 70 w/cm². The effects of heat flux and flow rate on thermal resistance, pressure drop and coefficient of performance were studied in detail. Significant effects of flow rate on the onset of nucleate boiling (ONB), thermal resistance and pressure drop were observed at much higher mass flux than in the work of Wang et al. [7]. The dissipated heat from the heater block was removed by a unique microchannel impinging cold plate design. The variation of coefficient of performance (COP, the ratio of power input Q and pumping power P) with heat flux is presented.

2. Experimental Apparatus

The schematic diagram of the setup is shown in Fig. 1. The dielectric coolant from the reservoir was pushed through fluid loop using a centrifugal pump. Flow rate in the loop was controlled by adjusting the DC power supplied to the pump. The flow rate of the single-phase liquid coolant entering the cold plate was measured using a flow meter (Omega FTB 336D) with a range of 0.2-2 LPM. There were two thermocouples and two pressure sensors located at inlet and outlet of the cold plate for measuring coolant temperatures and pressures. The details of the instrumentation are provided in Table 1. Base temperature is the temperature at the bottom of the cold plate. Two T-type thermocouples were used to measure the temperatures at the base of the cold plate. Two-phase liquid coolant after exiting the cold plate was passed through a liquid-liquid heat exchanger to condense the vapor. An external chiller Julabo LH40 was used to control the temperature of the condenser. Finally, the condensed single-phase liquid coolant came back to the reservoir to finish fluid cycle. The data acquisition systems (DAQ) from NI-Model 9219 and NI-Model 6001 was used for collecting data from thermocouples and pressure sensors of the setup. The details of DAQs are provided in Table 1.

A LabVIEW program was developed for data collection. A picture of the mock package is shown in Fig. 2. The copper block was placed on top of an insulating ceramic block - Sheffield Pottery-Model TCHTB (0.09 W/m.K, with stable properties up to 1260 °C). The heat input of the copper block was provided using four cartridge heaters (Omega CIR-3020/120V) installed in the bottom portion of the copper block. The cold plate was placed on the top of the copper block using a thermal interface material (TIM) in between. The cold plate was fixed onto the top-surface of copper block by a holder of the frame. Weights were placed on the weight seat for maintaining pressure and constant interfacial thickness. A gap pad (model number C128 of Aavid Thermalloy) was used as the TIM between cold plate and top surface of copper block with a thermal conductivity of 12.8 W/mK. The whole mock package was wrapped around by insulation layers of ceramic cloth and fiber glass, then a polystyrene box was used to contain the mock package.

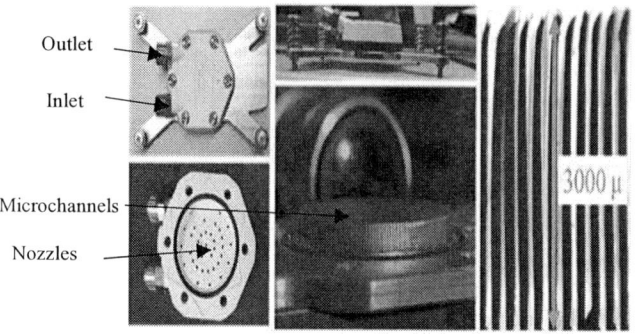

Figure 3. Cold plate with copper microchannel

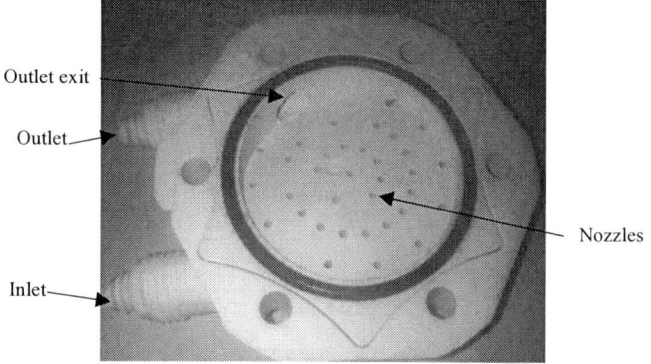

Figure 4: Cap with inlet, nozzles and outlet

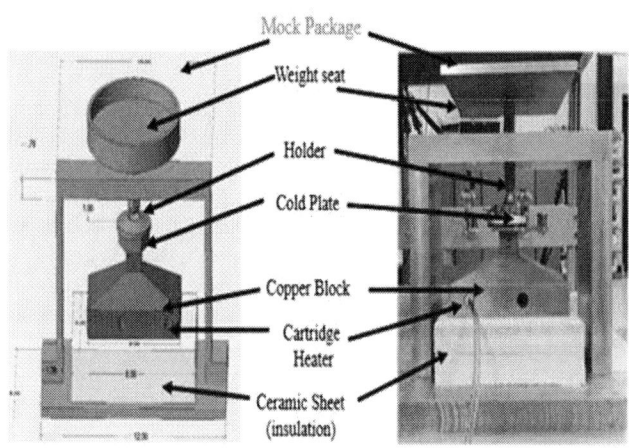

Figure 2. Mock package

The design of the microchannel cold plate can be found in Fig. 3. The coolant was pumped through inlet of the cap then spread to nozzles of the cap and impinged down to the microchannels before leaving the cold plate through outlet. Fig. 4 shows cap design with positions of inlet, nozzles and outlet exit. The fin height, fin thickness and channel width are 3000, 100, and 100μm respectively.

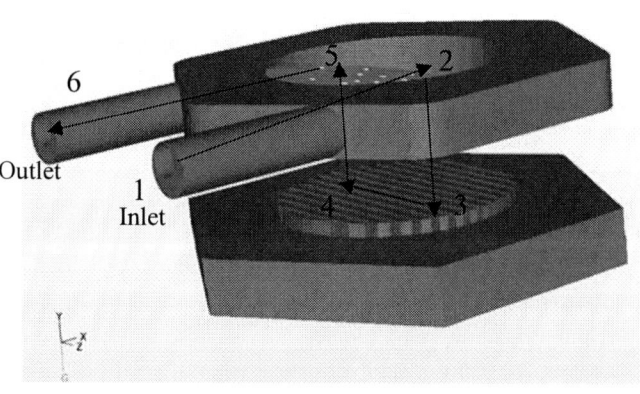

Figure 5: Flow field

Fig. 5 shows flow field of the cold plate. Fluid comes in the cold plate at inlet (path 1-2) then impinges down to channels through nozzles (path 2-3) and goes along the channel (path 3-4) before climbs up (path 4-5) and exits at the outlet (path 5-6).

3. Experimental Procedure

Before powering the heaters and collecting data, the setup was operated at a constant flow rate and a desired inlet coolant temperature for about 30 minutes to remove trapped air inside the loop. The inlet coolant temperature was controlled by adjusting temperature of the liquid in the loop of the Julabo

LH40 chiller. A constant flow rate was maintained by adjusting the supplied DC power to the pump. After 30 minutes of degassing, a ventilation valve on the reservoir was used to vent the trapped air inside the reservoir. An AC power supply was used to power the four cartridge heaters. In every step of increasing power applied to the copper block, the flow rate was adjusted in order to ensure that a constant flow rate was maintained throughout the experiments.

4. Experimental Uncertainties

The uncertainties of different sensors are listed in Table 1. The error analysis using root sum square method [8] showed the experimental uncertainty in the thermal resistance as presented in Fig. 6 for various heat fluxes. The details of uncertainty analysis can be found in Appendix A. The maximum uncertainties in the thermal resistance and pressure drop are 13 % and 12 % respectively, at the smallest heat flux of 5.4 W/cm^2. The uncertainties of the thermal resistance and pressure drop generally decrease with heat flux and reach the minimum of 1.3 % and 2.2 %, respectively at a heat flux of 70 w/cm^2. The experiment was performed in a single-phase state to determine the difference between sensible heat picked up by the coolant and the measured electrical power input. Using this method, the calculated heat loss was less than 9 % of electrical power input.

Table 1.

Instruments	Measurand	Uncertainty	DAQ Module
Pressure Gauges (Omega): PX309050A5V	P_{in}, P_{out}	±0.8 kPa	NI-USB-6009
T-type Thermocouple: Laboratory Made	T_{in}, T_{out}, T_b	± 0.2°C	NI 9219
Thermocouple (Omega): TJ36-CASS-116E-2-CC	Heat Flux Measurement	± 0.2°C	NI 9219

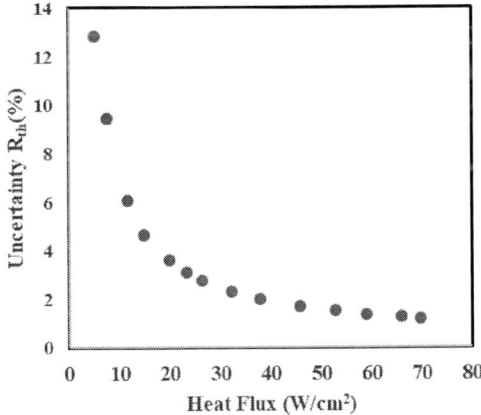

Figure 6. Experimental uncertainty of thermal resistance for various heat fluxes.

5. Results

Several tests for various heat fluxes were carried out when supplied coolant temperature was fixed at 25 °C. The boiling curve of the fluid is shown in Fig. 7 in the case of flow rate 1 LPM. The onset of nucleate boiling (ONB) is expected to occur when wall superheat reach values in the range 0-10 °C. The wall superheat was calculated by subtracting the saturation temperature from the wall temperature (i.e. cold plate's base temperature). Saturation temperature is defined from the thermodynamic table of HFE 7000 and the average of pressures measured at the inlet and outlet. It is observed in Fig. 7 that ONB appears when heat flux reaches around 23 W/cm^2 and wall superheat is very close to 8 °C. When boiling occurs, the slope of the curve increases significantly which elucidates the advantage of two-phase cooling as compared to single-phase cooling.

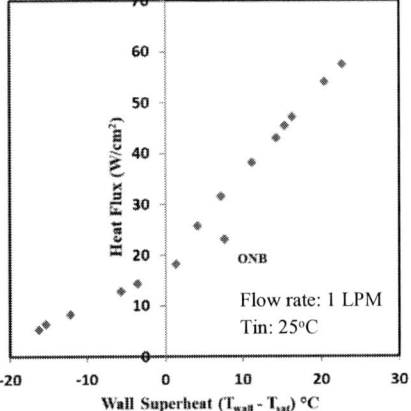

Figure 7. Boiling curve (the change of wall superheat with heat flux applied to cold plate)

The variation of pressure drop with heat flux is illustrated in Fig. 8 in the case of flow rate 1 LPM and inlet temperature 25 °C. Pressure drop first remains unchanged in the single-phase regime until a heat flux of about 23 w/cm^2; it then increases gradually when the boiling regime occurs. The reason for higher pressure drop is vapor production, which increases the frictional and acceleration pressure gradients along the microchannel.

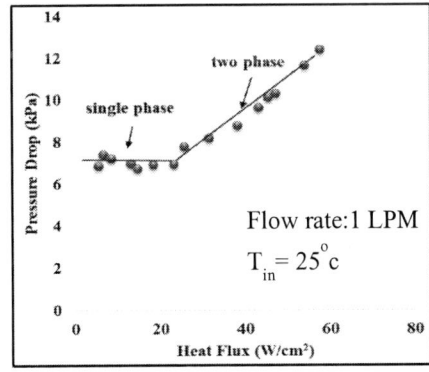

Figure 8. Pressure Drop Variation with Heat Flux

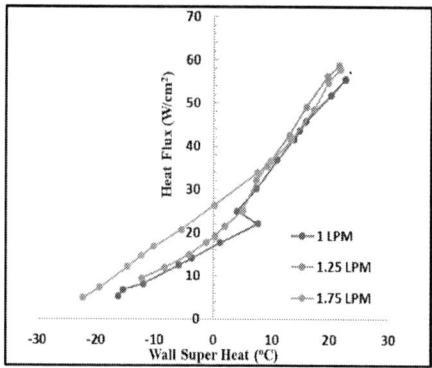

Figure 9. Boiling curves at different flow rates

The boiling curves at different flow rates are shown in Fig. 9. Increasing flow rate delays the onset of boiling and broadens the single-phase heat transfer region considerably [9]. At lower heat fluxes, forced convection heat transfer is dominant which results in a lower superheat at higher flow rate. At moderate heat fluxes, nucleate boiling contribution becomes large compared to single phase liquid convection. Therefore, all three boiling curves collapse to one curve. When heat flux is higher, forced convection become dominant again.

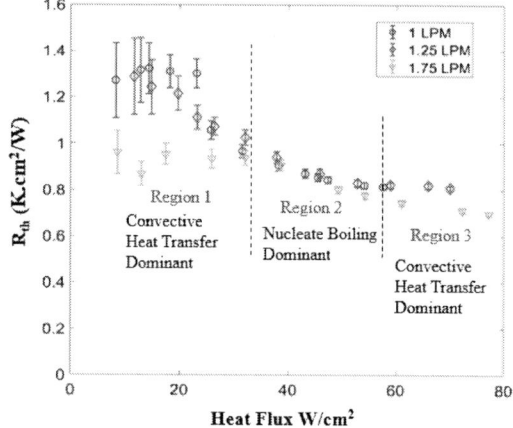

Figure 10. Variation of the specific thermal resistance of cold plate with heat flux at different flow rates

The effect of flow rate on the specific thermal resistance of cold plate is shown in Fig. 10. The specific thermal resistance is calculated by:

$$R_{th} = \frac{A(T_b - T_{in})}{Q} \qquad (1)$$

In which, T_b is base temperature, T_{in} is inlet temperature, A is the surface area of the heater block top surface and Q is the power dissipated to the copper block. The range of heat flux is divided into three regions in Fig. 10. In region 1 for low heat fluxes, thermal resistance decreases with increasing flow rate due to single-phase heat transfer behavior of the coolant. Increasing flow rate delays the onset of boiling incipience. The onset of nucleate boiling occurs at a heat flux of around 23 W/cm² for flow rate of 1 LPM and 40 W/cm² for a flow rate of 1.75 LPM. In region 2, when nucleate boiling heat transfer is dominant as compared to convection heat transfer, we observe similar thermal resistances in all three cases of flow rates 1 LPM, 1.25 LPM, 1.75 LPM. Due to the nucleate boiling dominance, the heat removal amount depends on the number of nucleate sites. Therefore, microchannel cold plate design is the main factor that decides thermal resistance value of heat sink. In region 3, as heat flux is further increased, it is observed that the thermal resistance decreases again with increasing flow rate. This can be explained by the dominance of convection heat transfer at higher heat flux.

Variation of pressure drop with heat flux at different flow rates is presented in Fig. 11. The pressure drop does not vary significantly with heat flux when the experiment is in the single-phase regime. When boiling occurs, the pressure drop rises considerably.

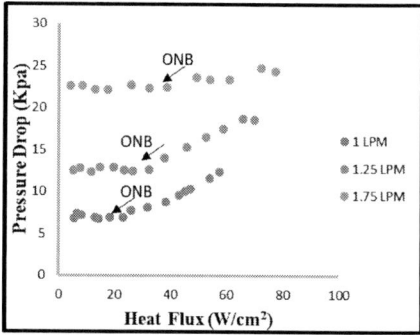

Figure 11. Variation of pressure drop with heat flux at different flow rates

Coefficient of performance (COP) is calculated via Eq. (2):

$$COP = \frac{Q}{P} \qquad (2)$$

Where Q is the power picked up by the dielectric coolant and P is the thermodynamic pumping power in the heat sink. The thermodynamic pumping power can be estimated as:

$$P = \Delta p \times V \qquad (3)$$

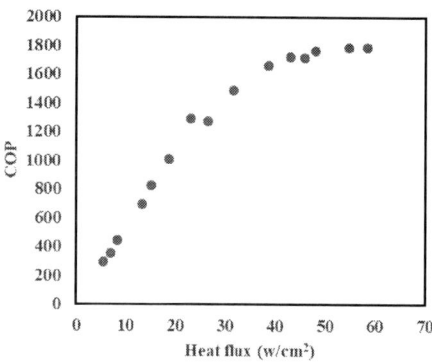

Figure 12. Coefficient of performance variation with heat flux at flow rate 1 LPM

Where V and Δp are volumetric flow rate and pressure drop through the cold plate, respectively. The variation of COP with heat flux is presented in Fig. 12. It is seen that the COP increases linearly in the single-phase regime. Then the slope decreases in the two-phase regime until the graph reaches a plateau. The reduction of slope is attributed to increasing of the pressure drop caused by vapor production. It should be noticed that when accounting for the pump efficiency and the rest of the system pressure drop, the COP will fall substantially below the values shown in Figure 12. However, the data in Figure 12 is the proper way to plot things and explain what happens to the COP.

6. Conclusions

An experimental apparatus was built to study two-phase cooling in a micro-channel heat sink using dielectric fluid HFE 7000. The experiments were carried out for a range of heat fluxes from 0 to 70 W/cm^2 and flow rates of 1, 1.25, and 1.75 LPM. Key findings from the study are as follows:

- Thermal resistance remains unchanged in the single-phase regime at lower heat fluxes but then reduce significantly when heat fluxes is higher and two-phase boiling occurs. Thermal performance in nucleate boiling state is independent on flow rate and is related to microchannel cold plate design only (eg nucleate site density, channel architecture, etc) .
- Pressure drop decreases slightly with heat flux in single-phase regime and then increases considerably with heat flux from the incipient of boiling phase. The coefficient of performance increases linearly with heat flux in single-phase regime and then the slope reduces when boiling occurs until it plateaus at higher heat fluxes.

The future plan of research is to optimize the cold plate design for enhancing heat transfer and hydraulic performance of the cold plate and improving the efficiency in two-phase cooling of electronic devices.

Acknowledgements

This work is supported by funding from the Semiconductor Research Corporation (SRC) and NSF IUCRC Award No. IIP-1738793

References

1. S. Alkharabsheh et al., "A Brief Overview of Recent Developments in Thermal Management in Data Centers," J. Electron. Packag., vol. 137, no. 4, Sep. 2015.
2. B. Ramakrishnan, "Experimental Assessment of Liquid Cooled Cold Plates Intended for Improving the Operational Efficiency of a High Density Data Center Environment." Order No. 13904547, State University of New York at Binghamton, Ann Arbor, 2019.
3. M. M. Ohadi, S. V Dessiatoun, K. Choo, M. Pecht, and J. V Lawler, "A comparison analysis of air, liquid, and two-

phase cooling of data centers," in 2012 28th Annual IEEE Semiconductor Thermal Measurement and Management Symposium (SEMI-THERM), 2012, pp. 58–63.
4. K. P. Drummond et al., "A hierarchical manifold microchannel heat sink array for high-heat-flux two-phase cooling of electronics," Int. J. Heat Mass Transf., vol. 117, pp. 319–330, 2018.
5. M. K. Sung and I. Mudawar, "Single-phase and two-phase cooling using hybrid micro-channel/slot-jet module," Int. J. Heat Mass Transf., vol. 51, no. 15–16, pp. 3825–3839, 2008.
6. F. Yang, W. Li, X. Dai, and C. Li, "Flow boiling heat transfer of HFE-7000 in nanowire-coated microchannels," Appl. Therm. Eng., vol. 93, pp. 260–268, 2016.
7. S. Wang, H. Chen, and C. Chen, "Flow Boiling Heat Transfer of HFE7000 in Manifold Microchannels through Integrating Three-dimensional Flow and Silicon Nanowires," in 2018 17th IEEE Intersociety Conference on Thermal and Thermomechanical Phenomena in Electronic Systems (ITherm), 2018, pp. 630–638.
8. R. J. Moffat, "Describing the uncertainties in experimental results," Exp. Therm. Fluid Sci., vol. 1, no. 1, pp. 3–17, 1988.
9. V.P. Carey, Liquid-vapor phase-change phenomena: an introduction to the thermophysics of vaporization and condensation processes in heat transfer equipment, 2nd ed., Taylor and Francis (New York, 2007).

Appendix A.

The uncertainty of thermal resistance in this paper is calculated using Root Sum Square method.

$$\delta R = \left\{ \sum_{i=1}^{N} \left(\frac{\partial R}{\partial Xi} \times \Delta Xi \right)^2 \right\}^{1/2} \quad (A1)$$

A1 is the basic equation of uncertainty analysis. Each term represents the contribution made by the uncertainty in one variable, Xi, to the overall uncertainty in the result, δR. Each term has the same form: the partial derivative of R with respect to Xi multiplied by the uncertainty interval for that variable.

Three thermocouples mounted to copper block (Fig. A1) were used for measuring power dissipated to cold plate. Power Q is calculated using Fourier's law (equation A2). In which, T_{bot} and T_{top} are the temperature readings using thermocouple 1 at the bottom and thermocouple 3 at the top as shown in Fig A1, respectively, and Δx is the distance between these two thermocouples. A stands for the top surface area of copper block where heat transfer goes through.

$$Q = \frac{KA(T_{bot} - T_{top})}{\Delta x} \quad (A2)$$

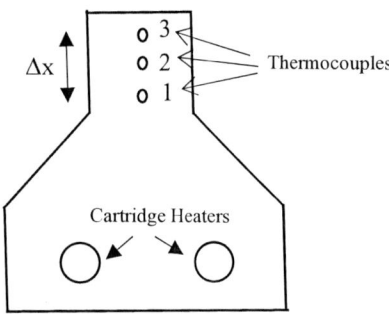

Figure A1. Copper block and position of thermocouples used for measuring heat flux

Substituting Q from equation (A2) to equation (1), we obtain the equation for thermal resistance:

$$R_{th} = \frac{(T_b - T_{in})\Delta x}{K(T_{bot} - T_{top})} \qquad (A3)$$

Applying basic equation of uncertainty analysis (A1) for the thermal resistance equation (A3) with 5 variables (N=5): T_b, T_{in}, Δx, T_{bot}, T_{top}. The Table A1 shows values of 5 variables, their uncertainty intervals and uncertainty of thermal resistance in the case of experiment with inlet temperature 25 °C and flow rate 1 LPM. It is found that uncertainty of thermal resistance reduces with power input from 12.7 % at power 34.8 W to 1.3 % at power 371 W.

Table A1. Values of 5 variables, their uncertainty intervals and thermal resistance uncertainty in the case of experiment with inlet temperature 25 °C and flow rate 1 LPM.

Δx (m)	T_{base} (°C)	T_{in} (°C)	T_{bot} (°C)	T_{top} (°C)	$\Delta(T_{base})$ (°C)	$\Delta(T_{in})$ (°C)	$\Delta(T_{bot})$ (°C)	$\Delta(T_{top})$ (°C)	$\Delta(\Delta x)$ (m)	R_{th} (°C.cm²/W)	Uncertainty of R_{th} (%)	Power (W)
0.0127	31.2	24.6	35.0	33.3	0.20	0.20	0.20	0.20	6.10^{-5}	1.22	12.78	34.88
0.0127	32.3	24.7	36.8	34.8	0.20	0.20	0.20	0.20	6.10^{-5}	1.18	10.70	41.74
0.0127	35.5	25.0	41.7	39.0	0.20	0.20	0.20	0.20	6.10^{-5}	1.27	8.34	53.34
0.0127	42.0	25.0	51.7	47.6	0.20	0.20	0.20	0.20	6.10^{-5}	1.32	5.36	82.98
0.0127	44.1	25.0	54.9	50.4	0.20	0.20	0.20	0.20	6.10^{-5}	1.32	4.79	92.95
0.0127	49.1	25.1	62.8	57.0	0.20	0.20	0.20	0.20	6.10^{-5}	1.31	3.80	117.78
0.0127	55.1	24.9	72.6	65.2	0.20	0.20	0.20	0.20	6.10^{-5}	1.30	3.01	149.40
0.0127	52.4	25.1	71.7	63.5	0.20	0.20	0.20	0.20	6.10^{-5}	1.06	2.75	166.30
0.0127	55.4	24.9	79.2	69.2	0.20	0.20	0.20	0.20	6.10^{-5}	0.97	2.28	203.44
0.0127	59.1	24.4	87.8	75.7	0.20	0.20	0.20	0.20	6.10^{-5}	0.91	1.91	246.60
0.0127	62.2	24.8	94.6	81.0	0.20	0.20	0.20	0.20	6.10^{-5}	0.87	1.72	277.56
0.0127	63.7	24.9	97.8	83.4	0.20	0.20	0.20	0.20	6.10^{-5}	0.85	1.64	292.57
0.0127	64.9	25.1	100.6	85.7	0.20	0.20	0.20	0.20	6.10^{-5}	0.84	1.59	304.30
0.0127	69.6	25.4	110.4	93.2	0.20	0.20	0.20	0.20	6.10^{-5}	0.82	1.41	348.37
0.0127	71.9	25.1	115.4	97.2	0.20	0.20	0.20	0.20	6.10^{-5}	0.81	1.33	371.09

CFD Investigation of Dispersion of Airborne Particulate Contaminants in a Raised Floor Data Center

Satyam Saini, Pardeep Shahi, Pratik Bansode, Ashwin Siddarth, Dereje Agonafer
The University of Texas at Arlington
701 S Nedderman Drive
Arlington, Texas
satyam.saini@mavs.uta.edu

Abstract

Modern data center facilities administrators are finding it increasingly difficult to lower the costs incurred in mechanical cooling of their IT equipment. This is especially true for high computing applications like Artificial Intelligence, Bitcoin Mining, Deep Learning, etc. Airside Economization/free air cooling reduces the mechanical cooling costs by using outside air to cool IT equipment under favorable ambient conditions. In this process, administrators risk their equipment to the exposure of fine particulate/gaseous contaminants that might enter the data center facility with the cooling airflow. Literature suggests that the nature of failures caused by particulate contamination is very intermittent which makes the failures tough to predict. While the recommended filters can remove $PM_{10-2.5}$, it's the fine and ultra-fine particulates like DPM (Diesel Particulate Matter), corrosive salts of high ionic content like sulfates and nitrates with low DRH (Deliquescent Relative Humidity) values that are the cause of concern.

The present investigation utilizes a 3-D CFD modeling of particle-laden flow in a rectangular flow domain, imitating the flow through floor tiles as in a raised floor data center. Literature was reviewed to study various numerical models that have been used for simulating particle dispersion and particle deposition in ventilated rooms, air ducts and particle behavior across physical obstructions of various geometries. A Discrete Phase Modeling approach was chosen using ANSYS FLUENT to calculate trajectories of the dispersed contaminants. 6SigmaRoom was used to predict accurate boundary and flow conditions of the fluid flow leaving the floor tiles.

Keywords

CFD, Particle Transport, Data Center, Servers, Contamination

Nomenclature

ACH	Air Change per Hour
CFD	Computational Fluid Dynamics
CBB	Confined Bluff Body
DPM	Diesel Particulate Matter
DRH	Deliquescent Relative Humidity
ISO	International Standards Organization
MERV	Minimum Efficiency Reporting Value
PCB	Printed Circuit Board
RANS	Reynolds Averaged Navier Stokes
Re	Reynolds Number
SIMPLE	Semi Implicit Method for Pressure Linked Equations
SIR	Surface Insulation Resistance

1. Introduction

To cope with rising computation and cloud storage demands, data center proliferation has been increasing unabatedly. Rising computational needs have also increased power densities at the chip level, causing a corresponding spike in the cooling demands. While novel cooling technologies like indirect and direct liquid cooling and immersion-cooling, evaporative cooling [27-29,36,39], have been shown to dissipate high heat fluxes, cloud providers like Google, Microsoft and Facebook have achieved PUE close to 1.1 using free air cooling at ideal locations. [1-3] Still, the majority of the data center administrators refrain from resorting to frees air-cooling methods owing to overhead costs of filters and dehumidification devices installation and the threat of introducing particulate and gaseous contaminants into the data center white space.

The ASHRAE T.C.9.9 subcommittee on Mission Critical Facilities, Data Centers, Technology Spaces and Electronic equipment has defined temperature and humidity ranges to ensure reliable operation of IT equipment [9-14]. They areaccepted by ITE manufacturers and their clients to be as follows: 18°C to 27°C (64.4°F to 80.6°F) dry bulb temperature, 5.5°C to 15°C (41.9°F to 59°F) dew point temperature, and less than 60% relative humidity. This recommended envelope was expanded by ASHRAE Thermal Guidelines for Data Processing Environments 2008, thereby, allowing short excursions into the allowable regions A1-A3, as shown in Figure 1, and an increase in the number of economizer hours. This exposes the IT equipment to the threat of gaseous and particulate contaminants. Two main failure modes associated with IT equipment failure because of surrounding the environment are: electrical open circuits resulting from corrosion of surface mount components due to gaseous contaminants and electrical short circuits due to copper creep corrosion, electrochemical migration and settled hygroscopic matter.

While much has been studied about the failure modes due to the presence of corrosive gases in the data center environment, less attention has been paid to failures because of particulate contaminants owing to the intermittent nature of the failures. 2011 The Gaseous and Particulate Contaminants Guidelines for Data Centers [23] recommends keeping the data centers clean as per ISO Class 8 by the following means of filtration:

• The room air may be continuously filtered with MERV 8 filters, as recommended by ASHRAE Standard 127 (ASHRAE 2007).
• For data centers using airside economizers, the air entering the data center may be filtered with MERV 11 or MERV 13 filters as recommended by ASHRAE (2009b).
The mechanism involved failure due to settled ionic particulate matter is moisture absorption from the surrounding humid air.

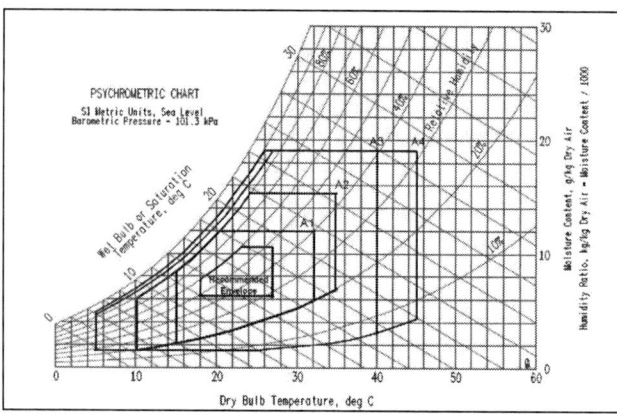

Figure 1: Psychrometric chart for ASHRAE recommended and allowable classes

Such hygroscopic matter forms a conductive aqueous solution above its DRH value, thus, reducing the Surface Insulation Resistance of the PCB and causing short-circuiting of closely spaced PCB features due to ion migration. The above-mentioned concerns point towards a planned effort in addressing these concerns through experimental and numerical studies of the impacts of particulate contamination at the server level.

The present investigation addresses the impacts of particulate contamination in data centers by utilizing a numerical approach to determine the flow patterns of sub-micron-particulate contaminants. Because of a lack of literature on particle transport studies in the data centers, an in-depth literature review was done on particle transport and particle dispersion studies in ducts and ducts with obstructions. Conclusions were made about the dominating forces involved and accurate numerical models to be used by correlating the flow and particle transport characteristics valid for the present investigation. 6SigmaRoom, a commercially available data center CFD code, was used to predict accurate pressure and velocity boundary conditions of the flow through a floor tile in a raised floor data center. The particle transport study was then conducted in ANSYS FLUENT where, for a rectangular flow domain representing volumetric flow rate through a floor tile. A transient Discrete Phase Modeling approach was used to track particle diameters between 1μm-10μm, and realizable k-e model was used to model flow turbulence as per the literature review. Particles were injected in the form of 2-D surface injections at the inlet and a low-pressure boundary condition was used to represent server inlet. Particle tracks and average volume fractions were obtained based on varying particle diameters and particle residence time in the flow.

2. Literature review

Data center contamination because of particulate matter has been addressed in the form of case studies and from a risk assessment point of view, where best practices to mitigate harmful particulates are addressed [30,31]. Existing empirical studies have investigated the failure modes and failure mechanisms, dominated by corrosion studies at PCB level and interconnect level [21,22]. Particle laden flow, in general, has been studied widely for buildings and indoor environments [24-26]. The challenging part in the case of particle-laden flow in

data centers is the presence of a multitude of inlets, outlets and varying pressure distribution in the room. To account for the lack of literature and accurate flow models for particle flow in the data centers, the authors constructed a simple model of the flow in a fluid domain, representing flow through a floor tile. The flow and particle transport models were studied from existing literature, as presented below, on particle transport in ducts and 2-D channels with/without obstructions. This enabled the authors to significantly simplify the problem and formulate a set of assumptions that closely matched the flow characteristics in a real-world data center.

An in-depth literature review was done for particle dispersion and deposition studies in ducts and across bluff bodies of various geometries. Several experimental [4,5] and numerical [6-9] investigations have studied the phenomena of particle dispersion in shear flows, where the flow regime is dominated by large vortex formations [18]. Recent examples of particle-laden channel and confined bluff body (CBB) flow can be found in, for example, Breuer and Alletto [10], Mallouppas and vanWachem [11], Sardina, Schlatter, Brandt, Picano, and Casciola [12] and Wang, Manhart, and Zhang [13], while turbulence modification in turbulent particle-laden channel flows can be found in Vreman [14, 19].

The deposition efficiency of liquid particles in 5.03–8.51 mm diameter circular cross-section tubes was investigated by Pui et al. in both turbulent and laminar flows [15]. Particle transport and deposition in vertical and horizontal turbulent square duct flows were studied by Zhang and Ahmadi for different gravity directions [16]. This study implemented Direct Numerical Simulation of the Navier–Stokes equation to provide that particle deposition velocity varies as per the direction of gravity. It was concluded that for horizontal ducts, because of gravity, particle deposition velocity on all four surfaces of the square duct is distinct. Most of the previous works have investigated particle sizes of less than 10μm. However, a study carried out by Zhao and Wu brought something new to the particle size selection [17]. Particle size distribution deposited on the ventilation duct was investigated in a room (4 m x 2.5 m x 3 m) with a typical mechanical ventilation system. The air supply volume rate was 240m^3/hour and the corresponding air change rate was 8 air changes per hour (ACH). The study reported that although the particle diameters smaller than 10 μm are the majorly airborne through the duct, the particles deposited on the duct surface are mostly larger than 10 μm [20].

3. Numerical Method

Particle-laden flow is a common phenomenon for many practical daily indoor and technical applications. CFD enables detailed prediction of complex fluid flows by discretizing complex geometries into smaller regions and numerically solving the desired flow characteristics in these individual discretized regions. Commercially available CFD codes have made it easier to visualize complicated flow phenomena like particle-particle interactions and particle-flow interactions.

Lagrange-Euler approach has been proven to solve multiphase particle-laden flows. This approach uses RANS equations to solve the continuous or carrier phase and the dispersed or particle phase is resolved by Lagrangian tracking. The CFD code was chosen based on its extensive abilities in resolving particle-particle, particle-flow interactions and accurate mathematical models in simulating turbulence involved in particle flow. As described in ANSYS FLUENT Theory guide [32,33], the continuous phase is calculated using the RANS equations as given below:

$$\nabla.\bar{u} = 0 \tag{1}$$

$$\frac{\partial u}{\partial t} + \rho\,(\bar{u}.\nabla)\bar{u} = -\nabla\bar{p} + \eta\Delta\bar{u} - \nabla.\overline{\tau^{RS}} + \vec{f}_D \tag{2}$$

Where $\bar{u}$ and $\bar{p}$ are the average velocities of continuous (air) and discrete (particle) phase. The second term on the left-hand side in equation (2) represents the Reynolds Stresses which are modeled using the eddy-viscosity approach. In this study, realizable RNG κ-ε model was used to model kinetic energy and turbulent dissipation. Equations (3), (4) and, (5) are solved to obtain the particle force balance and particle trajectories of particles of mass m_p.

$$m_p\frac{d\vec{u_p}}{dt} = \sum\vec{F_i} \tag{3}$$

$$\sum\vec{F_i} = \vec{F_D} + \vec{F_B} + \vec{F_G} \tag{4}$$

$$m_p\frac{d\vec{u_p}}{dt} = m_p\frac{\vec{u} - \vec{u}_p}{\tau_r} + m_p\frac{\vec{u}(\rho_p - \rho)}{\rho_p} + \vec{F} \tag{5}$$

The particles were assumed to be smooth and spherical, therefore, spherical drag law was activated, and default values of the coefficients were used for particles greater than 1μm. For sub-micron particles, as explained in theory guide, Stoke Cunningham drag law was used, which is given by equations 6, 7 and 8.

$$\vec{F_D} = \frac{3}{4}\frac{\rho}{\rho_p}\frac{d_p}{m_p}C_D(\vec{u} - \vec{u}_p)|\vec{u} - \vec{u}_p| \tag{6}$$

$$F_D = \frac{18\mu}{d_p^2\rho C_C} \tag{7}$$

$$C_C = 1 + \frac{2\lambda}{d_p}(1.257 + 0.4) \tag{8}$$

The particle relaxation time was used for predicting particle trajectories using the force balance on the particle in the Lagrangian time frame as given in equation (5). This describes the deacceleration of particles due to the drag force and was solved using equation (9).

$$\tau_r = \frac{\rho_p d_p^2}{18\mu}\frac{24}{C_d Re} \tag{9}$$

Here Re is the relative Reynolds number and is calculated by:

$$Re_p = \frac{|\vec{u} - \vec{u_p}|\rho d_p}{\mu} \tag{10}$$

As the particles considered in this study are of small diameter, the torque or particle rotation was not considered. For sub-micron particles, it has been concluded from the literature that their dispersion is dominated by Brownian Force which is calculated using equation (11). The particle lifts for particle diameters greater than 1μm can be solved using equation (12).

$$F_{bi} = \zeta_i\sqrt{\frac{216\rho v\sigma T}{\pi\rho_p^2 d_p^5 C_c\Delta t}} \tag{11}$$

$$\vec{F} = m_p\frac{2Kv^{\frac{1}{2}}\rho d_{ij}}{\rho_p d_p(d_{lk}d_{kl})^{\frac{1}{4}}}(\vec{u} - \vec{u}_p) \tag{12}$$

The Discrete Random Walk Model or eddy lifetime model can be used to model particle interaction with discrete fluid phase turbulent eddies which are classified by random velocity fluctuations given by u', v', w' and are calculated as given below in equations (13)-(15) where ζ is a normally distributed random number. The value of the RMS (Root Mean Square) fluctuating components on the right-hand side of these equations is calculated by equation (16) using known values of kinetic energy turbulence at each point in the flow.

$$u' = \zeta\sqrt{u'^2} \tag{13}$$

$$v' = \zeta\sqrt{v'^2} \tag{14}$$

$$w' = \zeta\sqrt{w'^2} \tag{15}$$

The characteristic lifetime of an eddy is given by equation (16). The same can be calculated using equation (17) as random variation about T_L, fluid Lagrangian integral time, r is a uniform random number greater than zero and less than 1 and C_L is the integral time scale constant.

$$\tau_e = 2T_L \tag{16}$$

$$\tau_e = -T_L\ln(r) \tag{17}$$

$$T_L = C_L\frac{k}{\epsilon} \tag{18}$$

4. Methodology

After an in-depth literature review of existing literature on particle transport in ducts, a set of assumptions was formulated that would simplify the current model giving near accurate results. A pressure-based solver was used for solving carrier fluid flow and pressure velocity coupling is achieved using the SIMPLE algorithm [38]. Based on the generated flow field, a defined number of particles were injected and were tracked as they traveled through the flow domain.

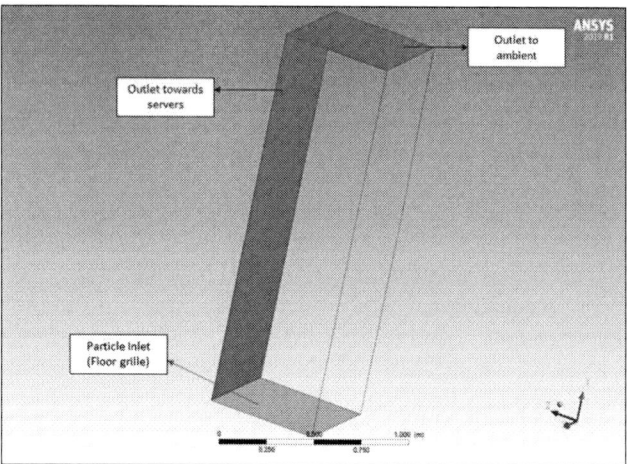

Figure 2: Boundary conditions used in the CFD study

The flow domain was designed in ANSYS DM as a rectangular channel extruded as a fluid representing the flow through a floor tile. Figure 2 shows the set of boundary conditions used in the CFD study, where a 0 Pa gauge pressure boundary condition was used for the outlet to ambient and -0.1 Pa for the outlet towards the servers. The meshing was done using an integrated mesher in the CFD code. As there were no obstructions curvatures within the flow domain, a fine quality mesh was homogenously generated consisting of 747,921 nodes. To visualize accurate boundary conditions, 6SigmaRoom was used to model airflow through floor tiles in a raised floor data center. This was done to realize the effect of neighboring floor tiles on the airflow through any specific tile. A server rack filled with 42 1U servers was modeled in front of a row of five floor tiles with an open area of 29%. 2-D contours of velocity through the floor tiles and pressure contours at the rack inlet were obtained. The obtained values of the velocity and pressure were then used in the CFD code.

Figure 3: Velocity profile through floor tile in 6SigmaRoom

As seen in Figure 3, a rather constant velocity contour can be seen in front of the server rack with a maximum outlet velocity of 4m/s from the floor tile. As seen from the pressure distribution in the front of the rack in Figure 4, the boundary condition at the outlet towards the servers in FLUENT was given as -0.1 Pa of gauge pressure.

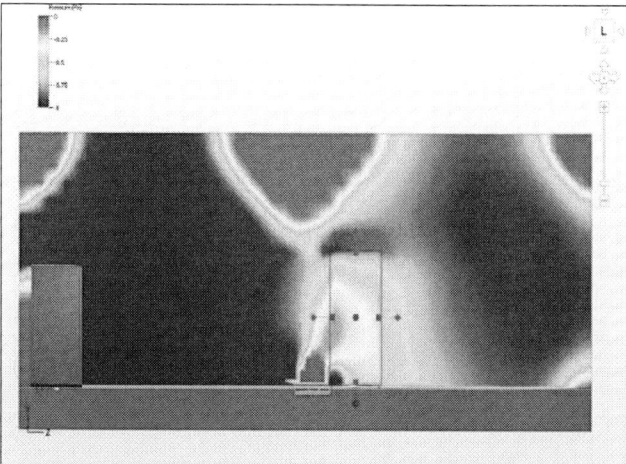

Figure 4: Pressure variation as seen in front of the rack in 6Sigma Room with a minimum pressure of -1 Pa

To validate the particle transport approach used, a validation study was conducted by simulating particle

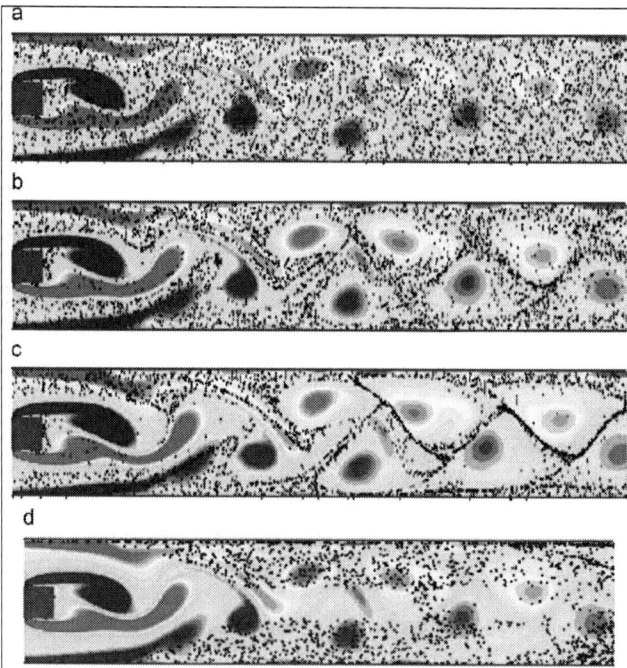

Figure 5: Validation case from the literature review [34]

dispersion across a 2D square cylindrical obstruction in a channel flow as described by Jafari et al [34]. As most of the particle transport studies in the reviewed literature use numerical techniques to model particle transport, it was necessary to determine if the models available in the CFD code could give near similar results to numerical exact solutions. The reviewed study numerically modeled particle transport and deposition by correlating the flow Stokes number for laminar flow. Similar mathematical models for forces and flow were activated in ANSYS and a transient simulation was run with a time step of 0.01 seconds for 1000 time steps. This allowed the flow to sufficiently fill the entire domain and aided in visualizing the vortex shedding across the bluff body as seen in Figure 5. This study concluded that particle dispersion is dominated by Brownian motion for sub-micron particles and by inertia forces for particle sizes of larger diameters.

The particle dispersion is also selectively distributed based on particle size. Smaller particles were unaffected due to the presence of the vortices, as can be seen in Figure 5. While the heavier particles are distributed on the periphery of the shed vortices. This is because inertial particles tend to move towards low vorticity regions because of the vortex generated centrifugal forces. Using the same design parameters and geometries, particle dispersion was obtained in the CFD code. The comparison of both the reviewed study and current study can be seen below, and a similar pattern is observed in present simulations as seen below in Figure 6.

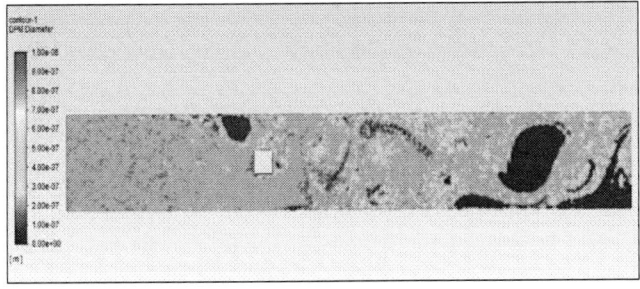

Figure 6: Particle traces as obtained for validation case

5. Results and Discussion

The final flow domain was imported to the CFD code and appropriate flow models and boundary conditions were activated as explained in Section 3 and Section 4. The particles were injected as surface injections where particles were then released from each facet of the surface. Here, the facet value of a variable is defined as the computed arithmetic average of the adjacent cell values of the variable. A total flow rate of 1e-04 kg/m^3 was used so that a sufficient number of particles can be generated and tracked. The maximum, minimum and mean diameters of the injections were specified

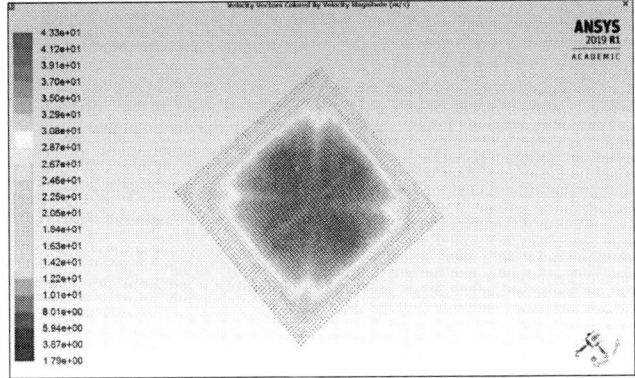

Figure 7: Velocity vectors of air at the outlet for base flow simulation

for size distribution. The particle properties were selected using anthracite (ρ=1550kg/m^3) as an inert particle from the software material library and relevant force balance laws were activated. Two-way coupling was used in which the continuous flow field was solved first, after which the discrete phase trajectories are calculated. After this, the continuous phase was solved again based on interphase momentum exchange (as no heat and mass transfer is considered in this study) and discrete phase trajectories were then recalculated for a modified flow field. This process was repeated until a converged solution was obtained. For the current study, a DPM iteration interval of 10 was selected, which means that a discrete phase iteration was performed every tenth continuous phase iteration. A time step size of 0.1 seconds was chosen with a total number of iterations equal to 100 and maximum iterations per time step equal to 10.

Figure 8: Particle concentration contour for various diameters

Multiple simulations were run after completing the validation case. The first case simulated in the present study was that of a rectangular channel with all four sides of the channel in a symmetry boundary condition. This was done to validate the approach that without any influence of gravity, all the particles must follow the flow streamlines of air/continuous fluid. Figure 7 shows the velocity profile of the flow as seen at the domain outlet. Comparing the velocity contour to the particle concentration, as seen in Figure 8, it can be concluded that the assumption made about particle following the flow streamlines holds. Also, from the knowledge of fluid dynamics, for a flow in a rectangular duct, the maximum flow velocity should occur at the center of the duct.

A similar flow and selective concentration pattern were observed for the particles, in this case, irrespective of their diameters. Unlike succeeding simulations, the effects of buoyancy and Brownian motion were ignored for the base simulation. Another base simulation with a particle diameter less than 1μm was also performed which showed the same flow pattern and was, therefore, not included in the results to avoid the repeated presentation of similar patterns.

The next and final set of simulations was performed by defining a second low-pressure outlet in the flow domain depicting the floor tile side facing the rack inlet. The value of the pressure was chosen from 6SigmaRoom by populating a rack with 42 1U servers and plotting a pressure contour on the

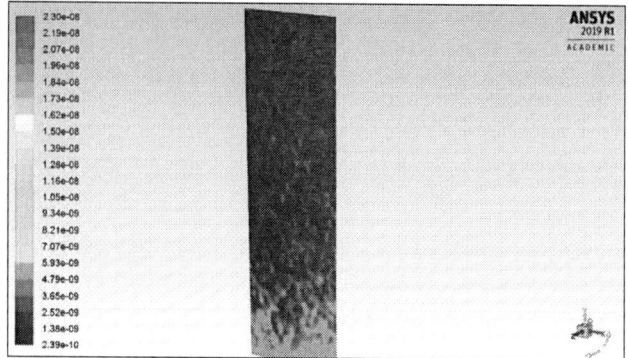

Figure 9: Particle concentration distribution on server side for low-density particles

server inlets. It was assumed that all the servers operated under a constant load, hence creating a constant pressure drop across the width of the rack. The other three sides of the flow domain were given a symmetry boundary condition. Physically, the symmetry boundary condition, in this case, can be interpreted as a floor tile present at the center of a cabinet row with no effect on its airflow from any of the neighboring tiles. Unlike the previous case, the effects of particle buoyancy and Brownian motion were also considered. Figure 10 shows the instantaneous distribution of particles based on particle diameters.

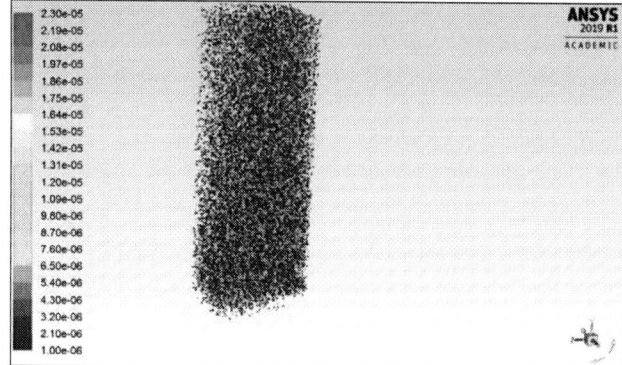

Figure 10: Real time Particle diameter distribution in flow domain after complete simulation

Figure 9 shows the concentration distribution of low-density particles as observed at the flow domain side facing the server inlets. Comparing this to Figure 11, the distribution of particles is more pronounced at the bottom and top ends of the

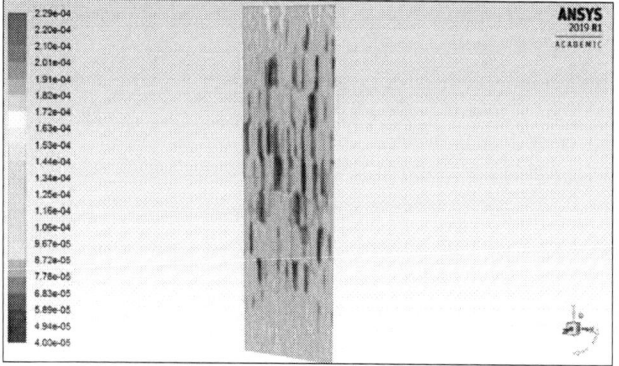

Figure 11: Particle concentration of dense particles on the server inlet side

flow domain faces. Finally, the density of the particles was increased to 8800 kg/m³ to see how a large variation in density of the inert particles will affect the particle tracks obtained under the same boundary conditions.

Figure 11 and Figure 12 show overall particle concentration by total mass and concentration variation by diameters on the domain face towards the rack inlet respectively. Most of the particle mass was retained at the bottom of the domain and escaped through the outlet facing the servers. The particle diameter distribution at the final time step shows that most of the particles that escaped the outlet facing the servers were of diameters less than the mean diameter considered in this simulation. The inequality in the parentheses represents mean diameter in the simulation bounded by minimum and maximum diameters (1e-07μm <5e-07 μm <1e-06 μm).

Property	Server side	Top outlet
Mass transfer	1.77e-04 kg	7e-04 kg
No. of particles escaped	6.4e+05	2.6e+08
Max. time before escaping	2.19 sec	2.15 sec

Table 1: Particle summary for low-density particles

The bulk distribution within the fluid domain of particle diameters was observed to be slightly different than that of lighter particles. This was observed from the fact that a greater number of particles escaped out from the inlet itself when compared to the particles of low density.

Property	Server side	Top outlet
Mass transfer	1.77e-04 kg	7e-04 kg
No. of particles escaped	1.6e+06	1.8e+06
Max. time before escaping	2.1 sec	1.83 sec

Table 2: Particle summary for high-density particles

Since a lot of elements in the flow domain were reported by the solver having reverse flow because of a low-pressure outlet, as concluded from the validation case, the heavier particles can be trapped in their vicinity and stay in the flow domain for longer times. Particle summary was obtained in post-processing which showed that a total of 3.4e12 particles were ejected from the floor tile for 10 seconds for low-density particles. Table 1 shows the number of particles and mass transferred from the two outlets considered in the flow domain. The total mass and the number of particles escaped from the top outlet is almost four times greater than the outlet facing the servers. This means that less than 10e-07 of the total particles injected and tracked by the solver went towards the IT equipment. Although this number might seem small, as the literature suggests the failures related to particulates are intermittent and happen after sufficient particulate accumulation and favorable environmental conditions. For high-density particles, the total particle count injected in the flow domain was around 1.9e+11. The fraction of particles escaping towards the server side was an order of magnitude more than that obtained for lighter particles. Also, the total number of particles escaping both the outlets were similar.

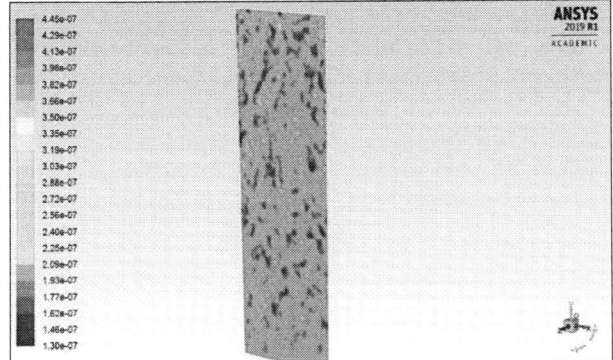

Figure 12: Particle diameter distribution at the side facing server inlets for high-density particles

6. Conclusions

Particle transport of particulate contaminants was studied for a simplified model of a raised floor data center using CFD. Particle distribution was obtained for various boundary conditions within the flow domain and two different particle densities. The approach was validated as per existing particle dispersion studies in the literature. It was observed that particle sizes or particle mass is the dominant factor that dictates particle dispersion. For lighter or lower diameter particles, the dispersion was rather random. Also, the particle concentration plots obtained show that particle deposition will be more pronounced around the bottom servers due to low-pressure regions. This observation might need actual experimental validation, where either outflow or pressure variation can be measured at the rack level in an actual data center. The particle escape regions were inferred to be located towards the bottom of the flow domain, implying that the maximum particle concentration or deposition should be found in the servers located at the bottom of the flow domain.

In a real flow inside IT equipment, when the vortices formed due to impedance inside the server eventually lose their energy, the particles carried by them will be deposited at that location. For servers, where the flow is enclosed and is generally incident on multiple obstructions, it can be predicted that the stagnation points of the flow on these obstructions will be most vulnerable to particle deposition. It can be further extrapolated that if some of these particles are ionic in nature, there can be a high probability of equipment failure due to short-circuiting. Further simulations of particle flow within the server will give a better interpretation of the deposition and dispersion phenomenon within the server. Various raised floor data center cooling strategies and local airflow delivery methods through floor tiles [37] will also affect the particle distribution. This necessitates the requirement of further detailed simulations of the floor tile itself to get a better idea of special distribution and dispersion of the particulate contaminants. Particle distribution will also be dependent on particle shape based on variation in drag and lift forces [40] which should also be taken into account for future studies.

7. Future Work

This study is preliminary work in conducting a detailed study of the particle-laden airflow through the server rack as well as individual IT equipment. A detailed model of the combined floor tile and server rack will be created by experimentally determining the pressure drops across the server rack and inlet and outlet velocities. The effect of changes in particle particulate characteristics like particle density, diameters, and particle coagulation will be seen. The objectives of these subsequent studies are to determine the percentage deposition occurring inside the server and using CFD, predicting the locations of these depositions based on flow characteristics inside the server. User-defined functions can be compiled and imported into the CFD solver to study particle deposition which is directly related to particle's Stokes number.

Acknowledgments

This work is supported by NSF I/UCRC in Energy-Smart Electronic Systems (ES2). The authors would also like to extend special gratitude towards Mark Seymour and Kourosh Nemati of Future Facilities for their guidance and feedback throughout the project.

References

1. https://www.opencompute.org/blog/cooling-an-ocp-data-center-in-a-hot-and-humid-climate
2. Google. Efficiency: How We Do It, 2014
3. Microsoft. Microsoft Shares Video Tour of its Cloud Datacenters,2011
4. Longmire, E. K., and Eaton, J. K. (1992). Structure of a Particle-Laden Round Jet, *J. Fluid Mech.* 236:217–257.
5. Lazaro, B. J., and Lasheras, J. C. (1992). Particle Dispersion in the Developing Free Shear Layer, Part 1—Unforced Flow, *J. Fluid Mech.* 235:143–178.
6. Crowe, C. T., Chung, J. N., and Trout, T. R. (1988). Particle Mixing in Free Shear Flows, *Progress in Energy and Combustion Science* 14:171–194.
7. Uthuppan, J., Aggarwal, S. K., Grinstein, F. F., and Kailasanath, K. (1994). Particle Dispersion in a Transitional Axisymmetric Jet: A Numerical Simulation, *AIAA J.* 32:2004–2014.
8. Aggarwal, S. K. (1994). Relationship Between the Stokes Number and Intrinsic Frequencies in Particle-Laden Flows, *AIAA J.* 32:1322–1325.
9. Chang, E., and Kailasanath, K. (1996). Simulation of Dynamics in a Confined Shear Flow, *AIAA J.* 34:1160–1166
10. Alletto, M., & Breuer, M. (2012). One-way, two-way and four-way coupled LES predictions of a particle-laden turbulent flow at high mass loading downstream of a confined bluff body. *International Journal of Multiphase Flow*, 45, 70–90.
11. van Vliet, E., Singh, M., Schoonenberg,W., van Oord, J., van der Plas,D., & Deen, N. (2013). *Development and validation of Lagrangian-Eulerian multi-phase model for simulating the argon stirred steel flow in a ladle with slag.* Proceedings of the 5th International Conference on

Modelling and Simulation of Metallurgical Processes in Steelmaking, STEELSIM. Ostrava, Czech Republic.

12. Sardina, G., Schlatter, P., Brandt, L. P., & Casciola, C. (2012). Wall accumulation and spatial localization in particle-laden wall flows. *Journal of Fluid Mechanics*, *699*, 50–78.

13. Wang, B., Manhart, M., & Zhang, H. (2011). Analysis of inertial particle drift dispersion by direct numerical simulation of two-phase wall-bounded turbulent flows. *Engineering Applications of Computational Fluid Mechanics*, *5*(3), 341–348.

14. Vreman, A. (2015). Turbulence attenuation in particle-laden flow in smooth and rough channels. *Journal of Fluid Mechanics*, *773*, 103–136

15. Pui DYH, Romay-Novas F, Liu BYH. Experimental study of particle deposition in bends of circular cross-section. Aerosol Sci Technol 1997;7:301–15

16. Zhang H, Ahmadi G. Aerosol particle transport and deposition in vertical and horizontal turbulent duct flows. J Fluid Mech 2000;406:55–80.

17. Zhao Bin, Wu Jun. Modeling particle fate in ventilation system—Part II: Case study. Build Environ 2009;44:612–20.

18. D. J. Brandon & S. K. Aggarwal (2001) A Numerical Investigation of Particle Deposition on a Square Cylinder Placed in a Channel Flow, Aerosol Science & Technology, 34:4, 340-352

19. Franziska Greifzu, Christoph Kratzsch, Thomas Forgber, Friederike Lindner & Rüdiger Schwarze (2016) Assessment of particle-tracking models for dispersed particle-laden flows implemented in OpenFOAM and ANSYS FLUENT, Engineering Applications of Computational Fluid Mechanics, 10:1, 30-43, DOI: 10.1080/19942060.2015.1104266

20. Gao, R., and Li, A. (2012). Dust Deposition in Ventilation and Air-Conditioning Duct Bend Flows. *Energy Convers. Manage.*, 55:49–59

21. Jimil M. Shah, "Characterizing contamination to expand ASHRAE envelope in airside economization and thermal and reliability in immersion cooling of data centers", PhD dissertation, The University of Texas at Arlington, May 2018

22. Jimil M. Shah, "Reliability challenges in airside economization and oil immersion cooling", UTA-THESIS, 2016-05-16, Shah, Jimil Manojbhai, 0000-0003-2297-7413.

21. Prabjit Singh, Levente Klein, Dereje Agonafer, Jimil M. Shah and Kanan D. Pujara, "Effect of Relative Humidity, Temperature and Gaseous and Particulate Contaminations on ITE Reliability, DOI: 10.1115/IPACK2015-48176, ASME InterPACK 2015, San Francisco, CA.

22. Shah, Jimil & Awe, Oluwaseun & Gebrehiwot, Betsegaw & Agonafer, Dereje & Singh, P & Kannan Mestex, Naveen & Kaler Mestex, Mike. (2017). Qualitative Study of Cumulative Corrosion Damage of ITE in a Data Center Utilizing Air-side Economizer Operating in Recommended and Expanded ASHRAE Envelope. Journal of Electronic Packaging. 139. 021002. 10.1115/1.4036363

23. ASHARE. 2011. 2011 Gaseous and Particulate Guidelines for Data Centers, Atlanta, GA, USA.

24. Seymour, M.J., A. A. M. A., and Jiang, J., 2000. "Cfd based airflow modelling to investigate the effectiveness of control methods intended to prevent the transmission of airborne organisms". Air Distribution in Rooms, (ROOMVENT 2000).

25. Jones, P., and Whittle, G., 1992. "Computational fluid dynamics for building air flow prediction- current status and capabilities". Building and Environment, 27(3), pp. 321–338.

26. Chen, Q., and Zhang, Z., 2005. "Prediction of particle transport in enclosed environment". China Particuology, 3(6), pp. 364–372.

27. Jimil M. Shah, Chinmay Bhatt, Pranavi Rachamreddy, Ravya Dandamudi, Satyam Saini, Dereje Agonafer, 2019, "Computational Form Factor Study of a 3rd Generation Open Compute Server for Single-Phase Immersion Cooling," ASME Conference Paper No. IPACK2019-6602

28. Dhruvkumar Gandhi, Uschas Chowdhury, Tushar Chauhan, Pratik Bansode, Satyam Saini, Jimil M. Shah and Dereje Agonafer, 2019, "Computational analysis for thermal optimization of server for single phase immersion cooling", ASME Conference Paper No. IPACK2019-6587

29. Pravin A Shinde, Pratik V Bansode, Satyam Saini, Rajesh Kasukurthy, Tushar Chauhan, Jimil M Shah and Dereje Agonafer, 2019, "Experimental analysis for optimization of thermal performance of a server in single phase immersion cooling", ASME Conference Paper No. IPACK2019-6590

30. Jimil M. Shah, Roshan Anand, Satyam Saini, Rawhan Cyriac, Dereje Agonafer, Prabjit Singh, Mike Kaler, 2019, "Development of A Technique to Measure Deliquescent Relative Humidity of Particulate Contaminants and Determination of the Operating Relative Humidity of a Data Center, ASME Conference Paper No. IPACK2019-6601

31. Gautham Thirunavakkarasu, Satyam Saini, Jimil Shah, Dereje Agonafer,2018, "Airflow pattern and path flow simulation of airborne particulate contaminants in a high-density Data Center utilizing Airside Economization", ASME Conference Paper No. IPACK2018-8436

32. *ANSYS® ANSYS FLUENT Theory Guide, Chapter 16, Release 2019 R1*

33. *ANSYS® ANSYS FLUENT User's Guide, Chapter24, Release 2019 R1*

34. Jafari, Saeed & Salmanzadeh, Mazyar & Rahnama, Mohammad & Ahmadi, Goodarz. (2010). Investigation of particle dispersion and deposition in a channel with a square cylinder obstruction using the lattice Boltzmann method. Journal of Aerosol Science. 41. 198-206. 10.1016/j.jaerosci.2009.10.005.

35. Patankar, S. V., & Spalding, D. B. (1972). A calculation procedure for heat, mass and momentum transfer in threedimensional parabolic flows. *International Journal of Heat and Mass Transfer*, *15*, 1787–1806.

36. Dakshinamurthy, H. N., Siddarth, A., Guhe, A., Kasukurthy, R., Hoverson, J., & Agonafer, D. (2019, January). Accelerated Degradation Testing of Rigid Wet Cooling Media to Analyse the Impact of Calcium Scaling.

In *ASME 2018 International Mechanical Engineering Congress and Exposition.* American Society of Mechanical Engineers Digital Collection.

37. Mohsenian, G., Khalili, S., & Sammakia, B. (2019, May). A Design Methodology for Controlling Local Airflow Delivery in Data Centers Using Air Dampers. In *2019 18th IEEE Intersociety Conference on Thermal and Thermomechanical Phenomena in Electronic Systems (ITherm)* (pp. 905-911). IEEE.

38. Patankar, S. V., & Spalding, D. B. (1972). A calculation procedure for heat, mass and momentum transfer in threedimensional parabolic flows. *International Journal of Heat and Mass Transfer, 15*, 1787–1806

39. Kumar, A., Shahi, P., & Saha, S. K. Experimental Study of Latent Heat Thermal Energy Storage System for Medium Temperature Solar Applications.

40. Sarker, M. R. H., Chowdhury, A. R., & Love, N. (2017). Prediction of gas–solid bed hydrodynamics using an improved drag correlation for nonspherical particles. *Proceedings of the Institution of Mechanical Engineers, Part C: Journal of Mechanical Engineering Science, 231*(10), 1826-1838.

General Guidelines for Commercialization a Small-Scale In-Row Cooled Data Center: A Case Study

Yaman. M. Manaserh[1], Mohammad. I. Tradat[1], Ghazal Mohsenian[1], Bahgat G. Sammakia[1], Mark J. Seymour[2],
[1]Departments of Mechanical Engineering, ES2 Center, Binghamton University-SUNY, NY, USA
[2]Future Facilities, London, UK and NY, USA
E-mail:yyaseen1@binghamton.edu

Abstract

As the world is moving toward edge computing, the need for small scale, widely distributed data centers became more critical. The traditional raised floor cooling techniques might not be applicable in these data centers and it could be more efficient to use in-row coolers. Recently, a small-scale data center was built at Binghamton University. This data center consists of ten cabinets and six in-row air cooling units combined into a single cold aisle enclosure.

In this study, a CFD model for this data center is built considering all components using Future Facilities CFD tool (6SigmaRoom). After that, this model is utilized to conduct a detailed computational heat transfer analysis.

These analyses include: the data center heat load, IT-equipment different configurations, in-row coolers supply air temperature, in-row coolers airflow rate, the data center running cost and some instructions to improve the data center performance. Moreover, behaviors that could harm the IT-equipment were identified and proposed solutions to ensure the IT-equipment reliability were presented.

Results showed that the data center heat load should be limited to 156 kW (15.6 kW per cabinet) and the supply air temperature should not exceed 23℃. In case of installing switches, cabinet load should be limited to 14 kW.

Keywords:

Numerical Study, Cold Aisle Containments, Small Scale Data Centers, In-Row Cooled Data Centers, Data Center Thermal Management.

Nomenclature

ACU	Air Cooling Unit
CAC	Cold Aisle Containment
CRAH	Computer Room Air Handler
DC	Data Center
FR	Flow Rate
HAR	Hot Air Recirculation
ITE	Information Technology Equipment
LC	Loaded Cabinet
P	Pressure (Pa)
RF	Raised floor
SAT	Supply Air Temperature
T	Temperature (℃)

1. Introduction

As the IT operations are necessary for running any business, the need for building a data centers that assures the IT-equipment (ITE) reliability, information security and has appropriate infrastructure become more crucial.

It is well known that operating data centers requires using a cooling system to get rid of the heat generated by the ITE and to maintain an environment that satisfies the required operating conditions. These conditions are clearly defined by the "ASHRAE guidelines for data center processing environments" [1].

Based on the fact that data centers cooling systems consumes (30-50) % of the total power input to the data centers [2] which is estimated to be 3% of the electricity generated around the world [3], a considerable amount of research on improving the performance of the existing cooling techniques and developing new ones was conducted.

However, air cooling-based methods are the most reliable, well developed, widespread and easy to employ cooling method. In the air-cooling techniques air is being delivered to the ITE in three ways which are: 1- using a cooling unit(s) to deliver the cold air to a raised floor and then to a cold aisle 2- using an actual compressors and chillers inside the rack itself, removing the heat without entering the data center 3- using cooling units (in-row coolers) between cabinets that delivers air directly to the cold aisle.

Among these three air cooling techniques, research focused mainly on data centers in which the air is being delivered to the cold aisle through raised floor.

H. Alissa et al. [4] characterized a raised floor data center laboratory using practical measurements methods, including tiles and the computer room air handler (CRAH) flow measurements. Then a full physics based CFD model is built to simulate/predict the measured data.

Mohsenian et al. [2,5] designed a control methodology that considers the differential pressure between CAC and room as a control parameter (unlike many researchers who have investigated the temperature-based control systems). The developed fuzzy control system ensures that just enough airflow is delivered to the aisle, regardless of model, generation and workload of ITE.

M. Tradat et al. [6] presented an experiment-based analysis and comparison of environmental and power data collection using two different approaches; one uses a discrete sensors network and smart PDUs and uses another available data from the installed ITE (IPMI data). They showed that containment allows us to raise the SAT up to 24°C without any negative impact on the ITE airflow demand.

Another study presented an experimental investigation of a facility-level cooling system failure scenarios in which chilled water interruption introduced to the data center [7]. Results show that the CAC helps in keeping the ITE cold for a longer time and its effects on the ITE performance and response could vary and depend on the server's airflow and heat.

Recently, in-row cooling systems is becoming increasingly popular since it is bringing the cooling units closer to the heat source. As a result, more studies that discusses using in-row coolers started to be conducted.

K. Nemati et al. [8] investigated the performance of in-row coolers in both opened and cold-aisle contained environments. T. Gao et al. [9] presented a hybrid cooling technology which utilizes in row coolers in existing raised floor air cooled data centers.

A. Bhalerao et al. [10] employed the in-house data center modeling tool Villanova Thermodynamic Analysis of Systems (VTAS) software package to ascertain the influence of hybrid liquid-air components on overall data center exergy destruction. The results showed that the exergy destruction decreases for a hybrid liquid-air system using only an in-row cooler.

Based on the literature review, numerical techniques results showed a good agreement with the experimental ones in many studies [11-14], even some other studies proposed methods to improve the precision of the numerical techniques in predicting the behavior of the data center for example: a study illustrated basic guidelines for integrating empirical measurements to state of the art modeling approaches to improve the accuracy of the CFD models [15].

As the numerical methods showed its capability of replicating the DCs performance, a considerable number of researches were carried out using numerical techniques. S. Alkharabsheh et al [16] demonstrated the effect of Cold Aisle Containment Systems (CACS) on the airflow and temperature distribution inside a representative data center using CFD.

X. Xiong et al [17]. Proposed a rack layout configuration called Vortex Flow Layout (VFL) and they simulated it under fully, partial and no containment cases to evaluate its performance using CFD modeling. K. Karki et al [18]. described a computational fluid dynamics model for calculating airflow rates through perforated tiles in raised-floor data centers.

In this work, a computational heat transfer model that replicates the Pharmacy School at Binghamton University data center was developed based on the measurements and data that have been gathered from the physical facility. 6SigmaRoom software was used to build the model and to conduct an energy-based analysis. Figure 1 shows the real data center and its model in the software.

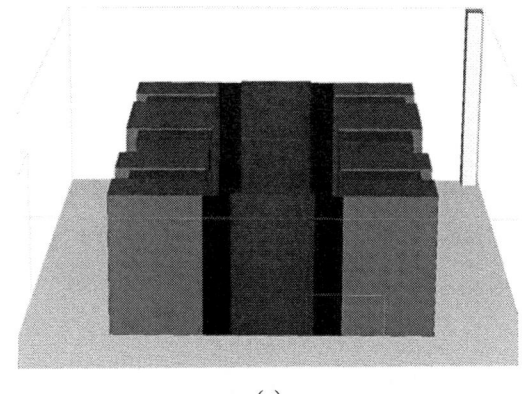

(a)

(b)

Figure 1. Front view of (a) the 6SigmaRoom model and (b) the actual DC.

This model was utilized to predict the behavior of the data center, identify saving opportunities, ensure IT-equipment reliability and providing general operation guidlines.

This study was carried out at a time when only few cabinets were occupied. A general method for defining the maximum cabinet capacity is introduced and the effect of installing the cabinets outlet near the wall is discussed.

2. Facility specifications

Binghamton University Pharmacy Building DC consists of a single cold aisle enclosure that has two rows of cabinets with 6 in-row coolers in between. Table 1 shows a detailed specification for the DC.

The data center is cooled by six Stulz CyberRow CRS-91 cooling units (80 H × 24 W × 42 D inch) each one is equipped with three blowers with a gross rate of 4927.1 m^3/h and a 33-kW cooling capacity.

Table 1. Pharmacy building DC specifications.

Property	Value
Physical	
Room area (m^2)	80.1
Room height (m)	3.05
Infrastructure	
Number of ACUs	6
Number of cabinets	10
Number of slots	420
Cooling capacity and heat load	
Total nominal cooling capacity (kW)	195
Total room heat load (kW)	16.8
Total ITE (kW)	1.69
Heat load	
Total nominal cooling airflow (m^3/h)	29562.8

3. Analysis Methodology

The data center performance was decided based on input and output variables. The input variables namely: ACUs SAT and flow rated (room level), cabinet heat load (rack level) and CPU utilization (server level). The output variable was the ITE inlet air temperatures.

The initial setpoint for the cooling units SAT was 23 °C with 100% maximum blowers speed, supplying 4927.1 m^3/h into the cold aisle containment (each). All ITE was assumed to dissipate 100% of the input power and concentrated in a fully sealed-contained cold aisle. Regarding the cabinets, a typical small leakage option was considered through them.

After that, a grid sensitivity analysis was carried out to determine the required number of cells that ensures the accuracy of the results.

In this study, initially a neutral pressure is maintained inside the CAC by using the full cooling capacity of the ACUs. SAT is varied to different values; namely 15, 17, 20, 23 and 25°C to investigate its effect on cabinet air inlet temperature.

Constant Turbulence Viscosity Models (k-ε) is used in the CFD solver [19]. The standard conservation equation for the turbulent kinetic energy k and rate of dissipation of turbulence energy ε are respectively:

$$\frac{\partial(\rho k)}{\partial t} + \frac{\partial(\rho k u_i)}{\partial x_i} = \frac{\partial}{\partial x_j}\left[\frac{\mu_t}{\sigma_k}\frac{\partial k}{\partial x_j}\right] + 2\mu_t E_{ij}E_{ij} - \rho\epsilon$$

$$\frac{\partial(\rho\epsilon)}{\partial t} + \frac{\partial(\rho\epsilon u_i)}{\partial x_i} = \frac{\partial}{\partial x_j}\left[\frac{\mu_t}{\sigma_\epsilon}\frac{\partial \epsilon}{\partial x_j}\right] + c_{1\epsilon}\frac{\epsilon}{k}2\mu_t E_{ij}E_{ij}$$
$$- c_{2\epsilon}\frac{\rho\epsilon^2}{k}$$

In such a small data center, the IT-equipment will be installed gradually so it might take years before occupying all the cabinets. For the current ITE load, it is not efficient to operate the data center as it will be operated when data center is fully loaded.

Based on that, the data center model was developed using 6SigmaRoom considering all components except the IT-equipment. Two levels of investigations were conducted which cover the current IT-equipment being used and then the cabinets were loaded to the data center maximum capacity of IT-equipment using servers from the design library of the CFD software. Each server power was set to 1 kW.

Up to this point, only one out of ten cabinets were occupied and here are the ITE that have been installed shown in table 2.

Table 2. Installed IT-equipment models.

ITE	Quantity
Dell M1000E blade server	1
M620 blade servers	3
M630 blade servers	4
Compellent SC4020 storage array controller	1
Compellent SC220 Enclosure (24 bay)	1

4. Results and Discussion
4.1. Case study 1: Current status
Meeting thermal load:

Current ITE thermal load is 1.69 kW which is considered a relatively low thermal load when it is being compared to the nominal cooling capacity of the in-row air cooling units that equals to 195 kW. Hence, it is not necessary to use all the air-cooling units to meet this thermal load.

Savings can be achieved by reducing the amount of energy consumed by the air-cooling units. By using the computational heat transfer model, three scenarios were investigated to meet the thermal load with minimal energy consumption. A description of those three scenarios were summarized in Table 3.

The first scenario studied running all the ACUs at full load, the second scenario considered running one cooling at full load and in the last scenario only one cooling unit was operated at 25% of its cooling capacity. Figure 2 Shows a top view of the DC model for these three scenarios.

Moving from scenario 1 to scenario 3 is limited by meeting the acceptable temperature range for operating a DC. Figure 3 Shows the operating temperature distribution of the loaded cabinet (LC) in the three scenarios.

Table 3. Description of the three scenarios.

Scenario	ACU units on duty	Cooling unit load (%)
1	6	100
2	1	100
3	1	25

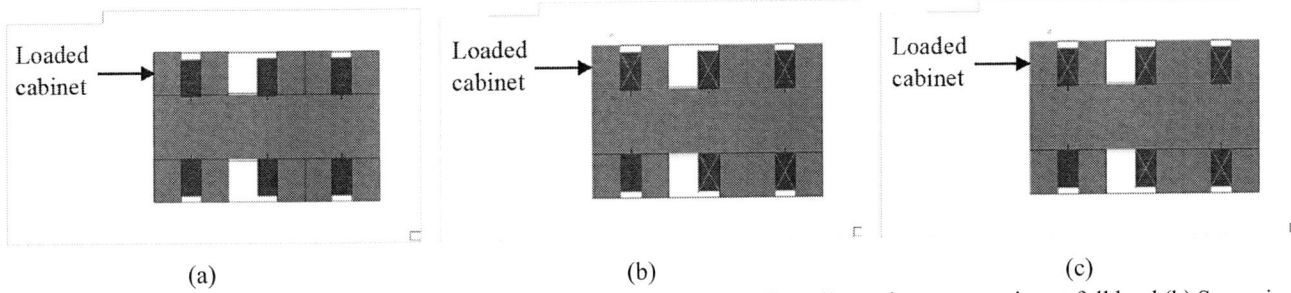

(a)　　　　　　　　　　(b)　　　　　　　　　　(c)

Figure 2. Top view of the data center for the three scenarios (a) Scenario 1: All cooling units are operating at full load (b) Scenario 2: One cooling unit is operating at full load (c) Scenario 3: One cooling unit is operating at partial load

Table 4. Effect of using the three scenarios on the cabinet temperatures, airflow rate, consumed energy and annual cost.

Scenario	Average cabinet inlet air temperature (℃)	Cooling airflow rate (m^3/h)	Fan operating power (kW)	Cooling power (kW)	Annual operating cost ($)
1	23	29562.8	16.8	1.68	32,200
2	23	4927.1	3	1.65	8,130
3	24.4	1231.8	0.75	1.6	4,120

In scenario 3 the temperature at the top of the cabinet seems to be high due to hot air recirculation. still, the ITE are safe since that the ITE are being installed in the bottom half of the cabinet.

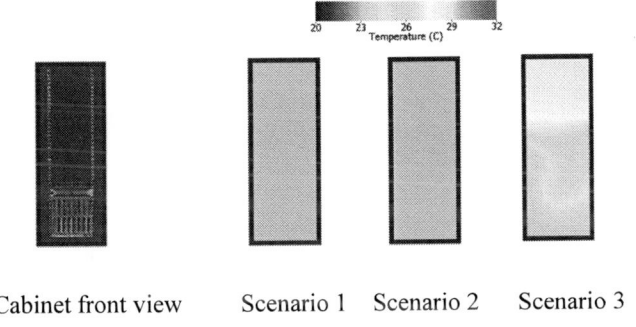

Cabinet front view　　Scenario 1　Scenario 2　Scenario 3

Figure 3. loaded cabinet front view and the temperature distribution of the loaded cabinet for the three considered scenarios.

Based on the above, in the three case scenarios all of the ITE operating temperature lie within ASHRAE recommended DC operating temperature range which is 18 ℃ to 27 ℃ and therefore savings can be achieved when moving from scenario 1 to scenario 3 by reducing the number of ACUs being used and by decreasing the operating energy of the ACU. Table 4 shows the details savings opportunities.

In table 4, the annual operating cost was calculated by considering that the 24h/day for the whole year and with average energy cost of $0.2/kWh. Based on these operating costs the maximum amount of annual savings that can be attained is $28k when comparing scenario 1 with scenario 3. These calculations are overestimating the annual savings since that the ITE load is relatively small in comparison with the cooling load. However, the over estimated saved cost between scenario 1 and scenario 3 is presented to qualify the wasted energy (money) when the pod is over cooled when no need.

General operation instructions: ACU location

Bringing the cooling units closer to the heat source may enhances cooling efficiency. However, in air cooling, the airflow pattern takes a place in determining the cooling efficiency.

When using only one cooling unit with one loaded cabinet, simulations results revealed that it is better to turn on the in-row cooler in the opposite row to the loaded cabinet than turning the one in the same row. Because hot air recirculation was reduced by running the ACU opposite to the loaded cabinet since the pressure distribution at the cabinet inlet is higher in this case.

Figure 4 shows the pressure distribution at the loaded cabinet inlet for the cases of running the in-row in the same row with loaded cabinet and running the in-row cooler in the opposite row.

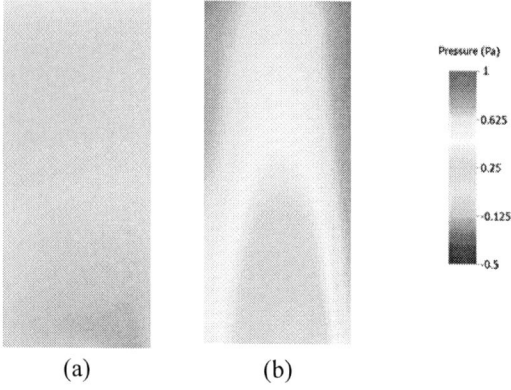

(a)   (b)

Figure 4. pressure distribution at the cabinet inlet when (a) Operating the ACU in the same of the loaded cabinet. (b) Operating the ACU in the opposite row of the loaded cabinet.

General operation instructions: Installing ITE

Because the number of ITE in the current stage is fewer than the capacity of each racks, it is important how to occupy the racks. Therefore, in the model, ITE were installed at the top, middle and bottom of the cabinets to achieve the most appropriate configuration. Figure 5 shows front view of the temperature contours of the left row cabinets when installing the ITE at these locations.

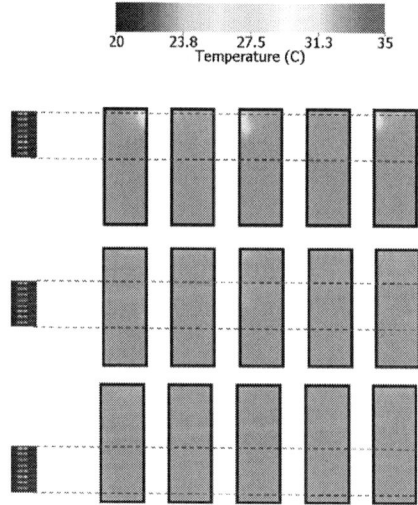

Figure 5. Temperature contours of the cabinets when installing ITE at different locations.

Simulation showed that hot spots started to appear if the ITE were installed at the top of the cabinet which caused by the hot air recirculation inside the cabinet. Hot air recirculation occurs due to the build of back pressure behind the cabinet when installed at the top. These hot spots temperature exceeds the ASHRAE recommended temperature range and can be harmful for the ITE. As a result, it is recommended not to install the ITE at the top of the rack.

General operation instructions: Installing blockage inside the cold aisle

As long as all the cabinets in the DC are not occupied, it is better to separate the settled cabinets using a blockage such as meat curtains which is shown in Figure 6. Installing such a blockage will help in guiding the airflow from cooling unit(s) to the ITE and eventually, energy saving can be achieved by reducing airflow demands and leakages.

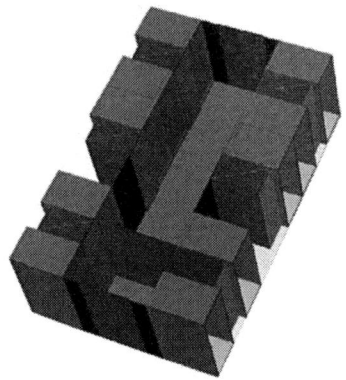

Figure 6. cold aisle containment when the obstruction being installed.

4.2. Case study 2: Future commissioning

With the passage of time, new ITE will be installed and these ITE thermal load should be less than the ACUs total cooling capacity because there are other components that emit heat inside the DC such as: uninterruptible power supply UPS, power distribution, air conditioning units and lighting. Based on that, a detailed analysis that showed the DC maximum capacity should be carried out.

DC capacity:

In a small typical raised floor DC (under 4,000 ft2) ITE thermal contribution is to be limited to 71% of the ACUs total cooling capacity to make sure that the DC is working within the recommended operation conditions [20].

Table 5. Effect of changing the SAT on the average cabinet inlet air temperature and the average room temperature.

Supply air temperature (℃)	Average ITE inlet air temperature (℃)	Average room temperature (℃)
15	16.1	24.9
17	18.1	26.8
20	21.1	29.4
23	24.1	32.7
25	26.6	34.6

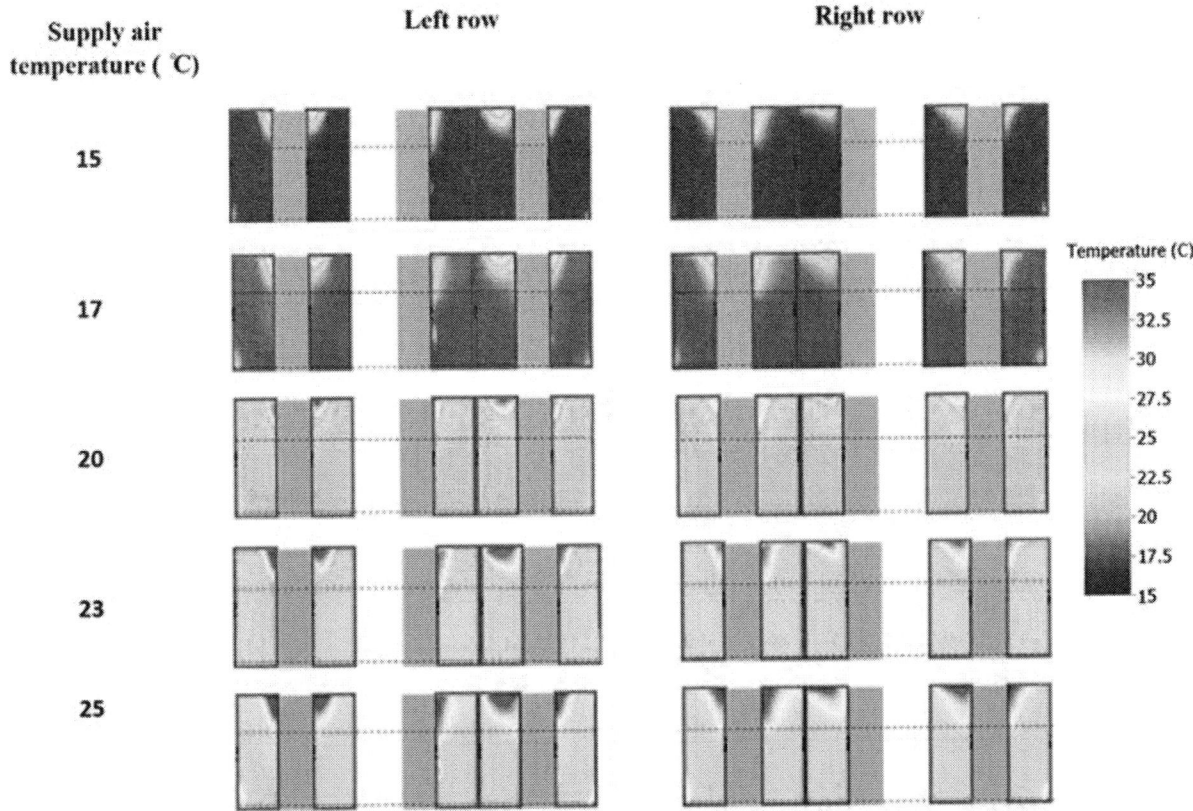

Figure 7. Cabinet inlet air temperature contours at different supply air temperature.

However, this percentage can be increased for the DCs that use unraised floor with in-row coolers, since that the cooling units are closer to the ITE and there are no leakages throw the raised floor.

Starting from this point, the ITE thermal load in this DC is to be limited to 80% of the nominal cooling capacity and by dividing this load equally on the cabinets to avoid hot spots formation, each cabinet capacity was limited to 15.6 kW. Later discussion will proof the validity of this assumption.

Identifying maximum SAT

Since that every DC has its own unique configuration, thereby, it could be considered as best practice of operation to identify its maximum allowable SAT. The constraints that limit the SAT are: first, satisfying ASHRAE guidelines and vendor specifications for environmental conditions. Second, achieving the comfort of humans who are occupying the room.

In order to specify the maximum SAT that can be achieved without violating those constraints, the effect of supplying air at different temperatures was investigated.

In this analysis, the cabinets were loaded to their maximum thermal load using 15 2U servers installed in each cabinet starting from the bottom of the cabinet. Results are summarized in Table 5 which shows the effect of changing the SAT on the average ITE inlet air temperature and the average room temperature.

Since that this DC will not be occupied by humans there are no restrictions on the room temperature which represents the temperature of the air outside the cold aisle containment. Therefore, the only restriction on increasing the air temperature is the ITE inlet air temperature.

For all of the various SAT, average ITE inlet temperature falls within ASHRAE recommended temperature zone. Anyway, the cabinet inlet air temperature contours should be considered to check if there are any hot spots were formed within the cabinets. Figure 7 shows the ITE inlet air temperature distribution when the air was supplied at a various temperature.

In Figure 7 the region between the dashed lines represents the ITE region and the upper region in the cabinets represent the blanked slots.

As can be noticed in the Figure 7, hot spots start to appear in this region when the SAT increased to 25 ℃. Therefore, the SAT should not exceed 23 ℃.

Regarding the blanked slots, this region encounters a high temperature due to the hot air recirculation anyway, because there is no ITE in this region, the it will not be harmful.

In a typical DC, it is not likely to have such a considerable amount of hot air recirculation at the top of the rack but in this DC, the distance between the cabinet outlet and the wall is relatively small.

When the air leaves the cabinet at a high speed it will run into the adjacent wall, as a result, a high-pressure region evolved near the wall at the back of the cabinet, which will push the air back into the cabinet.

Installing network switches

Usually, network switches are being installed at the top of racks. However, if hot air recirculation is considered to occur at the top of the cabinet that as stated before in the above section and giving the fact that the network switches have a weak fan, it may not overcome this recirculation. Hence, it will not be safe to install these switches at the top of the cabinet.

This issue can be solved by overprovisioning the cold aisle in a way that will eliminate the hot air recirculation at the top of the racks. This can be achieved by increasing the ACUs capacity or by reducing the cabinets thermal load. As it is obvious that the second solution is more applicable, the model was utilized to predict the cabinet capacity and the results shows that its capacity should be limited to 14kW in case of installing network switches.

5. Conclusions

For the current ITE it is enough to run one air cooling unit with 25% of its full flow rate.

In the case of running one cooling unit, it is better to run the one opposite to the loaded cabinet.

Avoid installing the ITE at the top of the cabinet.

Installing blockage inside the cold aisle containment will reduce the leakage and will improve cooling efficiency.

For the future as the DC grows, it is necessary to limit the room load to 156 kW (80% full cooling capacity) in order to keep the ITE safe.

It is recommended to distribute the ITE equally on the cabinets. Moreover, the cabinet load should be limited to 15.6 kW in order to keep environmental conditions within ASHREA limits and ensure ITE equipment safe functionality and reliability.

SAT should not exceed 23 ℃ for this data center.

In case in installing switches, cabinet capacity should be reduced to 14kW.

Acknowledgement

We would like to acknowledge Future Facilities Ltd . We would also like to thank the ES2 Partner Universities for their support and advice. This work is supported by NSF IUCRC Award No. IIP- 1738793.

References

[1] ASHRAE Technical Committee 9.9, Thermal Guidelines for Data Processing Environments, 4th ed. Atlanta: W. Stephen Comstock, 2015.

[2] G. Mohsenian, S. Khalili, and B. Sammakia. "A Design Methodology for Controlling Local Airflow Delivery in Data Centers Using Air Dampers." In 2019 18th IEEE Intersociety Conference on Thermal and Thermomechanical Phenomena in Electronic Systems (ITherm), pp. 905-911. IEEE, 2019.

[3] A. Marashi. "Power Hungry: The Growing Energy Demands of Data Centers". 2019. Retrieved from: https://www.vxchnge.com/blog/power-hungry-the-growing-energy-demands-of-data-centers

[4] H. Alissa, Kourosh Nemati, Bahgat Sammakia, Mark Seymour, Ken Schneebeli, and Roger Schmidt. "Experimental and numerical characterization of a raised floor data center using rapid operational flow curves model." In ASME 2015 International Technical Conference and Exhibition on Packaging and Integration of Electronic and Photonic Microsystems collocated with the ASME 2015 13th International Conference on Nanochannels, Microchannels, and Minichannels. American Society of Mechanical Engineers Digital Collection, 2015.

[5] Khalili, S., Mohsenian, G, Desu, A., Ghose, K., Sammakia, B., (2019). Airflow Management Using Active Air Dampers in Presence of a Dynamic Workload in Data Centers. SEMI-THERM 2019.

[6] M. Tradat, H. Alissa, K. Nemati, S. Khalili, B. Sammakia, M. Seymour, and R. Tipton., 2017, March. Impact of elevated temperature on data center operation based on internal and external IT instrumentation. In 2017 33rd Thermal Measurement, Modeling & Management Symposium (SEMI-THERM) (pp. 108-114). IEEE.

[7] M. Tradat, B. Sammakia, H.Alissa and K. Nemati. 2018, November. Experimental Analysis of Chiller Cooling Failure in a Small Size Data Center Environment Using Wireless Instrumentation. In ASME 2018 International Technical Conference and Exhibition on Packaging and Integration of Electronic and Photonic Microsystems. American Society of Mechanical Engineers Digital Collection.

[8] K. Nemati, H. Alissa, and B. Sammakia. "Performance of temperature controlled perimeter and row-based cooling systems in open and containment environment." In ASME 2015 International Mechanical Engineering Congress and Exposition. American Society of Mechanical Engineers Digital Collection, 2016.

[9] T. Gao, B. Sammakia, J. Geer, B. Murray, R. Tipton, and R. Schmidt. "Comparative Analysis of Different In Row Cooler Management Configurations in a Hybrid Cooling Data Center." In ASME 2015 International Technical Conference and Exhibition on Packaging and Integration of Electronic and Photonic Microsystems collocated with the ASME 2015 13th International Conference on Nanochannels, Microchannels, and Minichannels. American Society of Mechanical Engineers Digital Collection, 2015.

[10] A. Bhalerao, A. Ortega, and A. P. Wemhoff. "Thermodynamic Analysis of Hybrid Liquid-Air-Based Data Center Cooling Strategies." In ASME 2014 International Mechanical Engineering Congress and Exposition. American Society of Mechanical Engineers Digital Collection, 2014.

[11] K. Nemati, H. Alissa, B. Murray, B. Sammakia, and Mark Seymour. "Experimentally validated numerical model of a fully-enclosed hybrid cooled server cabinet." In ASME 2015 International Technical Conference and Exhibition on Packaging and Integration of Electronic and Photonic Microsystems collocated with the ASME 2015 13th International Conference on Nanochannels, Microchannels, and Minichannels. American Society of Mechanical Engineers Digital Collection, 2015.

[12] H. Alissa, K. Nemati, B. Sammakia, M. Seymour, K. Schneebeli, and R. Schmidt. "Experimental and numerical characterization of a raised floor data center using rapid operational flow curves model." In ASME 2015 International Technical Conference and Exhibition on Packaging and Integration of Electronic and Photonic Microsystems collocated with the ASME 2015 13th International Conference on Nanochannels, Microchannels, and Minichannels. American Society of Mechanical Engineers Digital Collection, 2015.

[13] E. Wibron, Anna-Lena Ljung, and T. Staffan Lundström. "Computational fluid dynamics modeling and validating experiments of airflow in a data center." Energies 11, no. 3 (2018): 644.

[14] Y. Fulpagare, Atul Bhargav, and Yogendra Joshi. "Dynamic thermal characterization of raised floor plenum data centers: Experiments and CFD." Journal of Building Engineering 25 (2019): 100783.

[15] K. Nemati, B. Sammakia, K. Ghose, and M. Seymour. "Innovative Approaches of Experimentally Guided CFD Modeling for Data Centers." In Semiconductor Thermal Measurement and Management Symposium (SEMI-THERM), 2015 31st Annual. 2015.

[16] S. Alkharabsheh, B. Sammakia, S. Shrivastava, and R. Schmidt. "A numerical study for contained cold aisle data center using CRAC and server calibrated fan curves." In ASME 2013 International Mechanical Engineering Congress and Exposition. American Society of Mechanical Engineers Digital Collection, 2014.

[17] X. Xiong, Y. Fulpagare, C. Sun, and P. Seng Lee. "Numerical Study of a New Rack Layout for Better Cold Air Distribution and Reduced Fan Power." In 2019 18th IEEE Intersociety Conference on Thermal and Thermomechanical Phenomena in Electronic Systems (ITherm), pp. 399-404. IEEE, 2019

[18] K. Karki, Amir Radmehr, and Suhas V. Patankar. "Use of computational fluid dynamics for calculating flow rates through perforated tiles in raised-floor data centers." HVAC&R Research 9, no. 2 (2003): 153-166.

[19] Versteeg, Henk Kaarle; Malalasekera, Weeratunge (2007). An introduction to Computational Fluid Dynamics: The Finite Volume Method. Pearson Education.

[20] N. Rasmussen. "Calculating Total Cooling Requirements for Data Centers". N.A. Retrieved from: https://www.apc.com/us/en/support/resources-tools/white-papers/calculating-total-cooling-requirements-for-data-centers.jsp

Self-Heating Investigation in SOI MOSFET Structures with High Thermal Conductivity Buried Insulator Layers

Konstantin Petrosyants and Dmitry Popov
National Research University Higher School of Economics (Moscow Institute of Electronics and Mathematics)
34 Tallinskaya, 123458, Moscow, Russia
kpetrosyants@hse.ru

Abstract

The self-heating of thin and short silicon on insulator (SOI) devices is a serious problem limiting operation and reducing device lifetime. The main source of this phenomenon is low thermal conductivity of SiO_2 buried oxide (λ_{SiO2}=1.4 W/m·K 100 time lower than that of silicon λ_{Si}=140 W/m·K). In this work three novel buried oxide (BOX) configurations in SOI MOSFET structure which suppress the self-heating effect are analyzed: 1) ultra-thin SiO_2 BOX; 2) buried layers with high thermal conductivity material AlN, Al_2O_3, Si_3N_4; 3) stack of traditional SiO_2 BOX with AlN plug. TCAD was used to study the 2D temperature distribution in device structures. The simulation results show that the thermal dissipation and self-heating were dramatically improved in all the devices. The UTB and UTBB ultra-thin SiO_2 BOXes reduce the maximum internal device temperature 25 K and 39 K lower than that of normal SiO_2 BOX; AlN, Al_2O_3, Si_3N_4 BOXes reduce T_{max} by 31-24 K lower than that of SiO_2 BOX. The advantages and difficulties in practical use of the SOI MOSFETs with novel buried insulator layers are discussed.

Keywords

SOI MOSFETs, buried oxide (BOX), buried insulator layers, thermal conductivity, self-heating, power dissipation.

1. Introduction

Due to the key advantages of SOI MOSFETs over bulk Si MOSFETs (perfect isolation on chip, higher speed, lower voltage/power operation, etc.) the SOI technology has been adopted by major companies for CMOS VLSIs fabrication.

However, in spite of many potential advantages over conventional bulk CMOS technology it has serious problem with self-heating in submicron SOI devices. The low thermal conductivity of the buried silicon-dioxide (SiO_2) λ_{SiO2} = 1.4 W/m·K (100 times lower than that of silicon λ_{Si} = 140 W/m·K) seriously limits the heat dissipation from active device region.

Several attempts were made to improve the conventional SiO_2 SOI MOSFET structure, and reduce the internal body temperature in operation regime. The following SiO_2 BOX configurations were developed to suppress the self-heating effect: SELBOX [1], Partial BOX [2], Quasi-SOI [3], Double-

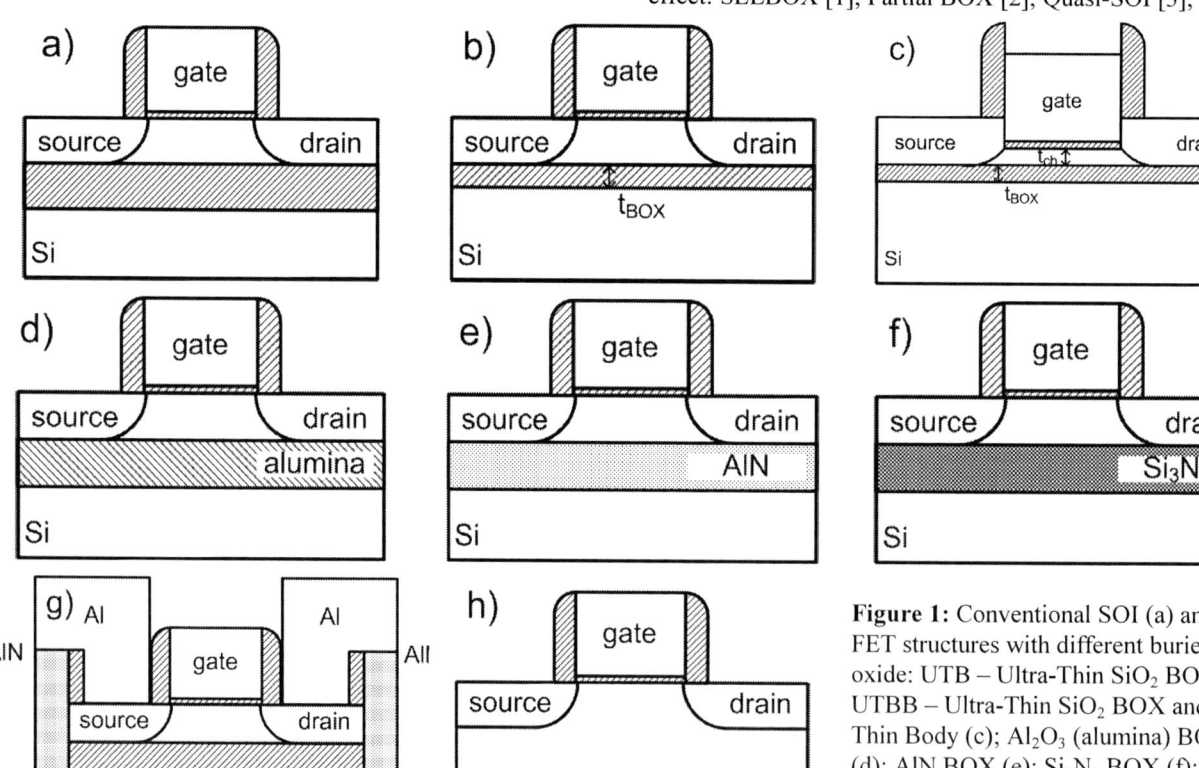

Figure 1: Conventional SOI (a) and SOI FET structures with different buried oxide: UTB – Ultra-Thin SiO_2 BOX (b); UTBB – Ultra-Thin SiO_2 BOX and Ultra Thin Body (c); Al_2O_3 (alumina) BOX (d); AlN BOX (e); Si_3N_4 BOX (f); stack of SiO_2 BOX with AlN plug (g) and traditional bulk MOSFET (h).

SOI [4], Ultra-Thin BOX (UTB) [5], Ultra-Thin Body and Box (UTBB) [6]. The effectiveness of these structures has been investigated in our previous works [7, 8]. It was shown that most perspective for deeply scaled VLSIs are UTB and UTBB structures. But the technology process of their fabrication is more complex than that of conventional SOI CMOS process.

Another method to solve this problem is to replace the traditional SiO_2 BOX layer with high thermal conductivity insulator (TCI) [9-12]. However, the effectiveness of these structures has not been quantitatively investigated. It is the purpose of this work.

2. SOI MOSFET structures with high TCI buried layers.

The device structures selected for consideration are presented in Fig. 1. The common n-channel SOI MOSFET structure (Fig. 1, a) was used with the following parameters: L=100 nm, $t_{gate,ox}$=2 nm, t_{Si}=70 nm, t_{BOX}=80 nm, N_{ch}=8·10^{17} cm^{-3}, $N_{d,s}$=1·10^{20} cm^{-3}, N_{LDD}=4·10^{19} cm^{-3}.

In Fig. 1, b, c the SOI MOSFETs with SiO_2 BOX are presented: b) Ultra-Thin BOX with t_{SiO2}=10 nm [5]; c) Ultra-Thin Body and BOX with t_{ch}=25 nm, t_{SiO2}=6 nm [6].

In Fig. 1, d the SOI MOS device with buried alumina (Al_2O_3) is presented. The thermal conductivity of alumina (λ_{Al2O3}=20 W/m·K) exceed by more than one order of magnitude that of SiO_2 (λ_{SiO2}=1.4 W/m·K).

In Fig. 1, e the SOI MOSFET structure with aluminum-nitride (AlN) buried layer [9-10] is presented. The thermal conductivity of AlN is about 100 times that of SiO_2 (136 W/m·K) and roughly equal to that of silicon itself (140 W/m·K).

In Fig 1, f the SOI device structure with silicon-nitride (Si_3N_4) buried layer with λ_{Si3N4}=30 W/m·K is presented [12].

In Fig. 1, g the SOI MOSFET structure with SiO_2 BOX and additional two-side high TCI plug of aluminum-nitride (AlN) is presented. When the source electrode is connected to the substrate, the internal temperature near the drain is not reduced enough, and the high TCI plug is necessary [13].

For comparison, in Fig. 1, h the traditional bulk MOSFET structure is presented as an example with excellent temperature dissipation.

3. SOI MOSFET electro-thermal model

The basical electro-thermal model built into Sentaurus Synopsys tool [14] was selected for thermal regimes description in the device structures Fig. 1. This model couples the equations for electron-hole transport and the heat transfer in the 2D/3D device structures:

$$j_{n/p} = f(n/p, \nabla(n/p), E, T, \nabla T) \qquad (1)$$

$$\nabla(\lambda \nabla T) - P(j_n, j_p, E, T) = \rho C \frac{\delta T}{\delta t} \qquad (2)$$

where:

- λ, C, ρ – thermal conductivity, capacity and specific gravity;
- n(x, y), p(x, y), E(x, y) – concentration of electrons and holes, electrical field distributions in device structure;

- T(x, y), j_n(x, y), j_p(x, y) – temperature and current densities distributions in device structure;
- P(x, y) – power density.

To take into account the special thermal effects in modern deep submicron and nanoscale devices the eq. (1)-(2) were complemented by the set of new models for the temperature-dependent physical parameters: thermal conductivities λ_{Si}(T), λ_{SiO2}(T) (see Fig. 2); effective carrier mobility μ_{eff}(T, E); oxide and trapped charge densities N_{ox}(T), N_{it}(T) (see Fig. 3).

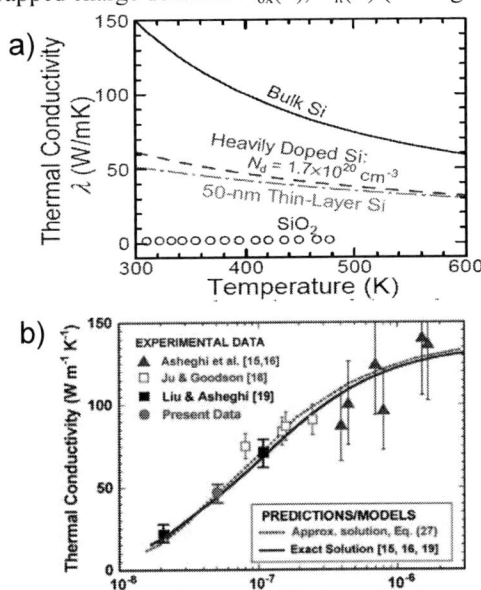

Figure 2: Temperature dependences of thermal conductivity of Si (λ_{Si}) and SiO_2 (λ_{ox}) on heavy doping concentration [6] (a) and on thickness of silicon channel layer [15] (b).

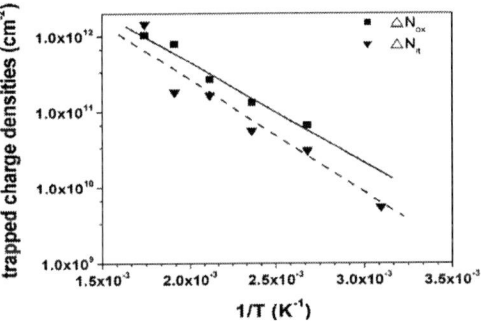

Figure 3. Oxide- and interface-trap charge densities vs. stress temperature for SiO_2 capacitors. [16]

4. Validation of electro-thermal model

In FETs the "hot spots" are located in the depth of the device structures. So the direct methods of temperature measurement (like IR thermography) are not applicable; the indirect thermal measurement methods are necessary for electro-thermal model validation. The four-terminal (4T) gate resistance measurement technique was used [17].

The test MOSFET structure with gate configured for four-point resistance measurements is shown in Fig. 4.

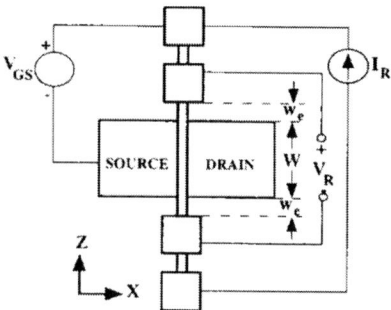

Figure 4: Top view of test MOSFET structure. [17]

The electrical resistance of the gate depends strongly on temperature. So the polysilicon gate plays the role of a temperature-sensing resistor for channel region, where the device power is dissipated. The details of the measuring method are presented in [17].

Model validation was carried out for two groups of SOI MOSFETs: submicron and nanoscale.

The submicron devices with W/L=10/0.3 um, silicon thickness t_{Si}=80 nm and BOX thickness t_{BOX}=360 nm were fabricated and measured using 4-T gate resistance technique. In Fig. 5 the 0.3 um SOI FET channel temperature vs power obtained by calibration measurement and TCAD modeling is presented. The maximum error is not more than 10 K at P_{tot}=10 mW. The good agreement is observed with experimental results published in [17]. As an example, the simulated 2D temperature distribution in MOSFET structure for P_{tot}=9.42 mW is presented in Fig. 6.

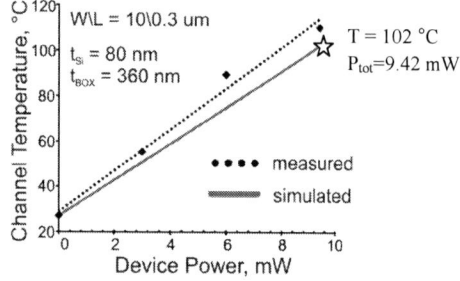

Figure 5: Channel temperature versus power for 0.3 um SOI MOSFET.

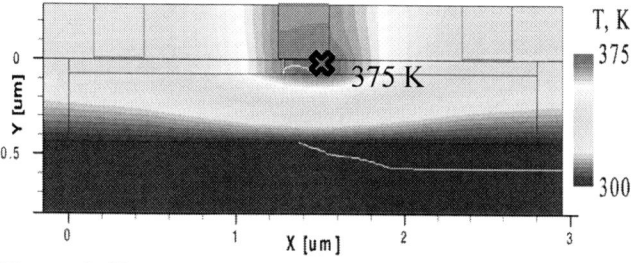

Figure 6: 2D temperature distribution in the test 0.3 um SOI MOSFET structure for dissipated power 9.42 mW.

For nanoscale SOI FET with W/L=0.45/0.045 um, silicon thickness t_{Si}=52 nm and BOX thickness t_{BOX}=147 nm the experimental data were taken from [6]. The comparison of measured and simulated channel temperatures is presented in Fig. 7. The good agreement with measured data was achieved.

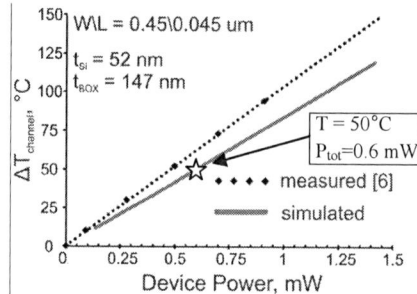

Figure 7: $\Delta T_{channel}$ versus power for 45nm SOI MOSFET; ($\Delta T_{channel}$=$T_{channel}$-27°C).

As an example, the simulated 2D temperature distribution in the 45 nm SOI MOSFET structure for P_{tot}=0.6 mW is shown in Fig. 8.

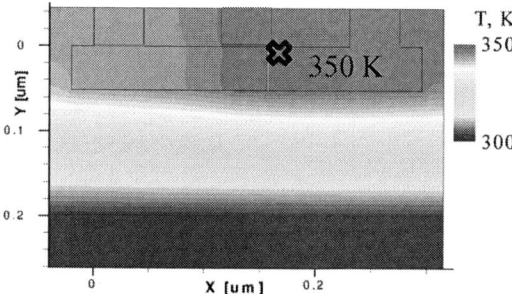

Figure 8: 2D temperature distribution in the test 45 nm SOI MOSFET structure for dissipated power 0.6 mW.

5. Analysis of thermal simulation results

The most critical parameter – maximal internal temperature T_{max} was selected for thermal regimes in device structures (see Fig. 1) analysis using TCAD modeling.

Temperature T_{max} for all the devices was simulated using TCAD for the operation regime V_{DS} = 2 V, V_{GS} = 2 V. The results were compared with the corresponding results for conventional SOI and bulk-Si FET structures. The 2D temperature profiles for the devices are presented in Fig. 9.

It is seen in Fig. 9 that for all the device structures the temperature peak T_{max} is located in the transistor body. The hot spot is shifted to the drain region.

Fig. 9, b, c show that the channel temperature can be reduced by as much as 25 K and 39 K for UTB and UTBB SiO_2 BOXes. These structures were recommended for deep nano-scaled SOI CMOS VLSIs [6], but the technology process of their fabrication is more complex than that of conventional SOI CMOS process.

If the Al_2O_3 (alumina) BOX is substituted to SiO_2 the internal device temperature is reduced by 26 K (see Fig. 9, d). On the other hand, alumina is a high-k dielectric and alumina layer (Al_2O_3) should not be too thin (lower limit 20-50 nm), otherwise the coupling between the front and back interfaces is reinforced and the subthreshold swing is degraded.

If the AlN BOX is substituted to SiO_2 the maximum device temperature T_{max} falls down up to 317 K (see Fig. 9, e) that is equal to the temperature of conventional bulk MOSFET. Thus, using AlN (λ_{AlN}=136 W/m·K is roughly equal to that of silicon itself λ_{Si}=140 W/m·K) as the buried insulator should essentially eliminate the self-heating penalty of SOI. But AlN is the material with high-k dielectric constant κ_{AlN}=8.7, and the coupling of drain and source terminals through the channel and BOX layers increases the off-state current I_{off} and threshold voltage shift ΔV_{th}. It was shown in [10] that the I_{off} in AlN SOI is 6.7 times higher than that in normal SOI MOSFET. So it is necessary for very small size devices to find the compromise between the heat dissipation path and the second order effects.

If the Si_3N_4 BOX is used (see Fig. 9, f) the T_{max} reduces by 24 K than that of conventional SiO_2 BOX. It can be compared with Al_2O_3 BOX.

Fig. 9, g shows the internal maximum temperature for the stack of conventional SiO_2 BOX with AlN plug. The plug provides effective thermal paths from source and drain to substrate. It successfully suppresses the lattice temperature rise throughout the whole device. T_{max} is reduced from 348 K for conventional SOI to 323 K for stack structure SOI. The novel SOI device structure has two very important advantages. It is almost free from the fabrication issues and performance issues in use of high-k material. It is easy to fabricate using current trench isolation techniques. The disadvantage of the structure Fig. 1, g is an increase of the device area.

6. Conclusions

The self-heating effect in SOI MOSFET structures with novel BOX configurations was analyzed using TCAD. The conventional SiO_2 BOX layer with low thermal conductivity (λ_{SiO2}=1.4 W/m·K) was replaced by three novel configurations: 1) ultra-thin SiO_2 BOXes; 2) buried layers

with high thermal conductivity materials AlN, Al_2O_3, Si_3N_4; 3) stack of traditional SiO_2 BOX with AlN plug.

The improved electro-thermal model built-into Sentaurus Synopsys tool is used to taking into account the special thermal effects in modern deep submicron and nano-scale SOI MOSFETs. The maximum internal temperature T_{max} was chosen as a criterion for device thermal stability analysis under hard operation conditions.

The simulation results have shown the serious improvement in the thermal dissipation and self-heating decrease for all the structures under consideration: T_{max} was reduced by 23-39 K in comparison with conventional SOI MOS structure (see Fig. 10). The minimum T_{max}=309 K was achieved in UTBB with SiO_2 BOX structure because of ultra-thin BOX thickness t_{BOX} =6 nm. It was confirmed that in the structure with the AlN buried insulator the self-heating penalty of SOI is eliminated, and T_{max}=317 K is the same as for bulk MOS device (T_{max}=316 K).

The advantages and difficulties in practical use of the SOI MOSFETs with novel buried insulator layers are discussed. The deep submicron SOI MOSFET structures with AlN, Al_2O_3 and Si_3N_4 BOXes could be recommended for use in high temperature applications (with reasonable constraints).

Acknowledgments

The research was supported by Basic Research Program at the National Research University Higher School of Economics in 2019, grant No. TZ-99 and by Russian Foundation for Basic Research, grant No. 18-07-00898 A.

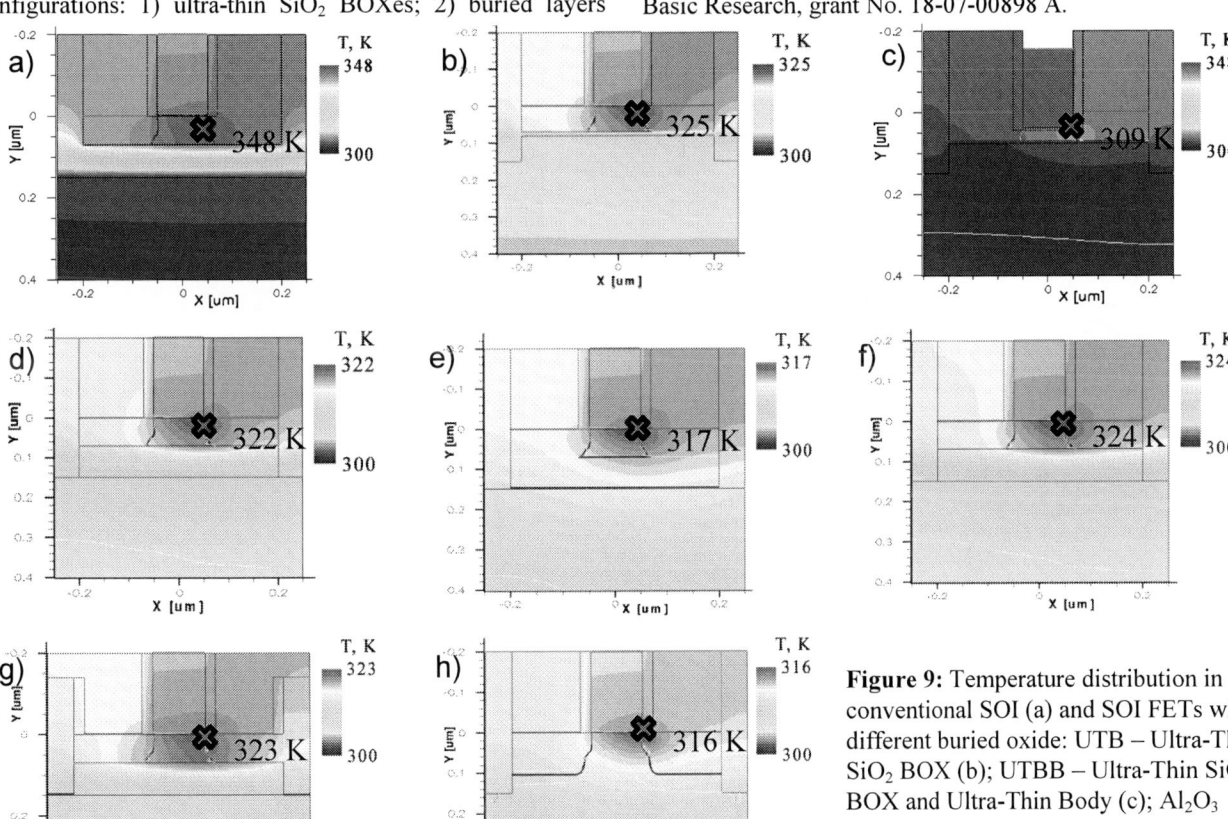

Figure 9: Temperature distribution in conventional SOI (a) and SOI FETs with different buried oxide: UTB – Ultra-Thin SiO_2 BOX (b); UTBB – Ultra-Thin SiO_2 BOX and Ultra-Thin Body (c); Al_2O_3 (alumina) BOX (d); AlN BOX (e); Si_3N_4 BOX (f); stack of SiO_2 BOX with AlN plug (g) and traditional bulk FET (h).

References

1. M. R. Narayanan, H. A. Nashash, Minimization of self-heating in SOI MOSFET devices with SELBOX structure, Proceedings of the 11th Int. Conference on Advanced Semiconductor Devices & Microsystems, 2016, pp. 61-64.

2. J. Cheng, B. Zhang and Z. Li, The Total Dose Radiation Hardened MOSFET with Good High-temperature Performance, IEEE, Proc. of ICCCAS, 2007, pp. 1252-1255.

3. Weikang Wu, Xia An, Taotao Que, Xing Zhang, Dongjun Shen, Gang Guo and Ru Huang, "Investigation of a radiation-hardened quasi-SOI device: performance degradation induced by single ion irradiation", Semiconductor Science and Technology, Vol. 31, No. 10, 105009 (6pp) (2016).

4. Huang Y., Li B., Zhao X., Zheng Z., Gao J., Zhang G., Li B., Zhang G., Tang K., Han Z., Luo J., An Effective Method to Compensate Total Ionizing Dose-Induced Degradation on Double-SOI Structure, IEEE Transactions on Nuclear Science, Volume: 65, Issue: 8 , Aug. 2018, pp. 1532-1539.

5. V.P. Trivedi, J.G. Fossum, Nanoscale FD/SOI CMOS: thick or thin BOX?, IEEE Electron Device Letters, Volume: 26, Issue: 1, Jan. 2005, pp. 26-28.

6. T. Takahashi, T. Matsuki, T. Shinada, Y. Inoue, K. Uchida, Comparison of self-heating effect (SHE) in short-channel bulk and ultra-thin BOX SOI MOSFETs: Impacts of doped well, ambient temperature, and SOI/BOX thicknesses on SHE, Electron Devices Meeting (IEDM), 2013, pp. 7.4.1-7.4.4.

7. Petrosyants K., Popov D., Simulation of self-heating effect of MOSFET with various configuration of buried oxide, Russian Microelectronics. 2019. Vol. 48. No. 7. pp. 487-493.

8. Petrosyants K. O., Popov D. Comparison of Self-heating Effect in SOI MOSFETs with Various Configuration of Buried Oxide, in: Proceedings of the 2nd International Conference on Microelectronic Devices and Technologies (MicDAT '2019). Barcelona : International Frequency Sensor Association (IFSA), 2019. pp. 24-28.

9. Biegel, B.A., Osman, M.A., Yu, Z. Analysis of aluminum-nitride SOI for high-temperature electronics, Proc. of HiTEC, June 2000, pp. 1-8.

10. Y. Yuan, G. Yong, G. Peng-Liang, A Novel Fully Depleted Air AlN Silicon-on-Insulator Metal-Oxide-Semiconductor Field Effect Transistor, Chin. Phys. Letters, 2008, Vol 25, №8, pp. 3048-3051.

11. Oshima K., Cristoloveanu S., Guillaumot B., Iwai H., Deleonibus S., Advanced SOI MOSFETs with buried alumina and ground plane: Self-heating and short-channel effects // Solid-State Electronics 48 (2004), pp. 907–917.

12. J. Roig, D. Flores, S. Hidalgo, M. Vellvehi, J. Rebollo, J. Millan, Study of novel techniques for reducing self-heating effects in SOI power LDMOS, Solid-State Electronics 46 (2002), pp. 2123–2133.

13. Komiya K., Kawamoto T., Sato S, Omura Y., Impact of High-k Plug on Self-Heating Effects of SOI MOSFETs // IEEE Transactions On Electron Devices, Vol. 51, No. 12, December 2004, pp. 2249-2251.

14. TCAD Sentaurus User Manual J-2014.09, Synopsys.

15. M. Asheghi, K. Kurabayashi, R. Kasnavi, K. E. Goodson, "Thermal conduction in doped single-crystal silicon films", J. Appl. Phys. Vol. 91, №8 (2002), pp. 5079-5088.

16. D. M. Fleetwood, M. P. Rodgers, L. Tsetseris et al., Effects of device aging on microelectronics radiation response and reliability, Microelectronics Reliability Vol. 47 (2007), pp. 1075-1085.

17. Lisa T. Su, James E. Chung, Dimitri A. Antoniadis, Kenneth E. Goodson, and Markus I. Flik, Measurement and Modeling of Self-Heating in SOI NMOSFET's, IEEE Transactions On Electron Devices, Vol. 41. No. 1. January 1994, pp. 69-75.

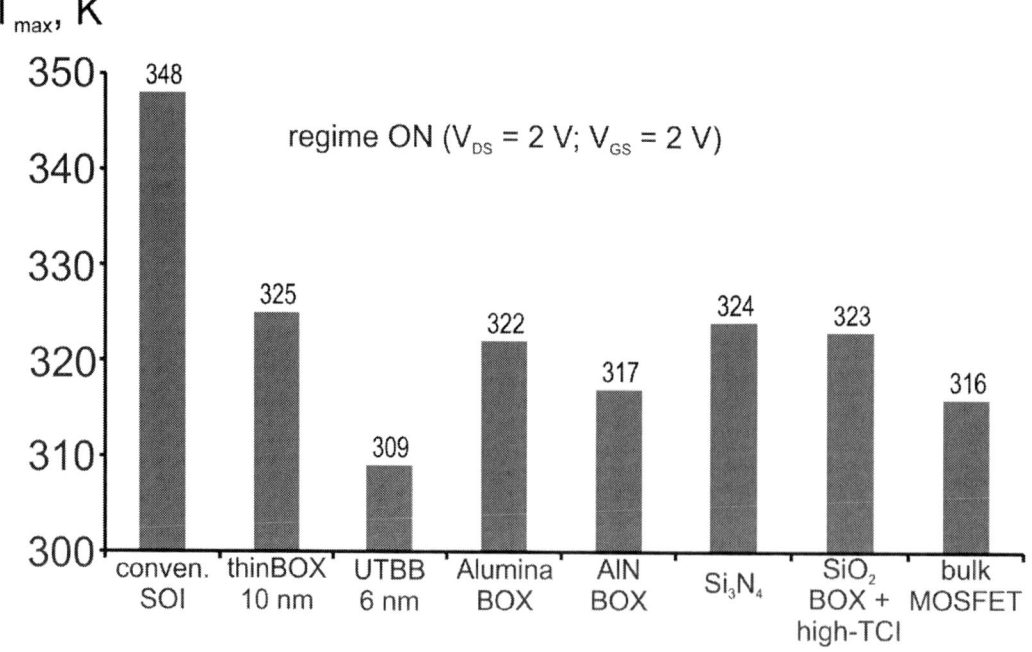

Figure 10: The simulated T_{max} for SOI MOSFETs with different buried oxide.

Transient Thermal Model for a Wearable Device in Contact with Human Skin

Bruce Guenin
Consultant
San Diego, CA USA
sdengr-bguenin@usa.net

Abstract

Makers of wearable devices, as in most areas of consumer electronics, try to provide as much performance and functionality in the use of these devices consistent with certain thermal limits. These thermal limits, of course, deal with peak temperatures in the silicon chips within the device. For wearables, the external temperatures of these devices are also critical from the viewpoint of user comfort and safety. To achieve the greatest accuracy in a thermal model for the wearable device, it is necessary to use a robust and accurate model for the transfer of heat into human skin. The typically used ad hoc assumption of an isothermal boundary condition representing the region of contact between a wearable device and human skin is no longer adequate.

Keywords

Biological, Medical, Thermal, Pennes Model, Wearable Electronic Devices

Nomenclature

MKS units are used throughout.

1. Introduction

In the medical and biological fields, the methodology for characterizing and modeling the transfer of heat into or out of living tissue, both in humans and other animals, is a mature area of study. The dominant methodology in this regard referred to as is the Pennes biothermal model.[1,2] It takes the form of Fourier's heat diffusion equation supplemented by the assumption of a mechanism for cooling the tissue by blood flow, which Pennes called "perfusion." Use of the Pennes model requires that certain specified material properties be measured for each of the different tissue types involved in the heat flow of interest, namely: thickness, thermal conductivity, specific heat, and perfusion rate.

2. Scope

The scope of the work is to explore the thermal interaction between a wearable device and human skin subject to various use conditions. The device is represented by a simple lumped element model consisting the aggregated heat capacity of the device and two thermal resistance paths conducting heat to its immediate environment: 1) the ambient air and 2) human tissue in the region of contact.

The representation of the skin includes heat conduction in the epidermis, dermis, and hypodermis skin layers and the effect of the perfusion cooling mechanism on each of these layers. Figure 1 depicts heat transfer from an electronic device in contact with human skin and heat removal due to the perfusion process.

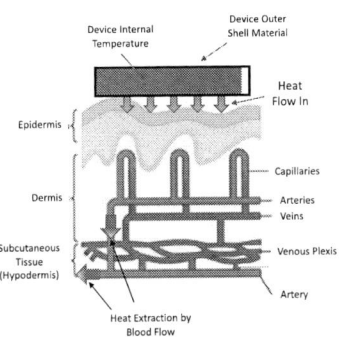

Figure 1: Heat Transfer from Device into Human Skin

Currently there are two distinct models that have been developed in the course of this work.

Model 1: the electronic device is not represented explicitly in the thermal network. The device simply provides a constant temperature boundary condition to the skin layers. The thermal network in Figure 2 is of this type.

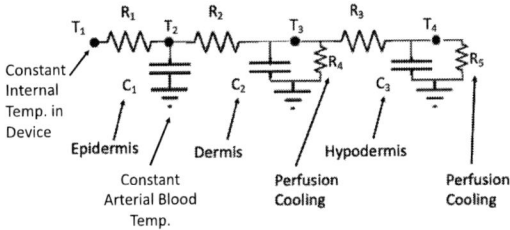

Figure 2: Thermal circuit representing the transient heat flow from a device into the various skin layers and then into the blood supply, the result of perfusion cooling.

Model 2: a wearable device is explicitly included in the thermal network. It is simplified in that its interior is assumed to be at a uniform temperature and to have the form factor and power dissipation of a modern smart watch.

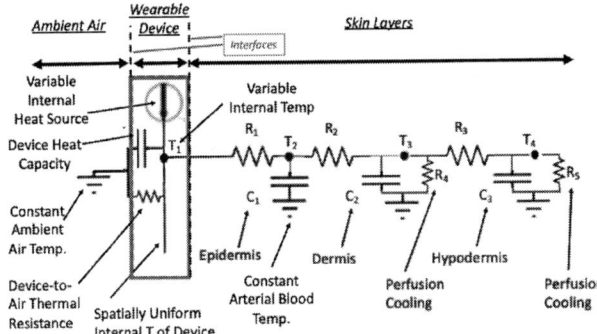

Figure 3: Thermal circuit representing the transient heat flow originating from the interior of a wearable device and then flowing out of its front and back faces by two paths: 1) out of its top face into the ambient air via convection and radiation and 2) out of its bottom face into the skin layers via conduction and out of them via the perfusion mechanism.

The following discussion focuses on Model 1 since it is the simpler of the two and the treatment of the heat transfer within the body is the same in either model. Figure 2, depicts a thermal circuit diagram representing heat flow from a device whose internal temperature, T1, was set at 49C. The temperature of the "ground" nodes was set to 37C (a typical value of arterial blood temperature for a healthy adult.

The following Tables provide the basic material properties as required by the Pennes model, the thermal resistance associated with each material layer, the lumped thermal resistance values calculated (R1, R2, and R3) the heat capacity values (C1, C2, and C3), and the perfusion equivalent resistances (R4 and R5). Note that all of these calculations assume 1-dimensional heat, flow perpendicular to the skin surface. This assumption is a good representation of the reality since the total thickness of the skin layers (~ 4 mm) is much less than the lateral dimension of a typical contact area (~ 25mm).

NOTE ON TABLES 1-4: The R and C values were calculated assuming 1-dimensional heat flow, normal to the top of each cuboidal volume, each representing a particular material layer. The cuboids were assigned thickness values from Table 1. The in-plane area of each cuboid was assigned a value of 1 cm². The material properties in Table 1 were measured on the anterior forearm of a human subject.[2]

Note that a highly detailed procedure for calculating values for all of these parameters has been published by this author and is available online.[3]

TABLE 1					
Material Properties					
Material Class	Material	Thickness	Th.Cond.	Spec. Heat	Perfusion Rate
		(m)	(W/mK)	(J/(m3·K))	(1/s)
Skin Layers	Epidermis	8.00E-05	0.1	3.44E+06	0
	Dermis	0.001	0.168	4.54E+06	0.0024
	Hypodermis	0.003	0.168	5.49E+06	0.0024
Device Shell Options	Metal (Al)	0.001	240	N/A	N/A
	Ceramic (Al2O3)		17		
	Plastic		0.2		

TABLE 2		
Thermal Resistance* per Each Material Layer		
LAYER	THERMAL RESISTANCE	
	SYMBOL	°C/W
Epidermis	R,epi	8.0
Dermis	R,derm	59.5
Hypodermis	R,hypo	179
Shell:Metal	R,shell	0.04
Shell: Ceramic		0.59
Shell: Plastic		50.0
*Representing thermal conduction		

TABLE 3			
Lumped Thermal Resistance Values			
SYMBOL	CALCULATION	SHELL MAT'L	RESISTANCE VALUE °C/W
R1	R,shell + 1/2 R,epi	Metal	4.04
		Ceramic	4.59
		Plastic	54.00
R2	1/2 R,epi + 1/2 R,derm	N/A	33.80
R3	1/2 R,derm + 1/2 R,hypoderm	N/A	119.00

TABLE 4				
Circuit Elements Associated with the Volume of Each Skin Layer				
LAYER	HEAT CAPACITY		R,PERFUSION	
	SYMBOL	(J/°C)	SYMBOL	°C/W
Epidermis	C1	0.0275	N/A	N/A
Dermis	C2	0.5	R4	1043
Hypodermis	C3	1.6	R5	348

3. Results

Results for Model 1 and Model 2 are discussed below.

3.1 Model 1 Results

Representative plots of calculated transient temperatures for the above model under two different assumptions for the device external shell material are shown: a metal shell, assumed to be aluminum, and a plastic shell. The results are plotted with two different time scales: 1) a short time scale appropriate for the rise time of the initial thermal transient and 2) a long timescale to capture the steady-state temperature behavior.

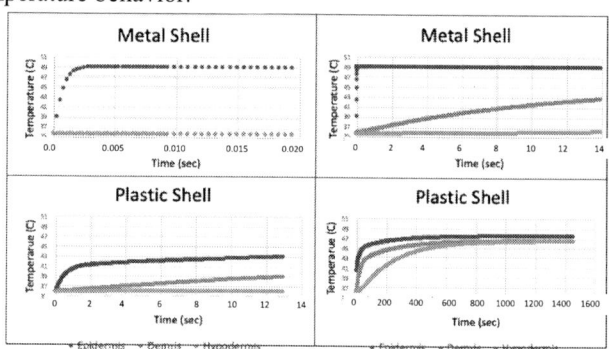

Figure 3 Thermal simulation results showing the transient behavior of skin layer temperatures due to contact with a powered handheld device having an outer shell made of a variety of materials having differing thermal conductivities.

A quick examination of these results shows that:
- The initial heating of the epidermis is much quicker with the metal shell (0.001 sec) than with the plastic shell (1 sec).
- The time to reach the steady state epidermis temperature is much shorter with the metal shell (0.01 sec) than with the plastic shell (~800 sec = 13 min).

The scope of this effort does not, at this time, include an experimental validation. However, a sanity check was performed by comparing the results of the model to the thermal behavior implicit in a widely used standards document relating burn risk to the temperature of a surface contacting the skin and the duration of the contact, as depicted in Figure 4.[5]

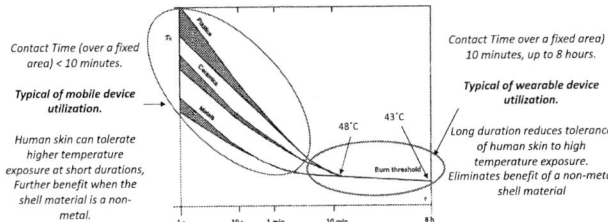

Figure 4: Burn threshold temperature vs contact time for different device shell materials.[5]

Consistent with the temporal predictions of the model, at short contact times (~1 sec) human skin can tolerate higher device temperatures with the plastic shell than with the metal shell. This is due to the much slower heating rate for the epidermis when the shell is plastic than when it is metal. As the model indicates, with the metal shell, the epidermis reaches the device temperature within 0.005 sec after the moment of contact.

On the other hand, once the contact time exceeds 10 min, the temperature of the epidermis is the same for both the plastic and metal device shell materials.

3.2 Model 2 Results

The use of Model 2 is still at the preliminary stage. However, results to date are very encouraging since they provide insight into the complicated interaction between a low-mass wearable device and the various skin layers.

Representative plots of calculated transient temperatures for the above model under two different sets of thermal conditions. Figure 5 demonstrates the effect of applying 0.25W to wearable device for 5 sec and then turning power off.

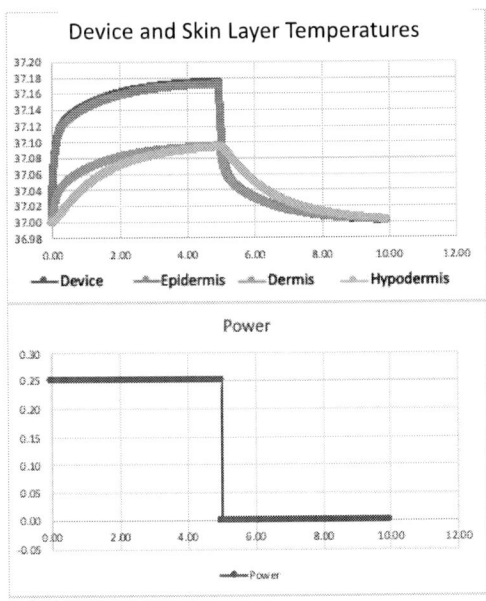

Figure 5: Thermal results using Model 2 showing the temperature behavior for a wearable device in contact with human skin during the application of 0.25 W of power for a duration of 5 sec.

Note that the epidermis temperature closely tracks that of the device. The dermis layer manifests half the temperature rise of the epidermis. The hypodermis layer shows a similar temperature rise as the dermis, but with a longer time constant.

Figure 6 shows the effect of a pre-heated wearable device being put in contact with human skin. Looking at the top graph, with the linear time scale, one sees that the device temperature quickly spikes up and then quickly cools. This is due to its small mass, 30 gm, and correspondingly small heat capacity. The bottom graph, with the logarithmic time scale, shows that the device temperature drops from 50C to 43C in only a millisecond. During this very brief temperature excursion, the dermis and hypodermis layers show almost no temperature change. The epidermis layer manifests a temperature rise of 5C, reaching a maximum temperature of 43C. However, it is at this elevated temperature for less than 0.1 second. Per the time-temperature guidelines in the IEC burn standard in Figure 4, at this very short heating duration, the device could have been at a much higher temperature without burning the wearer of the device.

The implication is that this widely-used standard, that has performed so well in determining the risk of burns with larger electronic devices, simply does not apply to wearable devices who are in the size range of smart watches.

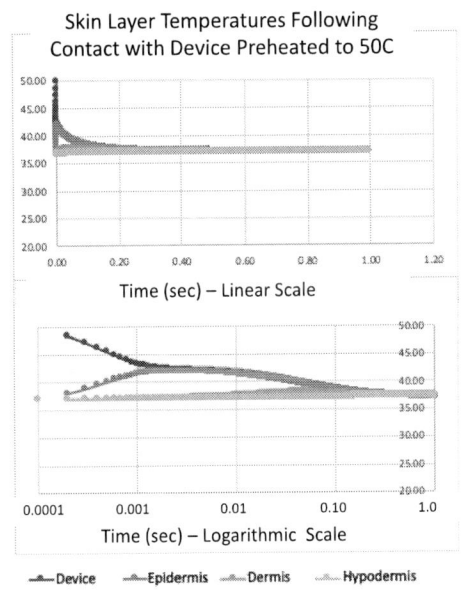

Figure 6

4. Conclusions

A numerical thermal network model has been successfully used to apply the Pennes bio-thermal model methodology to situations regarding skin contact with both larger electronic devices (using Model 1) and to wearable devices (Model 2). Future work will focus on the further development of Model 2 and the exploration of more situations involving wearable devices with respect to their detailed interactions with human skin. It will also be used to develop guidelines governing the burn risk in the use of wearable devices, given that the prevailing IEC standard seems to have limited applicability in this area.

Acknowledgments

This is to acknowledge the essential role of Alex Ockfen in encouraging me to submit this paper to Semi-Therm and for helpful discussions regarding the exploration of various scenarios of interest from the user point of view.

References

1. H. H. Pennes, "Analysis of tissue and arterial blood temperatures in resting human forearm," Journal of Applied Physiology, 1 (1948) pp. 93–122.

2. Maria Strąkowska, et al., "Evaluation of Perfusion and Ther- mal Parameters of Skin Tissue Using Cold Provocation and Thermographic Measurements," Metrol. Meas. Systems, Vol. 23 (2016), No. 3, pp. 373–381.

3. B. Guenin, "Transient Thermal Calculations for Skin Contact by Wearable Devices," Calculation Corner, Electronics Cooling, June, 2018, pp 9 - 13. https://www.electronics-cooling.com/wp-content/uploads/2019/06/ElectronicsCooling_2019_Summer_Edition.pdf

4. B. Guenin, "Transient Modeling of a High-Power IC Package, Part 1," Calculation Corner, Electronics Cooling, Dec, 2011. https://www.electronics-cooling.com/2011/12/transient-modelling-of-a-high-power-ic-package-part-1/

5. IEC GUIDE 117 -- Electrotechnical equipment - Temperatures of touchable hot surfaces Appendix A, 2010.

DNN-based Fast Static On-chip Thermal Solver

Jimin Wen, Stephen Pan, Norman Chang, Wen-Tze Chuang, Wenbo Xia, Deqi Zhu, Akhilesh Kumar,
En-Cih Yang, Karthik Srinivasan, Ying-Shun Li
ANSYS, Inc.
{jimin.wen,stephen.pan,norman.chang,wentze.chuang,wenbo.xia,deqi.zhu,akhilesh.kumar,
anita.yang,karthik.srinivasan,ying-shiun.li}@ansys.com

Abstract

Accurate prediction of on-chip temperature distribution becomes important for the performance and reliability of upcoming 5G, automotive, and AI chip-package-systems. In particular, a large thermal gradient (the temperature variation across a chip) accelerates electromigration and aging, and also impacts design performance and power. Furthermore, there are usually Tmax (maximum temperature) constraints on junctions of a chip, skin temperature concerns for mobile devices or wearables, and important placement considerations of on-chip thermal sensors for use in dynamic voltage and frequency scaling. However, obtaining an accurate and detailed thermal gradient on-chip is very time-consuming using the finite element method (FEM) or computational fluid dynamics (CFD) technology. Furthermore, there are many different functional scenarios for various applications that users need to identify possible Tmax locations on-chip. Therefore, there is an urgent need in the industry to provide a fast, yet accurate on-chip thermal solution in a chip-package-system or more complicated 3DIC design, which may include multiple chips. This paper proposes a method to use a data-driven DNN-based thermal solver that can be 100-1000x faster depending on the size of the chip compared to traditional FEM-based thermal solvers with the same level of accuracy.

Keywords

DNN, CFD, FEM, system-level thermal, DeltaT Predictor, power map, chip thermal profile

1. Introduction

As process technology advances toward 10/7/5nm and beyond, chip device and power density has increased significantly, resulting in more thermal challenges. For chip thermal considerations, both on-chip Tmax and thermal gradient are important.

On-chip thermal gradient is the temperature variation across a chip. Accurate prediction of Tmax and the on-chip thermal gradient becomes important for the performance and reliability of chip-package-systems in the upcoming 5G [1], Automotive [2], and AI (Artificial Intelligence) designs [3]. Large Tmax and on-chip thermal gradient accelerates electromigration (EM) and aging [4], and also impacts power/ground voltage drop [5] and timing [6,7]. Furthermore, there are usually Tmax (maximum temperature) constraints on junctions of a chip [8], skin Tmax of chip-package-system (CPS) for the maximum surface temperature of mobile devices [9], and also placement determination of on-chip thermal sensors [10,11] for DVFS (Dynamic Voltage and Frequency Scaling) design technique [12].

However, obtaining detailed Tmax locations and thermal gradient on-chip is very time-consuming using conventional finite element method (FEM) or computational fluid dynamics (CFD) technologies Adding to the complexity, there are many different switching scenarios for various applications that users need to analyze possible Tmax locations and on-chip thermal gradient. Therefore, there is an urgent need in the industry to speed up on-chip thermal analysis in a chip-package-system or more complex 3DIC designs consisting of multiple chips [13].

A system-level thermal analysis may need to account for multiple parts of the system level thermal environment including chips, packages, other heat sources on PCB, cooling elements, surrounding air volumes, fans, and chassis (Figures 1 and 2) [14]. In the figures, the fan-out wafer level package (FOWLP) design containing two chips is used as a popular low-cost package to replace a typical BGA substrate. The thin re-distribution layer (RDL) generated by the chip foundry is used directly along with the chip layer stackup without the need of a separate packaging process.

In the resulting on-chip thermal profile, the coarse mesh for the silicon chip usually has much less resolution than required in evaluating the reliability of devices and wires in each chip. Even with a detailed power map, the thermal results are largely smeared by the coarse mesh on-chip and cannot show the realistic Tmax and thermal gradient of the chip clearly. For example, in Figure 3, given a detailed tile-based power map on-chip (lower-left), the coarse mesh model smears the power details first (upper-left) and results in a coarse thermal gradient (upper-right). The fine grid response (lower-right) is desired to provide sufficient details in temperature distribution for accurate on-chip reliability evaluation. However, when the on-chip mesh density is increased to the system-level thermal analysis, the thermal analysis model will likely be much larger and prove challenging to be solved efficiently.

Applying numerical approaches like FEM, the best solution is definitely from the node-to-node continuous model. While this is hard in the case of system-level thermal problems, the common approach is to use constraint equations to connect coarse and fine sides of incompatible meshes or to use submodeling[15] for a chip in the system. Though the performance of these solutions is in general good, there could be accuracy losses depending on the modeling assumptions of the analysis. There are also other approaches like the multi-grid solution [16] to address the problems in simulation with large variations in physical scales. This is similar to constraint equations or submodeling methods and the accuracy may depend on the modeling assumptions.

The chip model in the system analysis was typically a silicon block. If the details of the thermally sensitive chip layers ae not included in the chip model, the accuracy of the resulting thermal profile would be questionable, even if a fine mesh on chip was used in the system analysis. Section 2 will discuss issues related to the realistic modeling of on-chip thermal behavior.

On the other hand, in a typical chip-package-system design flow of today's semiconductor industry, the system and chip designers have mostly independent workflows. When a chip is designed, it could be used by different system providers who may not have easy access to details like power map in sufficient granularity. A chip designer typically has an electronics background, while the system thermal expert may not have in depth knowledge of the chip. Addressing chip thermal reliability issues requires contributions from multiple groups and disciplines. The best thermal environment data available to the chip designer are commonly the approximate system-level temperature profile on the target chip. The more desired approach is for chip designers to be able to generate detailed on-chip Tmax and thermal gradient.

Figure 4 is a block diagram illustrating how on-chip thermal details could be created for chip designers. The shortcut we propose in this paper can bypass the costly field solution. The system-level thermal results were based on the full-chip power map or total power on-chip, but the power density everywhere was by average due to coarse mesh smearing when compared to some high-power density tiles in the power map. The localized thermal effects from the high-power tiles were missing in the system-level thermal analysis and the local temperature rise can be retrieved as a post-processing step as shown in the last step in Figure 4 for final T map generation at high resolution. The basic assumption is that the system-level thermal analysis is done with a very coarse grid for on-chip thermal analysis so there is no double-counting of solving the final Tmax or thermal gradient on-chip. The results may not replace the field solution directly, due to potentially complicated packaging, but it certainly is useful to chip designers in determining the hotspot locations, realistic temperature levels, and accurate thermal gradient for sensor placement applications. This should also be helpful for chip-level thermal-aware EM and stress reliability, voltage drop, timing, and other thermal-related issues of concern to chip designers.

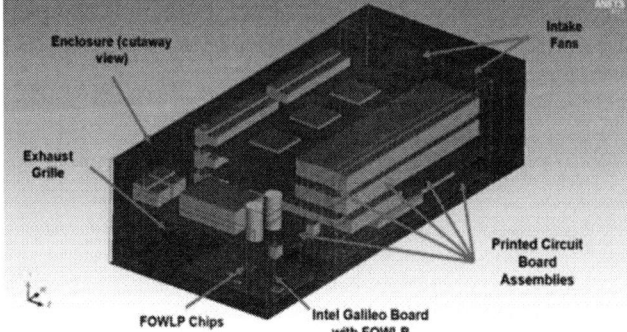

Figure 1: Typical thermal modeling components in system. There are two side-by-side chips in the FOWLP package [14].

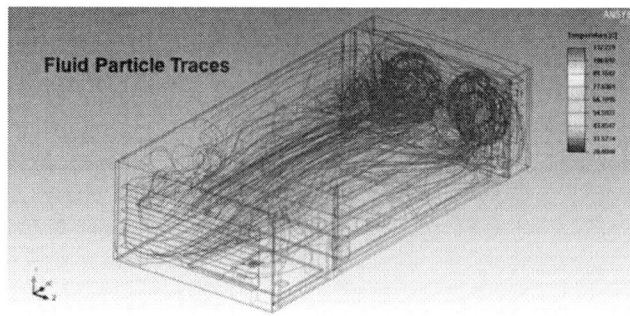

(A) Temperature on air partical traces in the system

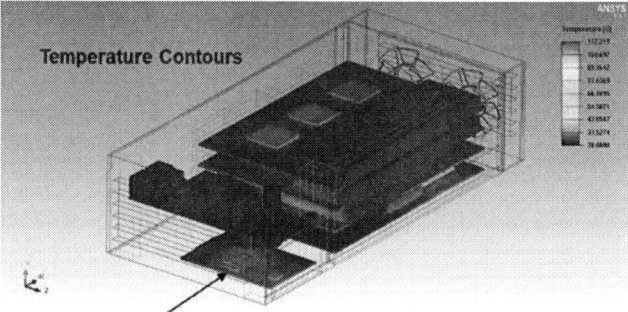

(B) Solid components in system

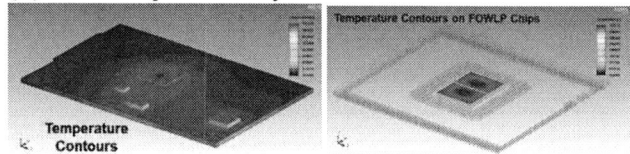

(C) Target pkg+board (left) and on chips in FOWLP (right).

Figure 2: System-level thermal responses, from (a) system air, (b) PCB's, to (c) chips [14]

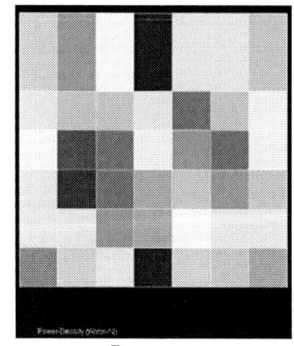

Power　　　　　Temperature

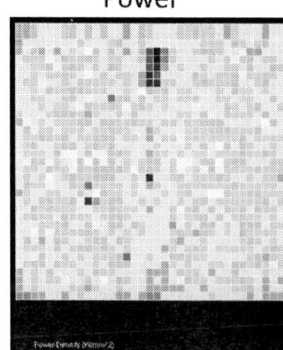
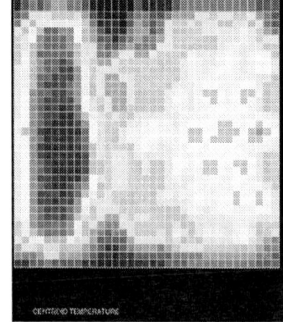

Figure 3: Example of coarse (100um tiles) and fine mesh (10um tiles) in a chip thermal analsysis.

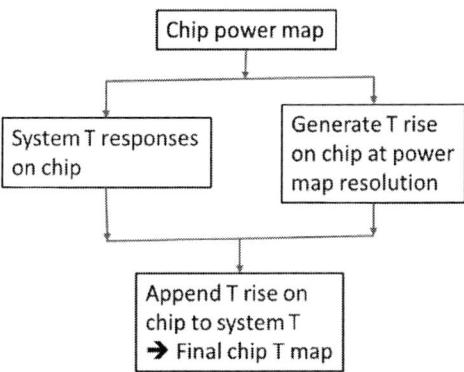

Figure 4: Block diagram for chip thermal map enhancement

The chip model at a system-level thermal analysis is typically treated as a silicon block with hierarchical modeling [17-20]. In the FEM or CFD method, the resolution enhancement approach is to make the chip mesh size comparable to that of the detailed power map from a chip power calculation tool [21]. For example, in the digital design of a chip, an instance realized by a standard cell, i.e., an abstract logic representation with multiple transistors generating heat, is considered as the basic element that generates the heat at the cell level. 10um x 10um is a typical grid or tile size that can capture enough details of the power variations at the cell level and is used in demonstrating the validity of our approach.

One important application of the detailed chip thermal profile is for determining the proper locations of thermal sensors in the neighborhoods of hot spots in design, as the thermal sensor is typically from 15 to 50um in size [10]. 10um in power map resolution and Finite Element (FE) size is sufficiently small for this purpose (lower-right picture in Figure 3). Note that the FE size should be even smaller to capture the detailed temperature variations. This certainly will increase the cost of generating results. This grid size of the power map is scalable as the technology advances. The detailed power map represented by 10um x 10um tiles can usually readily be obtained from a dynamic voltage drop analysis tool [21] and is referred to in the following discussions.

Our approach for resolution enhancement of a chip thermal profile is obtained by applying a Deep (learning) Neural Networks (DNN)-based DeltaT Predictor based on the detailed power map combined with the coarse on-chip thermal profile from the system-level thermal analysis. DNN allows computational models that are composed of multiple processing layers to learn representations of data with multiple levels of abstraction [22]. We use it to train and predict the temperature rise of a tile on-chip. The resulting temperatures on tiles include thermal interactions from the neighbors of each tile and the system-level thermal responses.   In the following sections, the chip thermal behavior will be reviewed in section 2 to understand how the chip should be modeled to get correct thermal responses for localized heating on a silicon chip. The description of the approach in using Delta T and the coarse temperature profile from system-level thermal analysis is described in section 3. The implementation of DNN through training with FEM results and the inferencing for the final temperature profile with the enhanced resolution are shown in

sections 4 and 5. Finally, section 6 covers the results and discussion, followed by the conclusion.

2.   On-chip Thermal Modeling

In system-level thermal analysis, chip(s) are usually modeled as pure silicon blocks with coarse meshes. Figure 5 shows an example of the fine grid power and thermal maps for a 10mm x 10mm chip at a resolution close to 10um in size. The chip has about 30 million instances. Using fine 10um mesh size on the silicon block, it took more than 12 hours to solve for the chip+package+PCB model using FEM.

Modeling the chip as a silicon block for thermal analysis ignores the details of many other structures on-chip (Figure 11). Although the main volume of a chip is the silicon substrate, the details in the chip structure have significant impact on the thermal responses on-chip. The important chip structures for thermal characteristics are the thin interconnection layers with embedded wires and insulator/BOX (Buried Oxide) layer under chip devices. There are non-uniform local temperature rises and decays everywhere in the layers, due to the existence of the dielectric material, e.g., SiO_2, with very low thermal conductivity. The effects of the detailed structures on the thermal response can be observed in the thermal field FE analysis using a sub-model of a chip including the chip design details [14]. An important observation is that local temperature rise will decay fast when the dielectric media exists. Hence, the conductive silicon model used in Figure 5 enables the thermal effects to reach far from the heat sources, even though this is not happening in reality. The predicted hotspot temperature will also be underestimated, due to fast heat dissipation in silicon materials. From our study, the realistic thermal decay distance for either device or wire segments is about 5um in the dielectric layers [17]. Also, the temperature increase on devices is expected to be higher than that in Figure 5 as the heat sources are buried in the dielectric materials in the CMOS chip configurations, as shown in Figure 11. As the device (self-heating) power is dominant ($\sim$ >90% of total power) in a chip [21], we will focus on the device power map and its thermal effects in this paper. Joule heating in wire segments is not in the scope of this study but covered in [17].

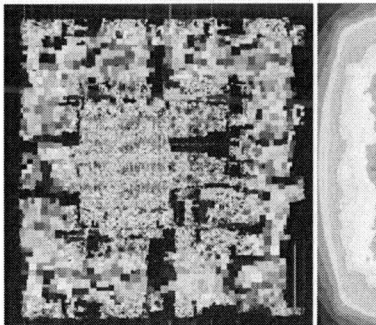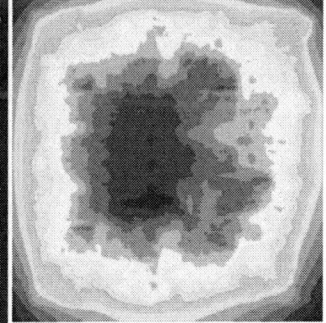

Figure 5: Tile-based power map (left) and thermal profile (right) of a 10mm x 10mm chip with 30 million instances, modeled as a silicon block at package+PCB level, was solved by FEM with elapse time exceeding 12 hours with mesh resolution about 10um in size.

Note that the thermal responses from a chip model of silicon block in Figure 5 are still valuable and can be treated as the thermal profile on the top face of the chip (i.e. at back-end or top metal layer near solder bump in Figure 11) as the results of the overall heat flow balances in the system. The real temperature profile on devices inside the chip remains unknown.

When the real device junction temperature is of concern, the dielectric media should be included in the analysis model, even in the system-level thermal analysis. Hence, two factors in analysis affect the thermal results of the chip, i.e., mesh fineness and dielectric media in the chip model. Results in Figure 5 address the mesh fineness, but not the effects from dielectric media in a chip.

Figure 6 [14] shows the temperature decay behavior in a chip model with as much detail in modeling as possible, i.e., via/wire layout, dielectric media, devices, and the silicon substrate. The temperature decay curve describes the temperature decreasing along the distance to a single heat source. The Y axis shows the normalized temperature change against the reference temperature without heat source applied. The decay curves are measured along different directions. For one of the directions along the metal layer – the X direction, its reverse direction also been measured. The decay behaviors are similar from either wire segment heating or device heating. Layout does have some impact to the thermal decay as shown in Figure 6, but the temperature impact will drop to near zero over the distance well before 5um. If a full chip model is to be constructed, the on-chip thermal decay behavior should be checked with those in Figure 6 for different technology nodes.

Instead of using fine mesh on-chip at system-level thermal analysis, which is impractical due to the long run time, we need an efficient way to get the temperature rise from devices across the whole chip.

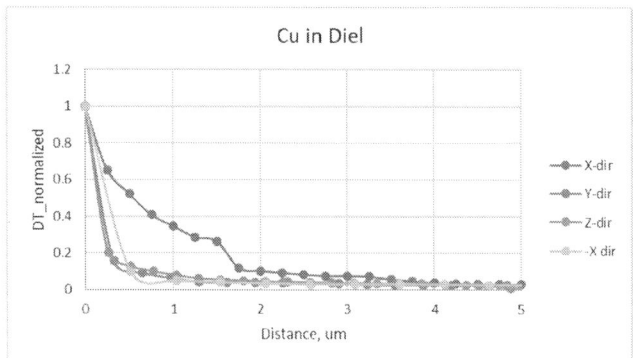

Figure 6: Temperature decay from a heated object in the device or interconnection layers of a deep submicron chip design [14]

3. Approach for Thermal Profile Enhancement Using DNN-based Modeling

This section explains how the localized heating effect (LHE) at multiple locations can be combined with the thermal response from a uniform average power on-chip to generate a DeltaT map at fine resolution. For an example of a multi-chip package, e.g., 3DIC, the system-level thermal response is shown in Figure 7A. The Theta-JA of each die can be extracted by the maximum temperature (Tmax) on-chip, as shown

schematically in Figure 7B, from the chip thermal profile of die (1), i.e., Theta-JA(1). For each die, Tmax depends on the heating of neighboring chips in the package or from the system, and the thermal profile may not be symmetric to the chip center. Note that the schematic thermal profile (bar chart) was plotted along a path through the Tmax location on-chip for illustration of the system-level temperature variation across the chip. The Theta-JA(i) of die(i) is calculated based on the coarse mesh on-chip modeling in the system. This Theta-JA and the coarse chip thermal profile represent the system response when LHE is not significant.

We assume that the coarse (base) thermal profile can be separated from the LHE effect of small tiles with a high power density as there was virtually no LHE in the coarse mesh model. In system-level thermal analysis, Theta-JA from a uniform or smeared power map is considered as the general thermal environment of a chip, regardless of the package details/types and board/system configurations. The maximum temperature rise on the chip can be calculated by the product of chip power and Theta-JA directly.

For a single chip package at the JEDEC test condition (Figure 8A) with the uniform power assumption on the chip, the Theta-JA was calculated from the Tmax on the chip with a symmetric thermal profile (Figure 8B). If the Theta-JA of the chip-package-system matches the Theta-JA(1) in Figure 7 by tuning the configurations of the package/system in Figure 8, we could use the model in Figure 8 to study or calculate LHE in the chip. This is considered to be package type and die size independent. The assumptions are that the chip size is much larger than the localized heating zone, e.g., 10mm vs. 50um (50um assumption of the LHE zone can be tuned for different technology nodes if needed), and die in any package type with the same Theta-JA has similar thermal responses to a uniform or smeared power distribution.

Figure 9 is an example of the thermal response plotted along a path through the Tmax location, i.e. the center of the chip in Figure 8 after adding a localized heating zone. The zone is with 10um x 10um power tile (different tile sizes for different technology nodes) at the chip center with high power density, with the remainder of the chip without power or with low and uniform power density, to extract the corresponding temperature rise (i.e. DeltaT). The DeltaT derived at the die center could be applied to the general locations inside a chip.

The mesh on-chip used in Figure 9 was with a fine grid such as 10um x 10um at the location of the localized heating, i.e., at the die center in this case, to capture the local heating accurately. (Note that the 10um element size is used to capture the power on the tile exactly, not a converged mesh size for thermal responses. Smaller element size e.g. 1um can be used for better accuracy as described in the following sections). The model with the localized heating effect is applicable to many locations "inside" the chip. The DeltaT is scaled by the local power in the full-chip power map for the system in Figure 7. Similarly, the localized heating effect at the die border (or edge) and corner can be extracted to represent those locations. Then, for a general system like the chip-package in Figure 7 with the detailed power map on-chip, the DeltaT map can be constructed by applying one localized heating at a time with high resolution for any die sizes with corresponding Theta-JA environment for any package type.

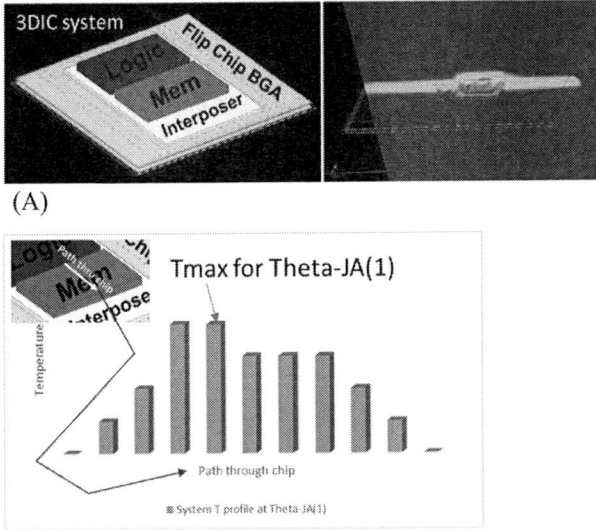

(A)

(B)

Figure 7: For multi-chips in a chip-package-system, the system-level thermal response is to the right (A) where each die has its own thermal profile and Tmax to calculate its own Theta-JA, e.g., Theta-JA(i) for the ith die. The schematic thermal profile was plotted along a path through the Tmax location on a chip (B).

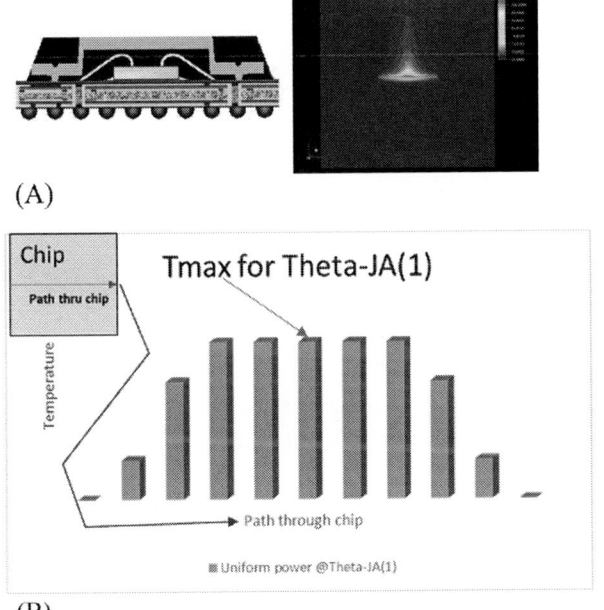

(A)

(B)

Figure 8: For JEDEC test of a single chip/package, the thermal response is to the right (A). Given uniform power on the chip, the thermal profile plotted along a path through the Tmax location can produce a Theta-JA effect similar to the Theta-JA (1) for die 1 in Figure 7B.

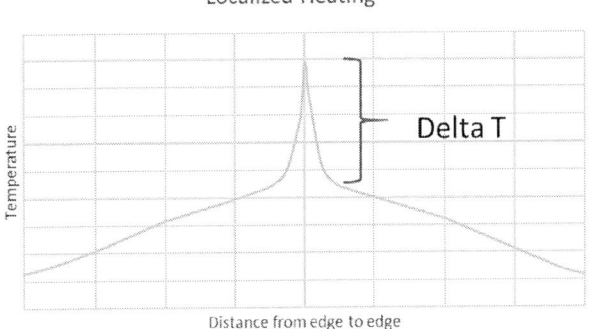

Figure 9: Example of temperature rise at Tmax location of a die due to localized heating on a chip modeled as a silicon block. Here, the "Distance" is the distance from edge to edge of the chip.

However, this approach with only one tile for localized heating ignored the heating from the neighboring tiles in the full-chip power map (Figure 5). A further enhancement is to extend the heating to an array of power tiles, e.g., 5x5 tiles with localized power pattern with two extra rows or columns to the center tile for proper thermal interactions. The left picture in Figure 10 is an example of the power pattern. The right picture is the corresponding thermal response. The purpose is to extract the DeltaT at the center tile.

Similar to the flow of the one-tile localized heating case discussed earlier, one could analyze the thermal response of a local power pattern and extract the temperature of the center tile which includes the thermal coupling of its neighbors. Just stride the power pattern one tile at a time and perform FE analysis for the temperature of the center tile. Then the full chip thermal map like the one in Figure 5 could be reconstructed piece-by-piece. But this will be very time consuming as one will need to repeat the FE analysis for each power template of 5x5 tiles across a given detailed power map in Figure 5. We need a fast calculation method to take the local power pattern and generate the thermal response of the center tile quickly to make this approach independent of the power map and be practical in usage. That fast calculation method can be a DNN-based DeltaT Predictor discussed in Section 4. Before moving on to the fast calculation method, we need to review the chip model of a silicon block commonly used in the system-level thermal analysis.

Figure 9 illustrates the LHE effect for a chip modeled as a silicon block. The dielectric layer structure of the chip was not included in this analysis. If included, the center spike in Figure 9 will have very sharp slopes as the device temperature rise will decay to zero within the short 5um range (Figure 6), which is very small compared to the typical die size, e.g., 4mm or larger. The localization of thermal response supports our assumption that the approach is independent of the die size and package type.

In order to include the effects of the dielectric layers, instead of the single silicon block model, an improved and more realistic 3-layer chip model with two dielectric layers on top of the thicker silicon substrate (Figure 11) can provide the thermal decay behavior similar to that in Figure 6, i.e., 5um in decay distance.

Zooming into the layer details (Figure 11), the CMOS chip structure stack-up from the bottom (silicon substrate) to top (solder bump) is basically layers of the silicon wafer, buried SiO_2, the front end of line (FEOL) devices, multiple via and Cu BEOL layers, and top seal layer/bumps. The silicon wafer at the bottom is typically very thick, e.g., 100um, as compared to the FEOL and the back end of line (BEOL) layers, not shown in Figure 11. Note that FEOL devices and BEOL Cu via/wire are all surrounded by low-K (conductivity) dielectric materials, e.g., SiO_2. The temperature decay in this kind of structure (Figure 6) is dominated by the presence of the low-K materials.

The thermal conductivity (k) for Si is close to 130 W-m/C while that of dielectric, e.g., SiO_2, is only around 1 W-m/C, which affects the thermal decay distance significantly. In this 3-layer model, the heating objects were buried between the upper dielectric layer as the equivalent interconnection layers and the lower layer as the insulation layer or BOX (Buried Oxide), with practical thickness, e.g., 6um and 0.05um (for example), respectively. Our studies showed that the existence of the insulation layer under the heating object is important in generating practical Tmax and keeping the decay distance realistic, and the thickness variations of the equivalent interconnection layer had secondary effects only. A ~30x temperature rise was predicted using fine grid FE analysis with the 3-layer model (T profile in Figure 21), as compared to that from the single silicon block model. The 3-layer model was used in the following studies in this paper.

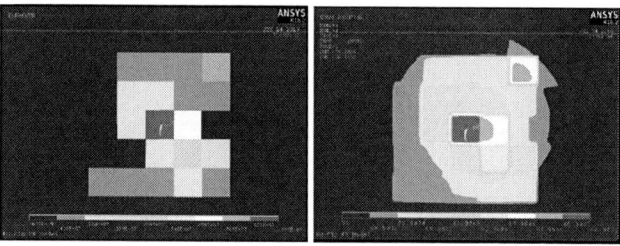

Figure 10: Example of the template of 5x5 tiles for localized heating analysis, power map (left) and T profile (right).

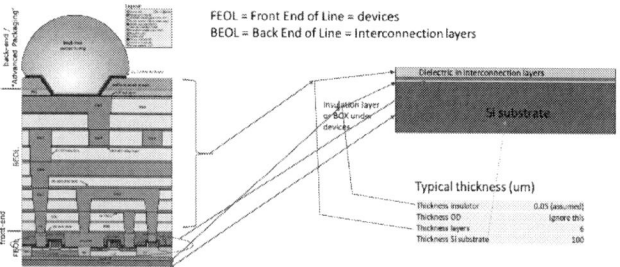

Figure 11: Power map of devices (FEOL) at interface of BEOL and Buried SiO2 (BOX). Typical CMOS structure (left) and the thermal simulation model (right).

The overall flow of the thermal profile resolution enhancement using DNN is as shown in Figure 12 and Figure 13 for both training and inferencing parts. A DeltaT Predictor

was created using the training cases with variations of power on the 5x5 tiles at a series of Theta-JAs. Given a new chip with a detailed power map with 10um x 10um resolution and the system Theta-JA, DeltaT Predictor can take the known system Theta-JA and step through each tile of a chip with its surrounding power map on a 5x5 neighboring power patterns to retrieve the proper DeltaT map. The combination of the coarse mesh system-level thermal profile and the DeltaT map gives the final thermal profile at enhanced resolution.

The detailed steps of the Fast Thermal Solver are the following:

(1) Create training cases for DNN-based DeltaT Predictor
(2) Run training cases with FEM and store DeltaT as ground truth for different 5x5 tile-based power patterns
(3) Template-based DeltaT Predictor Training using optimal Deep Neural Network (DNN) Architecture to generate a DeltaT Predictor model
(4) Inferencing step by applying Theta-JA and DeltaT Predictor for DNN-based fast thermal solver on every tile of a new chip
(5) Obtaining final Thermal map with the enhanced resolution by combining tile-based DeltaT and corresponding temperature from the coarse thermal profile of system-level thermal analysis

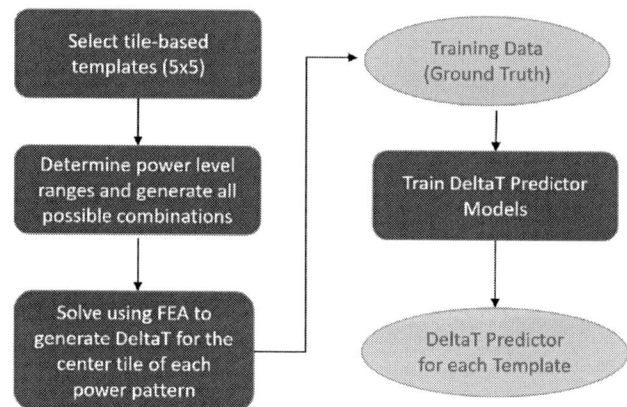

Figure 12: the model training flow for DeltaT Predictor on the three templates with several levels of Theta-JA

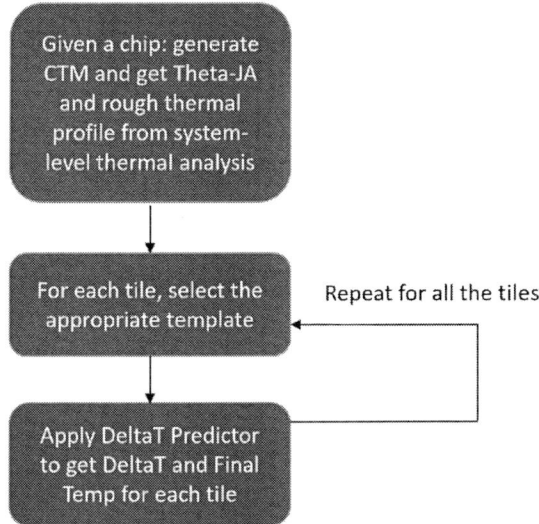

Figure 13: Inferencing flow given a new chip for obtaining full-chip DeltaT and Final T

4. DNN-based DeltaT Predictor Training

A typical BGA package was used with thermal board configurations (Figure 14) at a few Theta-JA's from low to high, e.g., 10 to 120 C/W, to cover the possible ranges of Theta-JA from the realistic system-level thermal analysis. These Theta-JA's are used for later interpolation to match the package/system Theta-JA from system-level thermal analysis at the inferencing step on a new chip.

$$\theta_{JA} = \frac{(T_J - T_A)}{P}$$

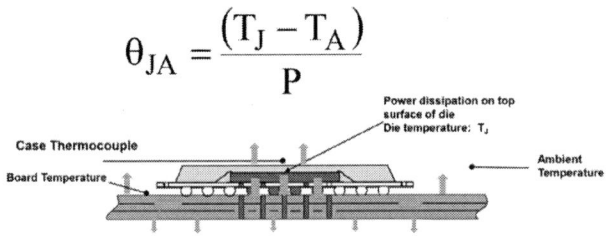

Figure 14: Package-system setup for extracting Theta-JA parameter

For each training case at a selected Theta-JA, one can create a fine grid pattern with a location template, e.g., 5x5 tiles at 10um resolution, within-the-die, along the border, or at the corner, and apply the localized power pattern in calculating the DeltaT at the center tile as shown in the flow (Figure 15). Note that for convenience of sketching, the template in Figure 15 was plotted as 3x3 arrays while in our test cases it was actually 5x5 arrays. The specific template is determined by the center tile distance to the nearest chip boundary and the tile size. The power values on the 5x5 tiles are based on the practical tile-base power density at different levels consistent with the technology node of the chip. For the 5x5 tiles example here, the possible power combinations could be as many as 3^{25}, and proper design schemes could be used to reduce the total number of power patterns to be solved, e.g., using symmetry and DOE (Design of Experiment) technique (Figure 16) to reduce the

number of training cases. Parallel processing of running training cases with various inputs of Theta-JA and power patterns on 5x5 tiles using FEM thermal analysis to generate the resulting DeltaT in the center tile as ground truth would be helpful to accelerate this step.

After all the training data are generated, the corresponding DeltaT Predictor based on the appropriate DNN architecture [21] can be trained as in Figure 12. The DNN architecture with 6 hidden layers, 64-128 neurons per layer, and with ReLU active element is used to create the DeltaT Predictor model.

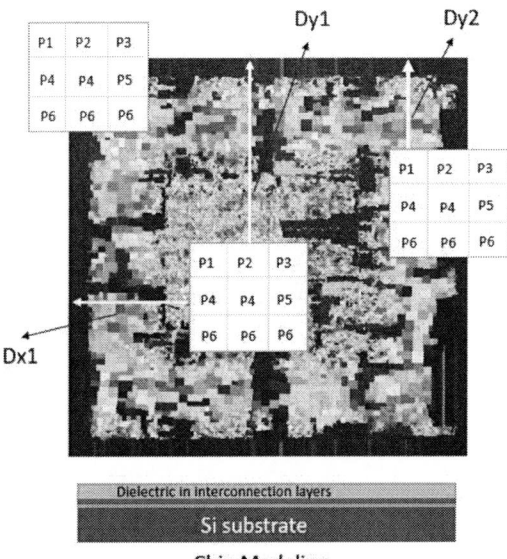

Figure 15: template-based 5x5 10um x 10um tiles example for within-the-die, along the border, and at the corner templates with LHE assumption that thermal coupling decay rapidly beyond 50-100um depending on the technology node.

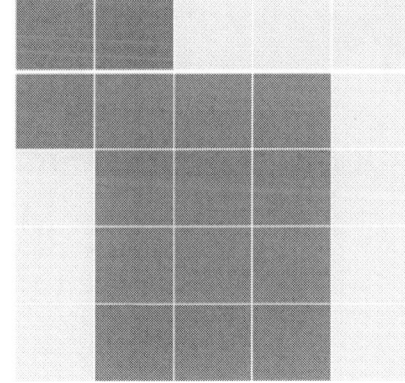

Figure 16: The Power-on-tile Patterns using DOE (Design of Experiment) for reduced input training cases for a Template, dividing the tiles into center and peripheral groups.

5. Inferencing Step

Based on the Theta-JA from the system-level thermal analysis and the tile-based power map from the chip power tool [21], the DNN-based DeltaT Predictor in Figure 17 efficiently extracts the DeltaT at the center tile of the template-based 5x5 tiles that has the corresponding power pattern at a tile location in the full chip power map. Given a new chip with a detailed power map and the system Theta-JA, the DeltaT Predictor can be applied by stepping through each tile of the chip. Each tile is with the 5x5 power pattern including its surrounding neighbors. And the DeltaT is predicted from the corresponding power pattern. Finally, the full-chip DeltaT map is generated (Figure 18).

Appending DeltaT map from the DeltaT Predictor to the coarse thermal profile from the system-level thermal analysis, the final thermal map with the enhanced resolution is generated for use in chip reliability analysis or determination of top-ranked Tmax locations for the placement of on-chip thermal sensor.

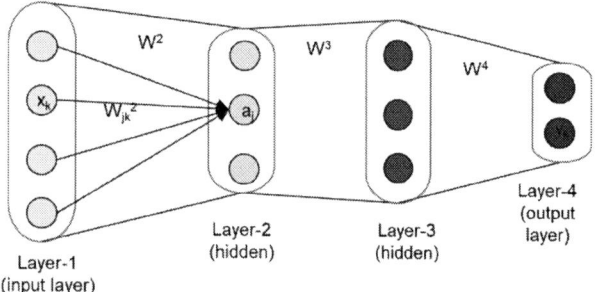

Figure 17: Template-based DeltaT Predictor training using appropriate DNN Architecture [22,25]

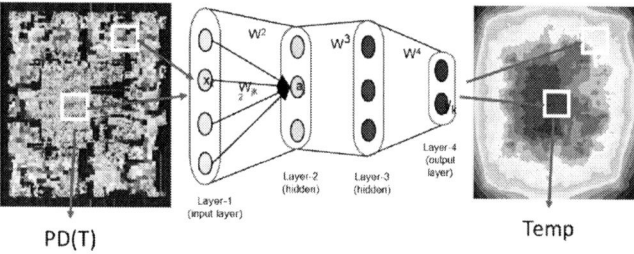

Figure 18: Inferencing step by applying DeltaT Predictor for DNN-based Fast Thermal Solver. With detailed power map on the left and resulting full-chip thermal profile on the right, using the DeltaT Predictor to calculate DeltaT with corresponding in-chip/border/corner template through striding with one tile size across the chip and appending to the coarse thermal profile from system-level thermal analysis.

6. Results and Discussion

About 80K training and 20K test cases with FE models were created and solved in the chip-package thermal tool [24] to generate the training data needed by DeltaT Predictor. Each FE model had 5x5 10um x 10um tiles with unique power patterns on each tile embedded in a 4mm x 4mm training chip for LHE.

It took about 8 seconds to solve in a Linux server with 32 cores using multi-threading. The number of cases was reduced significantly by considering factors such as symmetry and levels of power on tiles in power pattern. To generate a large number of training cases, large scale parallel processing was implemented to run in batches. The results were automatically collected for DNN/XGBoost model [23] training afterward.

For each training case, Theta-JA, power in the center 3x3 tiles and in the ring tiles (the remaining tiles of the 5x5 tiles) were the input features. For evaluating the model, we used 10-fold cross validation on the training set for model determination, and tested performance on the testing set. The statistics of target value DeltaT are shown in Table 1.

Table 1: Mean, standard deviation, maximum and minimum of DeltaT

Mean	7.2867
Standard deviation	2.0385
Maximum	12.106
Minimum	2.462

The power-level on each tile of the power template was assigned to three or more different levels to cover the practical ranges of power values for a particular node of advanced FinFET technology such as 7nm.

For different package/system environments, characterized as Theta-JA, the training cases were categorized into multiple Theta-JA values for the implementation of the DeltaT Predictor. The Theta-JA values for the training cases need to cover the realistic ranges of practical packages in a system.

The training cases of DeltaT Predictor with 5 Theta-JA values and about 100K power template patterns for each Theta-JA took about 11-14 seconds, while modeling and solving all the cases using FEA took more than 12 hours, even with the help of parallel processing on 32 cores.

Apart from the DNN-based DeltaT Predictor, another popular technique called XGBoost[23] was also used to generate the DeltaT Predictor as a part of this analysis. Below, we provide some details on the models that we assessed.

Table 2 shows the mean-squared-error (MSE), mean-absolute-error (MAE) and mean-relative-error (MRE) of 10-fold cross validation using the XGBoost [23] and DNN [22], as well as the training time. Note that the training time is dependent on hyper-parameters of the machine learningalgorithm and will be different when training on different hardware resources.

Table 2: MSE, MAE and MRE score of 10-fold cross validation and training time

Algorithms	MSE	MAE	MRE	MAX relative error	Training time (minutes)
XGBoost	8.38e-4	0.0227	0.0034	0.058	~25
DNN	5.61e-4	0.0105	0.0016	0.179	~19(GPU)

Table 3 and Table 4 show the hyper-parameters of XGBoost and DNN model used, which can be derived from manual tuning or automatically tuned algorithm in neural architecture search (NAS) [25].

Table 3: Hyper-parameters of XGBoost

Parameter	Value
No. of estimators	2500
Max depth	6
Column sample	0.7
Row sample	0.7

Table 4: Hyper-parameters of DNN

Parameter	Value
No. of hidden layer	4
No. of hidden units in each layer	32 - 128
Activation function	ReLU
Epochs	100

To verify the accuracy in inferencing, the 4mm x 4mm chip configuration was reused but with a randomly generated power pattern in a detailed FEA to obtain the golden results of temperature and delta T. The maximum differences found in DeltaT on tiles as compared to the "golden" was about 4%.

Similarly, for the same chip-package-system configuration, a tile-based power map of 4mmx4mm from a test chip is calculated from the Chip Thermal Model (i.e. 10um x 10um temperature-dependent power map) generated by a dynamic voltage drop analysis tool [21].

The total power on-chip was 3.377W, as shown in Figure 19, and the base Theta-JA (27.86 C/W) of the package/system was calculated from applying a uniform 1Watt on the chip with Tmax on-chip extracted from FEM analysis. The "golden" DeltaT on each tile was calculated by the difference of temperature on each tile and a reference temperature, e.g., Tmax of the uniform power case. Alternatively, the temperature profile of the uniform power case can also be used as the reference temperature in calculating the "golden" DeltaT map, as this could be more accurate at the border/corner locations of the chip.

The Ground Truth FinalT map using FEA is in Figure 20. The coarse thermal map from the system-level thermal analysis is in Figure 21 with 200um x 200um resolution and can be run efficiently. The DNN-based or XGBoost-based DeltaT Predictor generates the FinalT map as in Figure 22 with the inputs as the given randomly generated tile-based power map and a Theta-JA value of 27.86 C/W. The maximum differences found in DeltaT/FinalT on the tiles as compared to the "golden" was about <4%, as shown in Figure 23, which is a satisfactory result on accuracy.

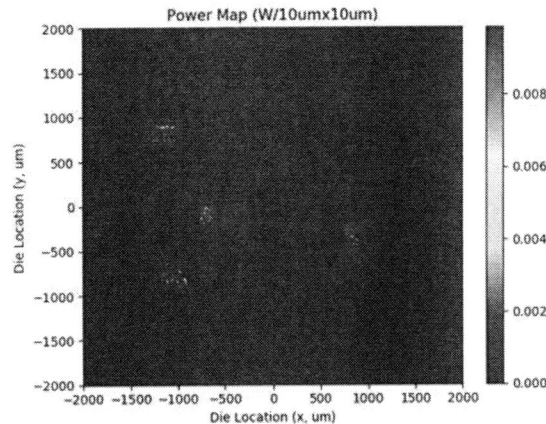

Figure 19: power map of the test case for the comparison between Ground Truth result from FEA and from DeltaT Predictor with rough thermal profile from coarse mesh FEA.

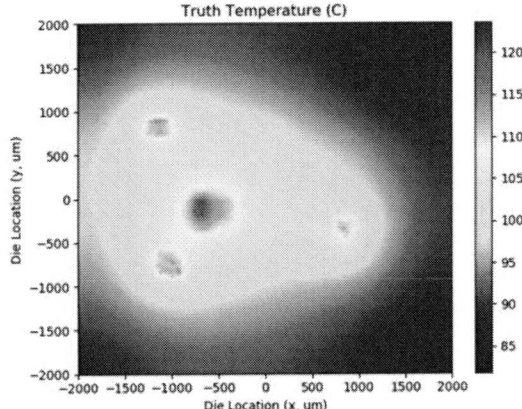

Figure 20: Final T map of the test chip using FEA

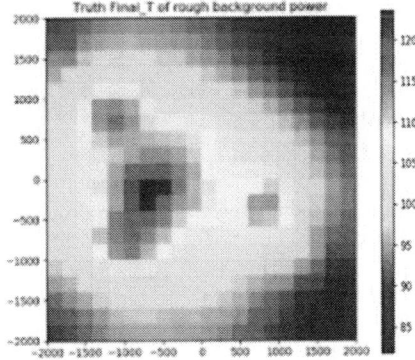

Figure 21: Coarse T map of the test chip using FEA from system-level thermal analysis and can be run efficiently

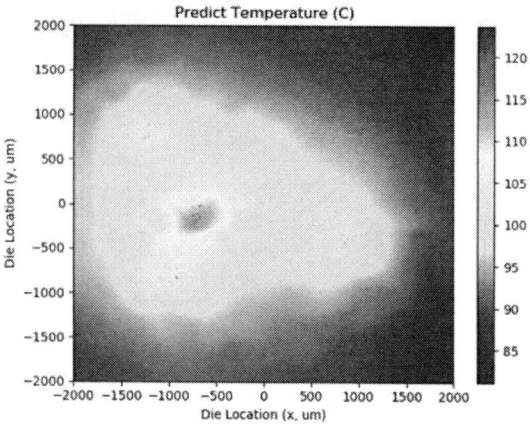

Figure 22: Final T map from the summation of DeltaT and Temperature map of rough thermal profile using coarse grid thermal analysis (FEA).

Chip size	Package size	Theta_JA
4mmx4mm	10mmx10mm	27.86

	Datasets	Run Time	CPU cores	Memory
Training	~100K	~12 hours	32	1.6 G
Rough profile	1	~14 s	1	75 M

	Max T	Delta T	Run Time	CPU cores	Memory
Ground Truth	123.58 C	12.09 C	42 mins	32	4 G
Predicted	123.60 C	12.98 C	10 s (~250x speed-up)	1	0.6 G

Figure 23: run time and memory statistics of detailed Ground Truth FEA run, rough thermal profile from coarse grid FEA run, and DeltaT Predictor

When new chip sizes become as large as 10mmx10mm or even 25mm x25mm such as V100 from Nvidia's latest GPU, the FE model with fine mesh on-chip will make the FEA approach less favorable as the run time could be up to 15 hours or more while the DeltaT Predictor run can be estimated 1000x faster.

7. Conclusion

A novel approach using a DNN-based or XGBoost model for efficient temperature rise prediction on modern chips was proposed and implemented. The approach starts with generating the training cases of each chip with LHE assumption in a Theta-JA environment. The LHE has a local power pattern that covers common power levels on a FinFET technology node and distributions in calculating temperature rise of a small tile (i.e. 10um x 10um) on chip at different locations of the chip, viz., inside, on the border, and at the corner of a chip. After training with a couple million cases, a DeltaT Predictor was generated that can predict the temperature rise at any location on any chip given the Theta-JA environment and the detailed power map. The technique traverses the full chip tile by tile, determines the appropriate template model to be selected for the current tile, and then uses the trained DeltaT Predictor to

predict the change in temperature of the tile, thereby generating a high resolution DeltaT map. The final thermal profile of the chip at high resolution is obtained by appending the DeltaT map to the rough on-chip thermal profile from the conventional methods of FEM or CFD on the coarse grid, which is usually available from system-level thermal analysis.

The analysis results showed that the DNN-based or XGBoost-based DeltaT Predictor can achieve runtime reduction of greater than 100-1000x with good accuracy compared to the traditional FEA based method.

References

[1] M. Emilio, "Hybrid Chips may Solve Thermal, Efficiency, and Integration Challenges in 5G Mobile Devices", Power Electronics News, Oct. 2019

[2] A. Mutschler, "New Thermal Issues Emerge", Semiconductor Engineering, Feb., 2018

[3] C. Peach and Y. Zhang, "Protecting AI Chips from Thermal Challenges during ATE Test", Evaluation Engineering Magazine, Jun., 2019

[4] Y. Sun, et al., "Localized Thermal Effect of Sub-16nm FinFET Technologies and its Impact on Circuit Reliability Designs and Methodologies", IEEE International Reliability Physics Symposium, 2015

[5] Y. Zhong, M. D.F. Wong, "Thermal-Aware IR Drop Analysis in Large Power Grid", 9th International Symposium on Quality Electronic Design, 2008

[6] S. Makovejev, S. Olsen, V. Kilchytska, and J. Raskin, "Time and Frequency Domain Characterization of Transistor Self-Heating", IEEE Transactions On Electron Devices, 2013

[7] R. Chandra, "It's Time To Consider Temperature Gradients In IC Design", Electronic Design, Feb., 2006

[8] B. Goel, M. Själander, "Techniques to Measure, Model, and Manage Power", 4.2.1 Temperature Effects, Advances in Computers, 2012

[9] M. Divakar, "Surface Temperatures of Electronics Products: Appliances vs. Wearables", Electronics Cooling, September, 2016

[10] C. Zhao, "CMOS On-chip Temperature Sensors for Power Management", PhD dissertation, Iowa State University, 2014

[11] S. Kim and M. Seok, "A Sub-50 um2, Voltage-Scalable, Digital-Standard-Cell-Compatible Thermal Sensor Frontend for On-Chip Thermal Monitoring", Journal of Low Power Electronics and Applications, May 2018

[12] R. Allen, "Explaining Adaptive Voltage Scaling And Dynamic Voltage Frequency Scaling", Semiconductor Engineering, July, 2018

[13] K. Tu, "Reliability Challenges in 3D IC Packaging Technology", Microelectronics Reliability, Volume 51, Issue 3, March 2011

[14] N. Chang, S. Pan, K. Srinivasan, Z. Feng, W. Xie, T. Pawlak, D. Geb, "Emerging ADAS Thermal Reliability Needs and Solutions", IEEE Micro, 2017

[15] M. Swenson, "Submodeling: Simple Solutions for Large Scale Problems" https://www.ansys.com/blog/submodeling-made-easy

[16] Wilson, J. S. and Raad, P. E., "A Transient Self-Adaptive Technique for Modeling Thermal Problems with Large Variations in Physical Scales," International Journal of Heat and Mass Transfer, Vol. 47, 2004

[17] S. Pan and N. Chang, "Fast Thermal Coupling Simulation of On-chip Hot Interconnect for Thermal-aware EM Methodology", ECTC, 2015

[18] S. Pan, Z. Feng, N. Chang, Chip/Package Co-analysis of Thermal-Induced Stress for Fan-Out Wafer Level Packaging, IWLPC, 2016

[19] K. Srinivasan, S. Pan, Z. Feng, N. Chang, T. Pawlak, An Efficient Transient Thermal Simulation Methodology for Power Management IC Designs, SemiTherm 2017

[20] S. Pan, N. Chang, and T. Hitomi, "3D-IC Dynamic Thermal Analysis with Hierarchical and Configurable Chip Thermal Model", IEEE 3D-IC Conference, 2013

[21] ANSYS RedHawk-SC Reference Manual, 2019

[22] Y. LeCun, Y. Bengio, and G. Hinton. "Deep Learning", Nature, 2015

[23] T. Chen and C. Guestrin. "XGBoost: A Scalable Tree Boosting System", Proceedings of the 22nd ACM SIGKDD International Conference on Knowledge Discovery and Data Mining, 2016

[24] ANSYS 3DIC CTA Reference Manual, 2019

[25] T. Elsken, J. Metzen, F. Hutter, "Neural Architecture Search: A Survey", Journal of Machine Learning Research, 2019

Inclination Angle Effects on Dual Cool Jet Heat Transfer

Sophia Brodish, Matthew Harrison, Ali Haider, Ted Brekken, Joshua Gess
Oregon State University
204 Rogers Hall
Corvallis, OR 97331

Abstract

Forced convection thermal management solutions require high heat transfer coefficients, electrically efficient operation, small form factors, reliability, and low noise. Piezoelectric synthetic jets capture all of these requirements. Using a high aspect ratio synthetic Dual Cool jet (DCJ) (AR = 44:1), heat transfer coefficients of nearly 200 W/(m²K) were achieved while the operational frequency was maintained at 170 Hz in the vertically oriented position above a heated surface. Normalized heights of y/D_h = 2, 5, and 10 were tested over an Amkor Thermal Test Vehicle (TTV) subjected to constant heat fluxes ranging from 0.2 to 1.4 W/cm². The TTV has a spatial resolution of 0.25" (6.6 mm), allowing for spatial temperature measurement along the major axis of the slit orifice synthetic jet. After validation of the DCJ performance over the flat plate, the effect of the inclination angle above the heated surface is examined with two DCJ's facing one another. Over the optimal operational frequency of 170 Hz, angles of 30°, 45°, and 60° were tested. Heat transfer coefficients over 250 W/(m²K) were achieved in this inclined dual DCJ configuration. Regression analysis shows a heat flux dependence on the heat transfer coefficient that exceeds that expected by laminar or turbulent free convection, suggesting that the periodic convergence and turbulent rise of the two coolant streams at the heated surface center exploits the increased energy density for improved thermal performance.

Keywords

Synthetic jets, piezoelectric fans, pulsed flow, forced convection, air cooling

Nomenclature

Symbol	Unit	Variable
A	m^2	Area
D	mm	diameter
h	$W/(m^2 \cdot k)$	Heat transfer coefficient
I	A	Current
q	W/m^2	Heat flux
S	-	Standard deviation
T	Kelvin	Temperature
t	-	Student t distribution value
u	-	uncertainty
x	m	Characteristic length
y	m	Distance from heated surface
θ	°	Angle of Inclination

Subscript	*Unit*	*Variable*
air	-	Air property
amb	-	ambient
ave	-	average
h	-	Hydraulic
i	-	Index of summation for diodes
local	-	Localized point
s	-	Surface
∞	-	Infinity

Introduction

Synthetic jets are used in a number of applications for low noise, reliable cooling of moderate energy densities. High heat transfer coefficients, over 100 W/m²K, are available at the mesoscale, ideal for handheld electronics and defense communications/power amplification. Synthetic jets distribute pulsed flow with maximum air velocities of 3 to 10 m/sec for orifices ranging from 1 to 6mm. Orifices vary in size and shape which can have minor effects on the overall heat transfer coefficient [1]. The variable orifice will result in an 8x expulsion velocity change, which impacts the overall out flow of the jet. These effects have been quantified using Particle Image Velocimetry (PIV) [2] and a more complete picture of the flow was captured using Tomo-PIV [3]. Changing the frequency at which the piezoelectric jet operates had a similar effect on the expulsion velocity. By increasing the operating frequency, mean velocity increases, which in turn has a direct effect on the enhancement of the overall heat transfer produced by the jet. At lower frequencies it has been shown that entrainment is facilitated and delays the overall dispersion of the cooling fluid [4].

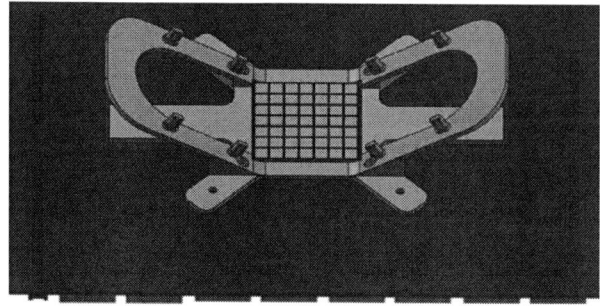

Figure 1 – Isometric view of dual jet test stand.

Synthetic jets have been used in various cooling applications due to their low profile and low noise production. Typical experiments commonly take place in cross flow orientation or impingement arrangements [5,6]. Due to the periodicity of the jet impingement, correlations do not correctly predict the overall heat transfer enhancement of the jets across variable heated geometries both in terms of scale and shape. So far, results on these effects have been inconsistent when compared to various authors' studies [7]. This shows that there is work to be done to fully understand how synthetics' configurations can be optimized.

Finally, other studies have explored how synthetic jets may be used in tandem with bulk fluid motion in both co-flow and cross flow arrangements. Heat transfer was enhanced by as much as 30% by adding a synthetic jet with bulk fluid flow in both a cross and co-flow arrangement. However, it was noted that careful consideration must be taken by the designer using cross flow due to areas of reduced heat transfer [7]. The current study documents the effects of inclination angle on dual synthetic jet configurations. The results show that the heat transfer coefficient increases with respect to heat flux beyond the expected superheat dependence noted in conventional laminar and turbulent free convection correlations.

Temperature Measurement Locations		
Vertically Oriented Single DCJ Configuration	Dual Inclined DCJ Configuration	
Diodes	4-8-9-10	Diodes
		1-2-3-4-5-6-7-8-9-10

Figure 2 – Regions actively measured for local temperatures are highlighted in orange. Symmetry is used to build a comprehensive thermal map for both configurations examined.

Experimental Facility and Testing Procedure

An Amkor Thermal Test Vehicle (TTV) with a 13.56 cm² square heated surface area was utilized for all experiments. The diodes in the board were calibrated in a Thermo-Scientific insulated oven, against a NIST traceable thermistor accurate to within 0.013K. The TTV diodes were measured in two different patterns: one for the vertical impingement configuration and another for the inclined dual jet configuration. Figure 1 shows the Dual Cool jet configuration mounting bracket installed on the TTV board. The illustrative squares highlighted on the TTV represent the regions where surface temperatures were recorded using centrally located embedded diodes. Figure 2 highlights the diode locations with details regarding which diodes were used for the single vertically oriented DCJ and the dual inclined-angle DCJ testing configurations. Symmetry is assumed for the dual DCJ configuration to create a complete board temperature profile. Symmetry was also invoked for the single vertically oriented DCJ configuration to more accurately compare these results with the literature and theoretical expectations. Average surface temperatures, which were used for heat transfer coefficient calculations, had uncertainties of ±0.55K and

±0.69K for the dual jet and single jet testing configurations, respectively.

Figure 3 displays the type of synthetic jets used in this study with dimensions of 1mm tall x 40mm wide for the orifice. This is GE Dual Cool Jet (DCJ). This type of jet was first used to cool microelectronics by Garg et al [8]. All synthetic jets were

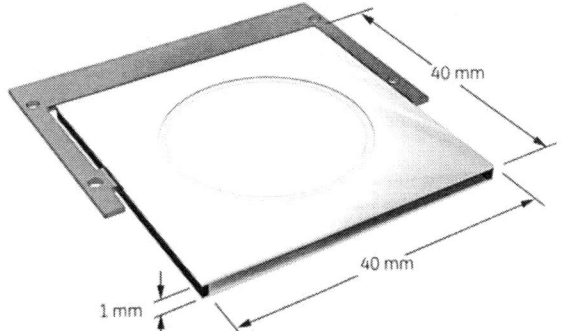

Figure 3 – Image of synthetic Dual Cool jet used in this study along with relevant dimensions.

operated at 52 Voltage peak to peak (Vpp) and 170 Hz. During single jet testing y/D_h varied from 1, 5, and 10 where y was the distance from the DCJ orifice to the heated surface, figure 5, and D_h is the hydraulic diameter.

Figure 4 show the side view of the custom 3D printed stands used in the dual jet configuration without the jets shown. Three stands were printed with angles, Θ, of 30°, 45°, and 60°. In the dual jet configuration y/D_h remained constant, where y refers to the distance from the jet opening to the leading edge of the TTV. During dual DCJ testing the nominal distance between the leading edge of the heated surface and orifice was 3 mm, shown in Figure 5. Figure 6 displays the experimental facility with the syn jets in the configuration used during testing.

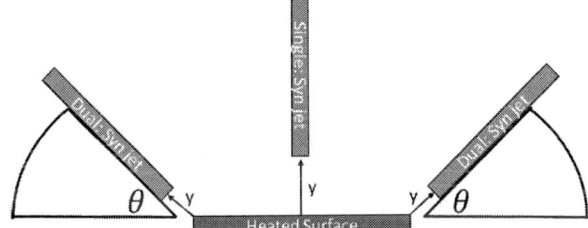

Figure 5 – Diagram of DCJ location in reference to heated surface and start and end measurement of length y

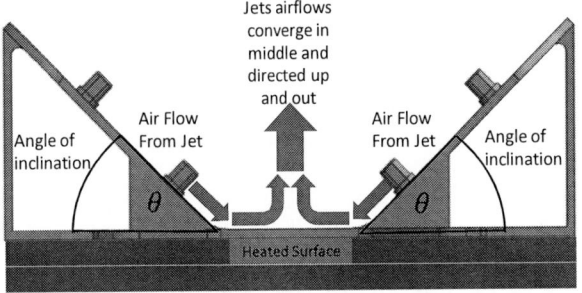

Figure 4 – Side view of the dual DCJ test stand illustrating airflow over the surface and how the inclination angle was implemented into the design. Jets not shown.

An in-house Labview Virtual Instrument (VI) was created for steady state determination during testing. "Steady state" here refers to being thermally stable as it is understood that flow is ejected periodically from the synthetic jet. Diodes were excited using a 1.05 mA current that provided adequate signal stability and minimal joule heating addition to the primary heat flux driven by epitaxial heaters embedded in the TTV. While temperatures were only measured at locations shown on Figure 1 and Figure 2, uniform heat flux was applied evenly across all regions of the TTV. In order to avoid signal interference from adjacent diode excitation, the VI drove a National Instruments (NI) solid state relay card that would close one relay and record the average temperature of the diode over a 6 second period at a continuous 1 kHz sampling frequency. The diode would then open, breaking the circuit, before moving forward to the next diode. In order to establish thermal steady state, diode temperatures were compared to one another from subsequent diode progressions. If the average temperature difference among all of the diodes were lower than 0.5K, thermal steady state was assumed.

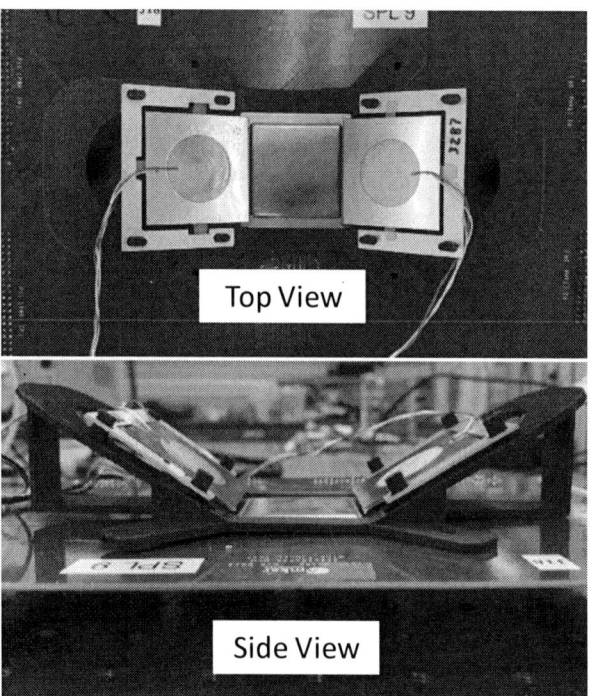

Figure 6 – Top View of Experimental Setup and side view to illustrate DCJ Orientation with the 30° angle of inclination bracket

Data Analysis and Results

Testing was first conducted with a single, vertically oriented DCJ, aligned over the center of the heated surface. Constant heat flux was applied to the entire surface, not just limited to diode regions directly under the DCJ. A plot of the heat transfer coefficient over various heat fluxes is provided in Figure 7. The heat transfer coefficient was calculated from equation 1. Figure 7 shows the average heat transfer coefficient (HTC), meaning each individual diode has a local HTC that was measured and the average from all the diodes are plotted.

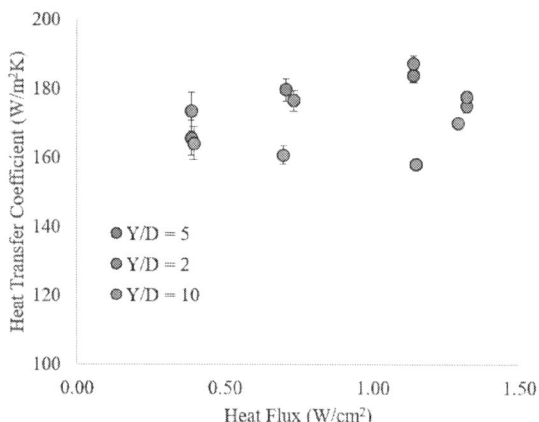

Figure 7 – Average heat transfer coefficients for the vertically oriented single DCJ configuration seem to be heat flux agnostic.

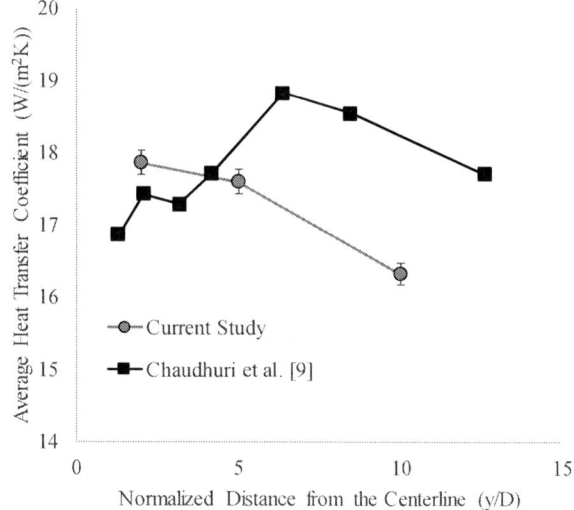

Figure 8 – Single DCJ configuration average heat transfer coefficient data relates well to a previous study under similar conditions.

$$h_{avg} = \sum_{i=1}^{4} \frac{q}{A_s(T_{si}-T_{amb})n} \tag{1}$$

From the data in Figure 7, there was no identifiable trend or dependency on heat flux for the average heat transfer coefficient, supporting theoretical expectations [9]. The data in Figure 8, however, does seem to indicate a precipitous drop in heat transfer coefficient at the normalized height $y/D_h = 10$. This trend is more evident in the local Nusselt number data over various normalized lengths along the heated surface. Figure 9 shows the local Nusselt number variation as the air flows away from the centerline of the heated surface. The local Nusselt number was calculated from equation 2. Where h, x, and k are the local heat transfer coefficient, distance from the centerline, and the thermal conductivity of the air, respectively.

$$Nu_{local} = \frac{h_{local} \cdot x}{k_{air}} \qquad (2)$$

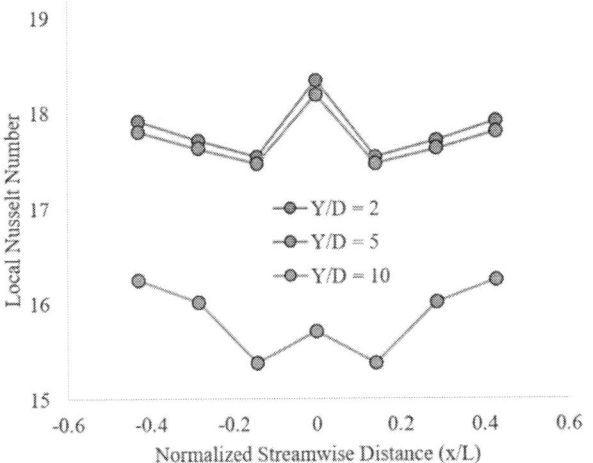

Figure 9 – Single DCJ configuration local heat transfer coefficient variation in the spent coolant streamwise direction

As expected, a local maximum exists at the stagnation point. In constant stream impingement cooling, the heat transfer coefficient is expected to decline in the direction of coolant flow as thermal boundary layer growth offers increased thermal resistance along the x-direction. However, periodic injection of fluid characteristics of synthetic jets serves to disrupt the thermal boundary layer development, driving increased heat transfer coefficients along the coolant exit direction. This path is parallel to the heated surface and can

Exponential Dependence of Heat Transfer Coefficient Data in Figure 10			
Inclination Angle	Exponential Dependence	Error	Improvement over Turbulent Free Convection Expectations
30	0.56	±0.10	39%-100%
60	0.58	±0.17	24%-127%
90	0.59	±0.18	24%-133%

Table 1 – Improvement of inclined DCJ over the best case turbulent free convection expectations.

result in several different Nusselt number minima and maxima [10]. The heat transfer coefficients decrease as the normalized height increases, another trend fundamentally expected. This is because reducing the distance to the heated surface leads to higher fluid pressure drops and tighter suppression of thermal boundary layer growth.

The average heat transfer coefficient over various normalized heights above the heated surface, y/D_h, was compared to a previous study [9] which used a similar configuration to the current work. The data in Figure 8 matched well with previous experimental work [9]. The aspect ratio (AR) of the slot jet used in the current study is 44:1 while that used by the previous work [9] was 20:1. The previous

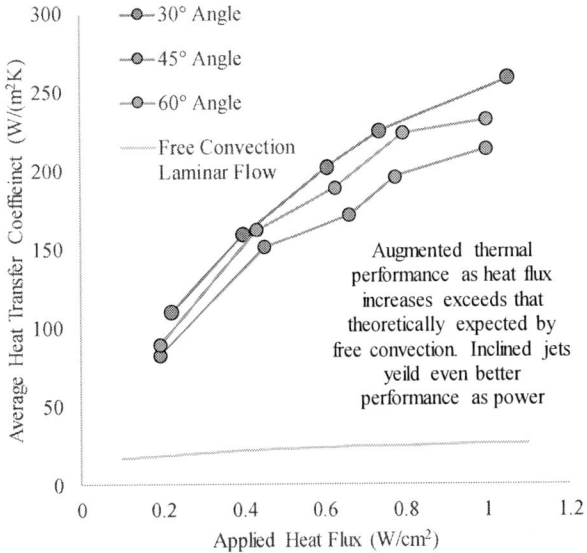

Figure 10 – Inclined DCJ configuration yields improved heat transfer coefficients with increased power density. Regression analysis shows that this improvement is beyond that expected by free convection from the surface.

work [9] showed a trend of reduced performance as aspect ratio grows; therefore, the lower performance, particularly at large y/D_h, is expected. At larger y/D_h, the impact of interference with the jet potential core is reduced. A comparison of the data in Figure 8 after the potential core is left unhindered, showed an identical slope in the heat transfer coefficient decay with respect to normalized height variation.

The dual DCJ configuration average heat transfer coefficient data with respect to varying heat flux is shown in Figure 10. The data showed an increase in heat transfer coefficient as heat flux applied to the surface increased. This trend was seen in various buoyancy driven heat transfer coefficient conditions such as boiling and free convection. Turbulent DCJ on inclined surfaces have shown a marked improvement by using various angles over the flat surface [11]. In order to prove that these effects exceed those expected by free convection, log-based regression analysis was used to determine the exponential dependence of each of the three curves in Figure 10. They were then compared to the 1/5 or 1/3 dependence on heat flux expected by laminar or turbulent Rayleigh numbers, respectively. The exponential dependence of heat transfer coefficients with respect to heat flux along with the error, as determined by the precision error of the slope in the regression analysis, is shown in Table 1. Even with the low-end value uncertainty prediction, the exponential dependence is still greater than either of the fractional dependencies mentioned previously. This means that the heat transfer augmentation, caused by improved buoyancy driven effects in the disrupted boundary layer by the pulsed synthetic DCJ flow, is beyond that of free convection. This suggests yet another heat transfer mechanism, like boiling and free convection, that is augmented by increased power density. Heat flux driven thermal performance is an asset in electronics thermal management as cooling performance improves as systems are asked to do more in smaller form factors.

The uncertainties in the heat transfer coefficients shown in Figure 10 are between 2%-5%, so there is clearly angular dependence on the resulting heat transfer augmentation. Additionally, the data crosses between the 45° and 60° inclination angles, suggesting that an angle exists where there is a minimum or potentially no improvement beyond free convection expectations. Ongoing work is driven towards finding these minima and maxima values through modelling the phenomena in conjunction with Particle Image Velocimetry (PIV) measurements for hydrodynamic characterization of the periodic and pulsed coolant flow.

Uncertainty Calculations

Uncertainties for the average surface temperature, heat transfer coefficient, and exponential dependence of the heat transfer coefficient on applied heat flux were all calculated with the Root-Sum Squared (RSS) method.

Each diode had a calibration curve that was acquired through a linear regression of the induced diode voltages over various oven temperatures measured with a NIST-traceable thermistor at an excitation current of 1.05 mA. The standard error of the fit, Equation 3, was used to calculate the calibration error while the thermistor error was minimal at 0.013°C across all temperatures examined. "yi" is the recorded value during calibration while "yci" is the predicted value from the curve established by linear regression analysis. "v" is the degree of freedom for the calibration curve which is the number of samples taken minus one (since the standard deviation is embedded in the defining equation). The calibration error is dominant as seen from the individual diode temperature sensing errors acquired from Equation 4 which are presented in Table 2. Equation 5 is used to acquire the error from the

average surface temperature. In the single DCJ case, "N" ranges from Diode 10, 9, 8, 4, 8, 9, 10 as the symmetry assumption is incorporated into the analysis. For the dual jet configuration, "N" spans all of the diodes shown on Figure 2. The resultant average temperature uncertainty was 0.55°C for the dual DCJ configuration and 0.64°C for the single DCJ configuration.

Heat transfer coefficient uncertainties were calculated using the RSS method applied to Newton's Law of Cooling, Equation 7. Equation 8 shows the uncertainties considered in the overall average heat transfer coefficient uncertainty analysis. The stated surface area of the TTV from Amkor was assumed to have little to negligible error in these calculations. Individual uncertainties are shown in Table 3. Heat transfer coefficient errors for all cases examined ranged between 1.2% and 2.6%. Arguably the most critical uncertainty calculation to this study is that of the heat flux dependence on the heat transfer coefficient seen in Figure 10. The disruption of the thermal boundary layer induced by the pulsed synthetic jet flow, corroborated by previous work [10], is augmenting the thermal performance beyond the expected natural convection dependence. The exponential dependence of the data in Figure 10 minus the low end of the uncertainty range must be larger than that expected by free convection in order to claim that the synthetic DCJ's are augmenting the performance beyond that

Diode Number	Local Temperature Error (°C)
1	1.76
2	1.86
3	1.86
4	1.75
5	1.72
6	1.64
7	1.71
8	1.68
9	1.64
10	1.72

Table 2 – Individual diode temperature sensing uncertainties

Value	Uncertainty
Current, I	±0.0005 A
Voltage, V	±0.005 V
Ambient Temperature, T_∞	±0.01°C
Average Surface Temperature, $T_{s,ave}$ (Dual DCJ)	±0.55°C
Average Surface Temperature, $T_{s,ave}$ (Single DCJ)	±0.64°C

Table 3 – Uncertainties used in heat transfer coefficient error calculations

$$S_{yx} = \sqrt{\frac{\sum_{i=1}^{N}(y_i - y_{c_i})}{v}} \tag{3}$$

$$u_{T,Diode} = \pm t_{0.95,v} S_{yx} \tag{4}$$

$$u_{\bar{T}} = \frac{1}{N}\sqrt{\sum_{i=1}^{N}(u_i)^2} \tag{5}$$

$$S_{a1} = S_{yx}\sqrt{\frac{N}{N\sum_{i=1}^{N}x_i^2 - (\sum_{i=1}^{N}x_i)^2}} \tag{6}$$

$$q'' = h\,(T_{s,ave} - T_\infty) = IV/A_s \tag{7}$$

$$u_{HTC} = \sqrt{\left[\left(\left(\frac{\partial h}{\partial T_{s,ave}}\right)u_{T_{s,ave}}\right)^2 + \left(\left(\frac{\partial h}{\partial I}\right)u_I\right)^2 + \left(\left(\frac{\partial h}{\partial V}\right)u_V\right)^2 + \left(\left(\frac{\partial h}{\partial T_\infty}\right)u_{T_\infty}\right)^2\right]} \tag{8}$$

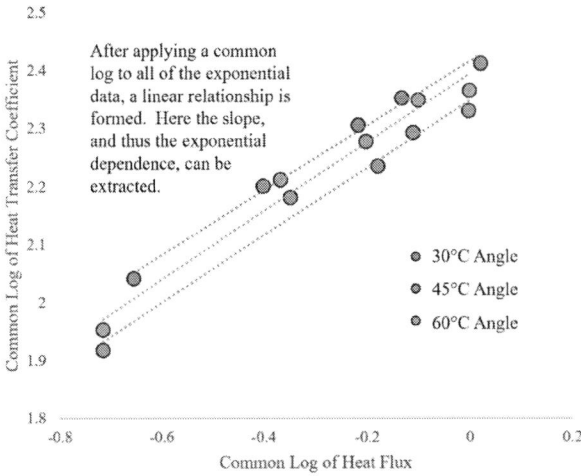

Figure 11 – Exponential data post common logarithmic manipulation shows a linear trend. The slope of this linear trend is the exponential dependence of the data based on conventional logarithmic rules.

expected by the natural convection phenomenon itself. In order to extract this exponent, a common logarithmic manipulation is applied to all of the data. Logarithmic rules then dictate that the exponent may be assigned as a leading coefficient resulting in a conventional "y = mx + b" linear form. The results of this common logarithmic manipulation are shown in Figure 10. The negative values on Figure 11 have no physical meaning as they are an artifact of the natural log of a value between zero and one yielding a negative value. All three of the angles tested have been overlaid, and the linear form is clear from the data in Figure 11. Equation 6, which is a precision estimate for a linear slope, is then applied to these common logarithmically adjusted results. The resulting errors are those shown in Table 2. Since temperature is included in Equation 6, the propagated errors from individual temperature measurements were incorporated using an RSS manipulation of Equation 6. These propagated errors were found to add approximately 10% more error to the exponential dependence beyond that predicted solely by Equation 6. In Equation 6, the Syx, standard error of the fit, is applied to the linear data post common logarithmic manipulation, i.e. the data seen in Figure 11.

Conclusions and Future Work

A vertically oriented single DCJ and an inclined dual DCJ configuration were tested over heat fluxes between 0.2 and 1.4 W/cm². Heat transfer coefficients above 180 W/m²K have been measured for the single DCJ configuration, matching previous published work and theoretical expectations. Two DCJ's were inclined at angles of 30°, 45°, and 60° from a centrally located flat surface, yielding heat transfer coefficients of nearly 250 W/m²K. Higher heat transfer coefficients could be yielded with the same input power to the air movers as it was determined that the thermal performance was augmented by the heat flux at the cooled surface. Regression analysis showed a 0.56-0.59 exponential dependence between heat flux and heat transfer coefficient for the inclined dual DCJ configuration. Even with uncertainty analysis on this exponent, ±0.10-±0.18 across the various angles was examined, this dependence exceeds the expectations of turbulent free convection by at least 24% for all

cases examined. The data showed there exists an operating angle which should be avoided as the heat transfer coefficient values are minimized, and future work is aimed at the determination of this value through modelling of the pulsed flow in addition to validation of hydrodynamic predictions using PIV measurements.

Acknowledgments

A special thanks to General Electric Company for allowing Oregon State to borrow the Dual Cool Jet's used in this study. Also their guidance and support for proper set up and operation.

References

[1] Zhang, Jing-Zhou, et al. "Convective Heat Transfer on a Flat Plate Subjected to Normally Synthetic Jet and Horizontally Forced Flow." *International Journal of Heat and Mass Transfer*, vol. 57, no. 1, 2013, pp. 321–330., doi:10.1016/j.ijheatmasstransfer.2012.10.035.

[2] Bock, H. Peter De, et al. "Particle Image Velocimetry Study on Dual Cooling Jet Flows." *2016 15th IEEE Intersociety Conference on Thermal and Thermomechanical Phenomena in Electronic Systems (ITherm)*, 2016, doi:10.1109/itherm.2016.7517708.

[3] Wernet, P. Mark. "Comparison of Tomo-PIV Versus Dual Plane PIV on a Synthetic Jet Flow". *National Aeronautics and Space Administration,* Glenn Research Center. May 2017

[4] Mongia, Rajiv K., et al. "Heat Transfer Enhancement Using Synthetic Jets for Cooling in Low Form Factor Electronics in Presence of Mean Flow." 2007 9th Electronics Packaging Technology Conference, 2007, doi:10.1109/eptc.2007.4469758.

[5] Bock, H. Peter De, et al. "Investigation and Application of an Advanced Dual Piezoelectric Cooling Jet to a Typical Electronics Cooling Configuration." *13th InterSociety Conference on Thermal and Thermomechanical Phenomena in Electronic Systems*, 2012, doi:10.1109/itherm.2012.6231582.

[6] Buchberger, Gerda, et al. "Simple Synthetic Jet Actuators for Cooling Applications Using Soft or Rigid Magnets." *Procedia Engineering*, vol. 168, 2016, pp. 1541–1546., doi:10.1016/j.proeng.2016.11.456.

[7] Persoons, Tim, et al. "A General Correlation for the Stagnation Point Nusselt Number of an Axisymmetric Impinging Synthetic Jet." *International Journal of Heat and Mass Transfer*, vol. 54, no. 17-18, 2011, pp. 3900–3908., doi:10.1016/j.ijheatmasstransfer.2011.04.037.

[8] Garg, J., Arik, M., Weaver, S., and Saddoughi, S., "Micro fluidic jets for thermal management of electronics", *Proc. ASME Heat Transfer/Fluids Engineering Summer Conference, Charlotte, North Carolina*, FED F-346, July 11-15, 2004.

[9] Chaudhari, M., Puranik, B., and Agrawal, A., 2010, "Effect of Orifice Shape in Synthetic Jet Based Impingement Cooling," Experimental Thermal and Fluid Science, Vol. 34, No. 2, pp. 246-256.

[10] Silva, L.A. and Ortega, A, 2003, "Convective Heat Transfer in an Impinging Synthetic Jet: A Numerical

Increased System Performance and Reduced Surface Touch (Skin) Temperature in Mobile Electronics Utilizing Composites of Graphite with Ultra-High Spreading Capacity and Insulation with Ultra-Low Thermal Conductivity

Mitchell Warren[1], Julian Norley[2], John Allen[1], Jonathan Taylor[2], Lindsey Keen[1]

1. W. L. Gore & Associates, 201 Airport Rd,
   Elkton MD, 21921 USA
   mwarren@wlgore.com, jallen@wlgore.com, lkeen@wlgore.com

2. NeoGraf Solutions, LLC 11709 Madison Ave.,
   Lakewood, Ohio 44133 USA
   jnorley@neograf.com, jtaylor@neograf.com

Abstract

Graphite foils with ultra-high spreading capacity and insulation sheets with ultra-low thermal conductivity were combined in a thermally stressed Google Pixel 3XL to reduce steady-state surface touch (skin) temperatures (T_S) by up to 3.2 °C with < 1 °C increase in max junction temperature (T_J) as compared to single-component thermal solutions of graphite, insulation, and air. An axisymmetric conduction model was simulated in COMSOL to determine trends in surface temperature reductions of five unique thermal solutions of comparable thickness (~350 μm). Four of these solutions were fabricated, tested and validated experimentally in Google Pixel 3XL thermal stress testing. The composite yielding the greatest T_S reduction was utilized to demonstrate an increase in steady-state system performance while maintaining a surface temperature suitable for user comfort and safety. The steady-state *3DMark – Slingshot Extreme* benchmark score increased from 3401 to 3823 resulting in a 12.4% increase in steady-state system performance. The enhanced device performance was linked with material properties by means of steady-state heat flow and thickness testing for through-plane thermal conductivity of insulation, and thermal diffusivity testing for in-plane thermal conductivity of graphite. In-plane conductivity of graphite was validated experimentally in a steady-state heat spreading test where 100 μm foils of high-performance graphite measured ~30% higher spreading capacity than 100 μm foils of synthetic and natural graphite.

Keywords

Graphite, ultra-high spreading capacity, insulation, ultra-low thermal conductivity, composite, heat spreader, thermal conductivity, thickness, surface touch (skin) temperature, hot spot, junction temperature, ambient temperature, steady-state, Google Pixel 3XL (Pixel), system on chip (SoC), *3DMark – Slingshot Extreme*, benchmark score, system performance, user comfort

Nomenclature

k	thermal conductivity (W/m·K)
t	thickness (mm, μm)
T_S	surface touch (skin) temperature (K, °C)
T_J	device junction temperature (K, °C)
ΔT	change in temperature (K, °C)
q"	heat flux (W/m²)
R"	thermal resistance (K·m²/W)
t·ΔT	intrinsic heat spreading capacity (μm·K)

Introduction

Thermal spreaders (graphite) and insulators (air, polymers) have been widely and commonly used to address heat challenges in the mobile electronics industry. As the trends for higher power processing and thinner form devices become standard requirements, mobile electronics continue to face a more pressing issue of user safety and comfort by means of the surface touch (skin) temperature (T_S).

The Underwriters Laboratories (UL) guidance for T_S is based on direct skin contact for specific temperatures and durations,[1] and is accepted across the mobile electronics industry. Where passive thermal solutions have previously been able to reduce the T_S below specification, many of the commonplace materials such as air and synthetic graphite are facing technical limitations.[2] In the absence of a thermal solution that maintains system performance, one widely practiced solution is power throttling of the processor, which may reduce system power by up to 50%.[3]

In thin mobile electronics with relatively low temperatures (< 100 °C) and no active cooling, conduction is the primary mode of heat transfer inside the device[4]; internal convection and radiation are considered negligible in comparison and not discussed further in this work.

Fourier's Law of One-Dimensional Conduction Heat Transfer, shown in Equation (1), states that the theoretical change in temperature (ΔT) is directly proportional to the thermal resistance (R") of the heat transfer medium.

$$(1) \quad q'' = \frac{\Delta T}{R''} \quad \left(\frac{W}{m^2} \right)$$

Assuming heat flux (q") in a given system is constant, ΔT is driven by R", which is defined as the ratio of thickness (t) to conductivity (k).

$$(2) \quad R'' = \frac{t}{k} \quad \left(\frac{K \cdot m^2}{W} \right)$$

Combining and rearranging Equations (1) and (2), T_S can be viewed as a function of the junction temperature (T_J), t, k, and q", which is shown in Equation (3) and the accompanying one-dimensional resistance network (Figure 1). In a constrained system with constant q" and t, T_S can be reduced by lowering k.

$$(3) \quad T_S = T_J - \frac{q'' \cdot t}{k} \quad (K, °C)$$

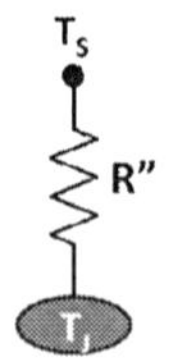

Figure 1: 1D thermal resistance network. Heat flows from T_J to T_S through R"

When a system is expanded into three dimensions of heat transfer (Figure 2), planar heat spreading can be an integral contributor to the resulting T_S. Both in-plane and through-plane conductivities deliver significant contributions to the resultant spreading of heat in a material of given thickness and area. Combining ultra-low (through-plane) conductivity insulation with ultra-high spreading capacity graphite yields a thermal composite solution with exceptional heat spreading performance compared to existing materials used for thermal management in thin mobile electronics.

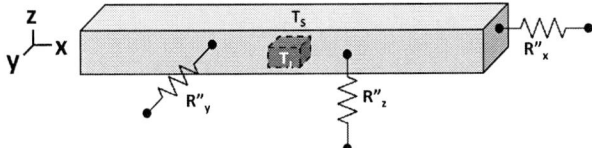

Figure 2: 3D thermal resistance network. Heat moves from T_J (center of device) in multiple directions including toward the surface of interest, T_S.

Material Selection

GORE® Thermal Insulation (W. L. Gore & Associates, Inc.) is an insulating material ("the insulation") exhibiting ultra-low thermal conductivity, below that of air, in thin sheet form (100 μm and 250 μm). NeoNxGen™ Thermal Management Solutions (NeoGraf Solutions, LLC) includes a thick foil graphite (70 μm to 270 μm) displaying ultra-high intrinsic heat spreading capacity ("high-performance thick graphite").

Individual layers of insulation and graphite may separately provide a reduction in T_S when placed between a heat source and the surface of interest. Insulation alone is an optimal solution when the ratio of available area to area of the surface hot spot is approximately one-to-one. While insulation is relatively isotropic, graphite exhibits highly anisotropic behavior, favoring thermal conduction in the plane of the material. This utility becomes impactful for T_S reduction when the ratio of available area to area of the surface hot spot approaches two-to-one or greater; in these system architectures, insulation can be combined with graphite to enhance its effective heat spreading capacity. A schematic of the ratio of available area to area of the surface hot spot is illustrated in Figure 3.

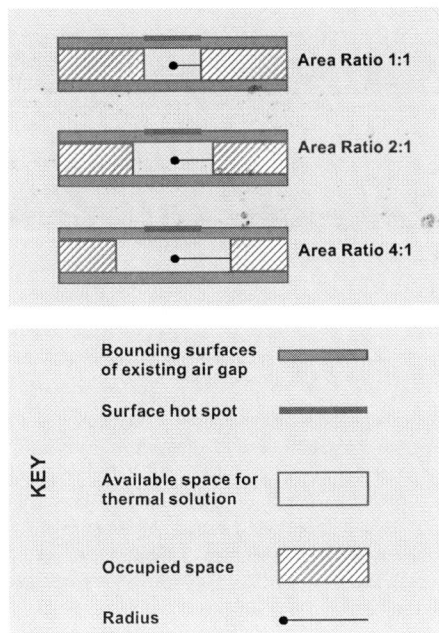

Figure 3: Schematic showing a cross section view of the ratio of available area to area of the surface hot spot. Area is proportional to radius squared.

Insulation Thermal Conductivity Characterization

The insulation is characterized by its distinctively low thermal conductivity, < 0.020 W/m·K, due to a conduction heat transfer phenomenon known as the Knudsen Effect. The Knudsen Effect explains that when the pore diameter in a medium is smaller than the mean free path of air (approximately 70 nm), the path of heat transfer through this medium is disrupted, relative to the path of heat transfer through air in free space.[5] This principle is often applied through the use of aerogels due to their morphology of high porosity with small pore diameters. The uniqueness of this insulation appears in the form of a homogeneous aerogel structure with ultra-low (and consistent) thermal conductivity and precise thickness resulting in a reliably high thermal resistance. Comparatively, the thermal conductivity of free air at room temperature is 0.026 W/m·K and it increases non-linearly with temperature (0.028 W/m·K at 50 °C),[6] which can result in variable and insufficient thermal resistance at elevated temperatures (> 50°C) in mobile electronics.

The through-plane thermal conductivity of this insulation is determined by measuring thermal resistance using a heat flow method and material thickness using a precision thickness method. Both tests are conducted with a pressure set point of 6 psi. A heat flow meter (TA Instruments, Model FOX 50), modified from ASTM C518-17, is used to measure thermal resistance under steady-state thermal transmission.[7] A thin and thick (layered) sample are both tested for thermal resistance. Thickness is then tested for each sample (Instron, Model 5565) using a modified ASTM F36-15 method.[8] A two thickness resistance procedure is used to calculate through-plane thermal conductivity, shown in Equation (4); this method is used to eliminate any effects of contact resistance in the heat flow method.[9]

$$(4) \quad k = \frac{t_2 - t_1}{R''_2 - R''_1} \quad \left(\frac{W}{m \cdot K}\right)$$

Graphite Thermal Conductivity Characterization

Graphite is used for spreading heat due to its inherently high conductivity in the planar direction and relatively low conductivity in the through-plane direction. Synthetic graphite thickness ranges from less than 25 μm (~1500 W/m·K) up to 100 μm (~600 W/m·K), with in-plane thermal conductivity trending inversely to thickness. Layering thin sheets of high conductivity graphite is a potential way to improve heat spreading capacity at higher thicknesses, though this often leads to inconsistencies in thermal performance as well as challenges in manufacturing. High-performance thick graphite foils prove to have the thermal conductivity benefits of thin synthetic graphite, up to 1100 W/m·K in-plane, at single-layer thicknesses similar to that of natural graphite. The through-plane conductivity is comparably ~3.5 W/m·K for each grade of graphite.

Two instruments were qualified to test the thermal diffusivity of high-performance thick graphite foils. The first, Angstrom instrument, was developed by Wagoner et al. to measure graphite fibers and named after the inventor of the technique.[10] In this instrument, the temperature of a long, thin specimen is varied sinusoidally at one end and measurements are taken of the resulting heat wave as it propagates along the specimen in a vacuum environment. One end of the specimen is affixed to a heat source while the other end is maintained under light spring tension. Two thermocouples contact the specimen along its length and measure the amplitude and time delay of the temperature wave as it propagates. The amplitude, time delay, and spacing of the thermocouple are used to calculate the thermal diffusivity of the specimen. A second instrument, the TA-33 Thermowave Analyzer, manufactured by Bethel Co., Ltd. irradiates the top surface of a square specimen with a modulated laser beam heat pulse and detects the changes in amplitude and phase of the heat pulse using an infrared detector at the center of the bottom side of the specimen. The laser frequency as well as the horizontal distance between the laser beam and the infrared detector can be varied. The frequency of the laser beam, the change in signal amplitude, and change in phase can be used to calculate thermal diffusivity. In-plane thermal conductivity can then be calculated from thermal diffusivity (α), density (ρ), and specific heat capacity (c_p), shown in Equation (5).

$$(5) \quad k = \alpha \cdot \rho \cdot c_p \quad \left(\frac{W}{m \cdot K} \right)$$

It has been demonstrated that the Angstrom instrument can reliably measure thermal diffusivity on the widest range of graphite specimen thickness, at least 32 µm to 940 µm thick.[11] However, the Bethel TA-33 instrument demonstrated similar results and less variation than the Angstrom instrument in the thickness range of 32 µm to 168 µm. Given the smaller specimen size, the non-contact measurement technique, and shorter test cycle time for the Bethel TA-33, it is the preferred thermal diffusivity test instrument for graphite specimen thicknesses up to 168 µm thick. The Bethel TA-33 test results were used to calculate the thermal conductivity of the 100 µm high-performance thick graphite samples in this paper.

Experimental Tests and Simulation

A series of experiments were conducted to measure the intrinsic heat spreading capacity of graphite, along with steady-state surface temperatures and performance responses for insulation-graphite composites in mobile electronics. Testing results were benchmarked against air and single-component solutions where applicable.

Steady-State Heat Spreading Test

The Steady-State Heat Spreading Test consists of 3-in. x 1-in. graphite strips, heated from one end with an electrical resistance heater applying constant power (4.16 W). Both ends of graphite were fixed in place and solidly in contact with thermocouples via thermal interface materials (TIMs). The temperature drop along the strip was measured at steady-state. A schematic of the test setup is shown in Figure 4.

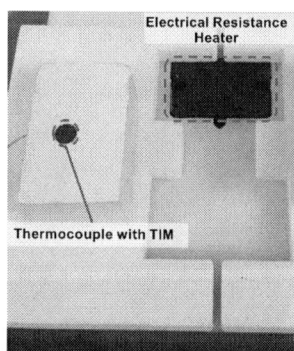

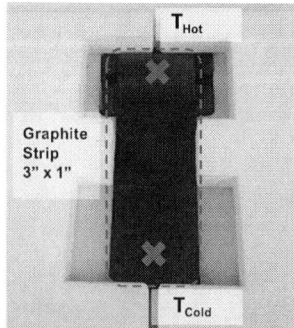

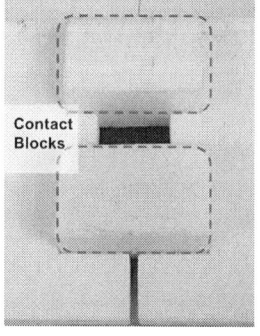

Figure 4 (a, b, c): 4a (top) shows the empty test setup with electrical resistance heater and one contact block with thermocouple-embedded TIM. 4b (lower left) shows the graphite strip placed in the test setup, designating the hot and cold thermocouple locations. 4c (lower right) shows both contact blocks in place, creating solid contact between the thermocouple TIMs and graphite strip.

Test results were analyzed, using Equation (6) to compare intrinsic heat spreading capacity of graphite samples. Temperature drop is multiplied by thickness of individual strips (as measured by compression test with Instron, Model 5565) to account for variations in thickness.

$$(6) \quad t \cdot \Delta T = t \cdot \left(T_{Hot} - T_{Cold} \right) \quad (\mu m \cdot K)$$

A lower temperature drop implies greater heat spreading, as the graphite surface temperature is more uniform from end to end. For a material that does not spread heat, the T_{Cold} thermocouple would approximately equal the ambient temperature, resulting in a high value for $t \cdot \Delta T$.

Simulation – Thermal Conduction Model

An axisymmetric thermal conduction model was created in COMSOL to simulate the impact on steady-state T_S and T_J for various thermal solutions in a representative smartphone architecture. The model consists of a constant power heat source, individual material layers, and a device cover; heat transfer coefficients and emissivities can be applied to external surfaces of the heat source and all individual layers. In-plane and through-plane thermal conductivities are defined for all layers and constant across temperatures. The system geometry is defined by a radius and thickness for heat source, material layers, and device cover. Critical model outputs are T_S, displayed in a radial profile along the cover, and maximum T_J on the heat source. Figure 5 shows a schematic of the general model setup and outputs.

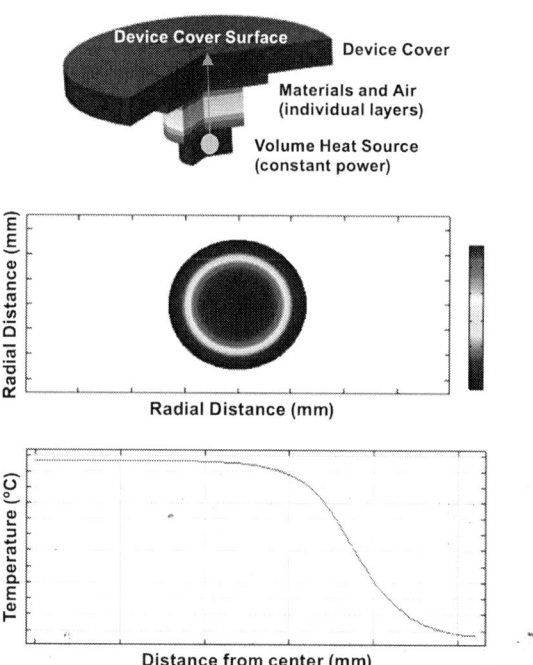

Figure 5 (a, b, c): 5a (top) shows a schematic of the axisymmetric thermal conduction model setup in COMSOL with critical components labeled including volume heat source, material and air layers, and device cover. 5b (middle) and 5c (bottom) show the simulation output in a heat map and radial temperature profile on the device cover surface, respectively.

This simulation focuses on a representative geometry of the Google Pixel 3XL back cover located over the system on chip (SoC); a cross section is shown in Figure 6. Measured thermal conductivity values were applied for insulation (through-plane) and high-performance thick graphite. Literature and data sheets were used to approximate thermal conductivity values for air and glass. Thermal conductivity values used are shown in Table 1. The total thermal gap was fixed at 500 µm for all configurations tested; air was used to fill the remainder of total thickness not filled by materials. All material solutions were modeled at 350 µm thickness to be consistent with physical testing materials. Simulation configurations are detailed in Table 2.

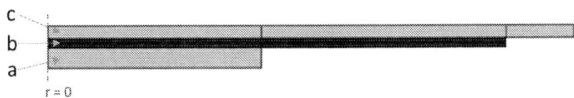

Figure 6: 2D schematic of the axisymmetric thermal conduction model before it is revolved around the "r=0" axis. Block "a" represents the heat source with 11.3 mm radius and 1 mm thickness. Block "b" represents an available thermal gap with 24.1 mm radius and 0.5 mm total thickness. Block "c" represents a glass cover with 27.6 mm radius and 0.65 mm thickness.

Table 1: In-plane and through-plane thermal conductivities values used in simulation

Material	In-plane Conductivity (W/m·K)	Through-plane Conductivity (W/m·K)
Air	0.028	0.028
Glass	15	15
High-performance thick graphite	1000	3.5
Insulation	0.018	0.018

Table 2: Configurations simulated in available thermal gap (Block "b") from Heat Source to Device Cover

Configuration	Material Thickness (mm)	Configuration Depiction
S1 (control)	Air, 0.500	
S2	Insulation, 0.350 Air, 0.150	
S3	Graphite, 0.350 Air, 0.150	
S4	Insulation, 0.175 Graphite, 0.175 Air, 0.150	
S5	Graphite, 0.175 Insulation, 0.175 Air, 0.150	
S6	Graphite, 0.117 Insulation, 0.116 Graphite, 0.117 Air, 0.150	

KEY

Cover	
Heat Source	
Air	
Insulation	
Graphite	

Google Pixel 3XL 3DMark Stress Test

An off-the-shelf Google Pixel 3XL ("Pixel") was purchased and modified to allow for constant power stressing without thermal throttling. UL's *3DMark – Slingshot Extreme* was chosen for testing as it is a widely-accepted benchmark used to score the physics (CPU) and graphics (GPU) of high-end smartphones.[12] In order to achieve steady-state test results, the Professional Version of 3DMark was purchased and installed on the Pixel to enable infinite looping of the 90-second *Slingshot Extreme* benchmark test. All testing was conducted in a still air environment with tightly controlled ambient temperature and humidity. Parameters available for measuring include: surface point temperatures via thermocouples, images via IR camera (Fluke, Model Ti55), internal component temperatures (CPU, GPU, etc.) via built-in thermistors, CPU and GPU clock frequencies, and system performance via *Slingshot Extreme* benchmark score.

An initial stress test was run in the out-of-box condition with IR imaging (Figure 7). Hot spot locations were identified and chosen for placement of thermocouples via TIMs (Figure 8).

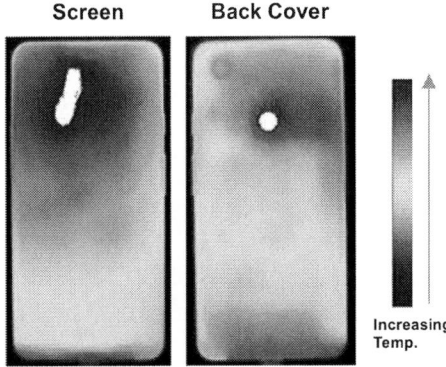

Figure 7: IR images of screen (left) and back cover (right) on the Google Pixel 3XL. A numberless temperature scale is shown to indicate directional trends between color and temperature. Surface hot spots are represented by the white areas.

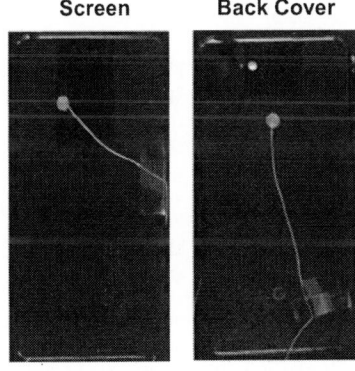

Figure 8: Screen (left) and back cover (right) with thermocouples attached via TIMs on the Google Pixel 3XL. Thermocouples were placed precisely to measure temperatures at the surface hot spot locations.

The Pixel back cover was removed by means of heating and breaking adhesive. A conformable polymer was placed inside the back cover at seven different locations near the SoC (Figure 9) to determine the space available for a thermal solution; the back cover was then replaced to compress the polymer into the existing air gap at each location. The back cover was removed again and thickness at all locations was measured via snap gauge on the compressed polymer. This process was repeated twice more and all thickness measurements per location averaged. Thickness means are detailed in Table 3.

Figure 9: Google Pixel 3XL with back cover removed. Existing air gap thickness measured by conformable polymer at seven locations shown.

Table 3: Air gap measurements near SoC in closed Pixel device

Location	Mean Gap Measurement (mm)
1	0.900
2	0.625
3	0.520
4	0.520
5	0.440
6	0.450
7	0.640

In order to avoid mechanical compression in Locations 5 and 6, a nominal thickness of 350 μm was chosen for all thermal solutions. Physical materials for testing include 110 μm insulation sheets, 110 μm graphite foils and 5 μm acrylic double-sided tape. Materials and example configurations are illustrated in Figure 10.

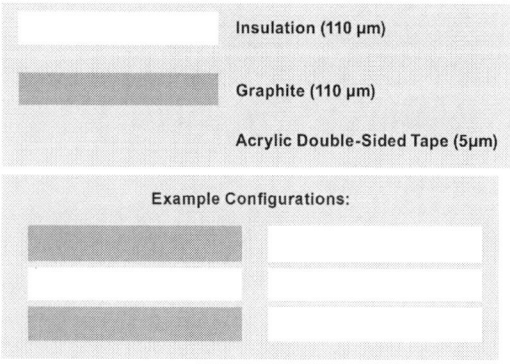

Figure 10: Depiction of physical materials for testing and example configurations of materials layered with adhesive.

The part geometry, shown in Figure 11, was chosen to maximize area with no or minimal disruption to internal components. For simplicity, only configurations with uniform thickness and layers with identical shape and area were considered. Further optimization in layer thicknesses and sizes are possible to achieve form, fit, or functional goals. A cross section schematic through the thickness of the phone is depicted in Figure 12. Simulation results were analyzed to inform material configurations chosen for Pixel testing.

Figure 11 (a, b): 11a (left) shows placement of the part inside the back cover. 11b (right) shows a composite sample cut to fit the designated geometry. Part area measured to be 1825 mm².

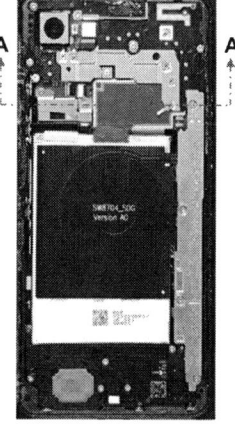

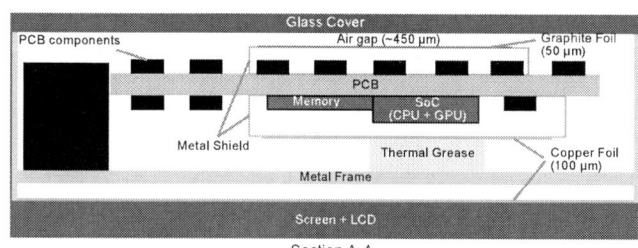

Figure 12 (a, b): 12a (top) denotes the location of cross section A-A in the Pixel. 12b (bottom) shows a schematic of section A-A through the thickness of the device.

Results

Steady-State Heat Spreading Test

Synthetic, natural and high-performance graphite grades were tested, all at 100 μm nominal thickness; t·ΔT values were obtained using Equation (6). Six individual samples of each graphite were tested in a randomized experiment. Results are shown in Figure 13.

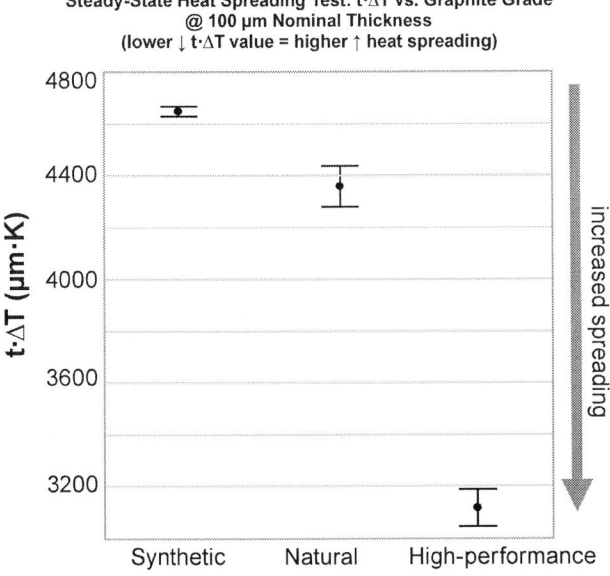

Figure 13: Graph (means and standard deviations) of Steady-State Heat Spreading Test t·ΔT, n=6 per graphite grade.

High-performance thick graphite exhibited the lowest t·ΔT value with a mean of ~3100 μm·K. This value is 29% lower than the mean t·ΔT value for natural graphite (~4350 μm·K), and 33% lower than the mean t·ΔT value for synthetic graphite (~4650 μm·K).

Simulation – Thermal Conduction Model

Power and heat transfer coefficients were iterated to achieve cover and heat source temperatures relevant to Pixel device testing. Surface emissivity was neglected for this simulation. Parameters chosen for all test configurations are shown in Table 4.

Table 4: Simulation inputs for all test configurations

	Power (W)	Device Cover Heat Transfer Coefficient (W/m²·K)	Heat Source Heat Transfer Coefficient (W/m²·K)	Material Layers Heat Transfer Coefficient (W/m²·K)
Set Point Value	1.5	20	25	1

Configurations S1 through S6 were simulated and outputs displayed in Figure 14 with results detailed in Table 5. All configurations are compared to the control scenario, Configuration S1 (air only). A zoomed in graph of cover surface temperature for graphite and graphite-insulation composite configurations (S3, S4, S5 and S6) is displayed in Figure 15.

Simulation: Device Cover Surface and Junction Temperature Study
Steady-State Surface Temperature Profile and Heat Source Max Temperature vs. Configuration

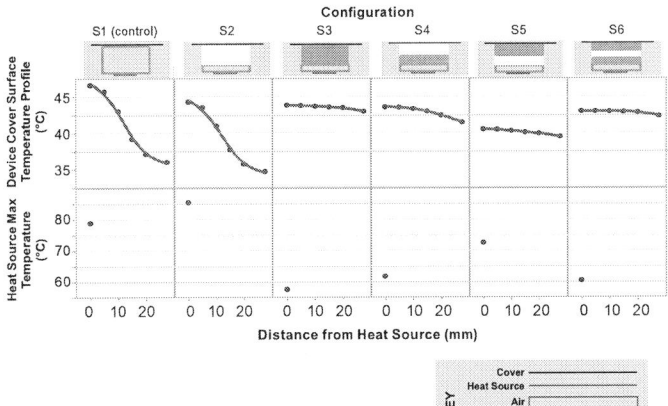

Figure 14: Simulation results by configuration. The top graph displays a radial temperature profile along the device cover surface from device center (r = 0 mm) to device edge (r = 27.6 mm). The bottom graph displays a single value for the device heat source max temperature.

Table 5: Simulation results for max temperatures on cover surface and heat source

Configuration	Cover Surface Max Temperature (°C)	Heat Source Max Temperature (°C)
S1 (control)	46.55	78.82
S2	44.28	85.43
S3	43.79	57.58
S4	43.54	61.73
S5	40.52	72.48
S6	42.96	60.42

Simulation: Device Cover Surface and Junction Temperature Study
for Configurations including Graphite
Steady-State Surface Temperature Profile vs. Configuration

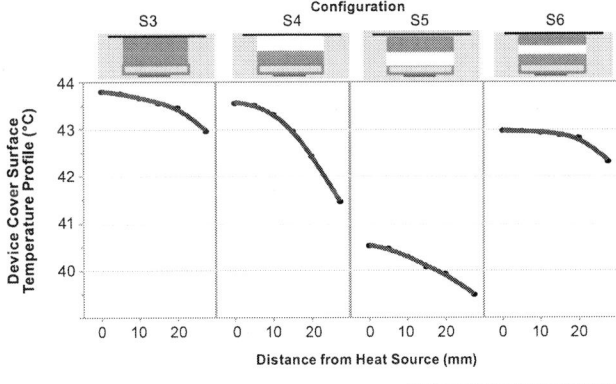

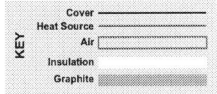

Figure 15: Zoomed into top graph of Figure 14 for the graphite only and insulation-graphite composite configurations (S3, S4, S5, and S6).

Configuration S5 yields the greatest reduction of max surface temperature compared to the control (Configuration S1). For all configurations tested, the max temperatures on device cover surface and heat source occur at the device center (r = 0). As heat travels radially from the device center, the temperature decreases. When insulation is introduced into the system (Configuration S2), the temperature profile along the surface looks similar to that of the control, though the magnitude is shifted down at each respective location along the surface. This effect is a result of the insulation's ultra-low conductivity and propensity to redirect heat toward the heat source, causing an increase in T_J. When graphite and insulation-graphite composites are introduced into the system (Configurations S3, S4, S5, and S6) the max surface temperature is reduced while the radial temperature profile is increased relative to the control. This result occurs due to graphite's preferential planar spreading of heat, producing a more uniform heat distribution along the device surface.[13] The simulated T_J is maintained or reduced for these four configurations relative to the control.

Google Pixel 3XL 3DMark Stress Test

Back Cover Touch Temperature Study

Configurations S1, S2, S3, S5, and S6 from simulation were selected for Pixel device testing and constructed with physical materials described in Figure 6 above; device test configurations are titled D1, D2, D3, D5, and D6 with D1 as the control scenario. The CPU and GPU frequencies were set at 2169.6 MHz and 675 MHz, respectively. Frequencies were recorded and verified at the end of each test run. Benchmark scores were recorded to show performance consistency across all test runs. Ambient temperatures in the still-air environment were held between 21.6 and 21.8 °C for all testing. All configurations were tested three times to steady-state (> 90 minutes) in a randomized experiment. After each test run, the Pixel was cooled down to idle operating temperature and opened up to setup the next test run. The steady-state back cover hot spot touch temperatures and GPU max temperatures are shown in Figure 16. IR images of the back cover are shown in Figure 17. Depictions, thicknesses, and measured outputs (means and standard deviations) for all tested configurations are detailed in Table 6.

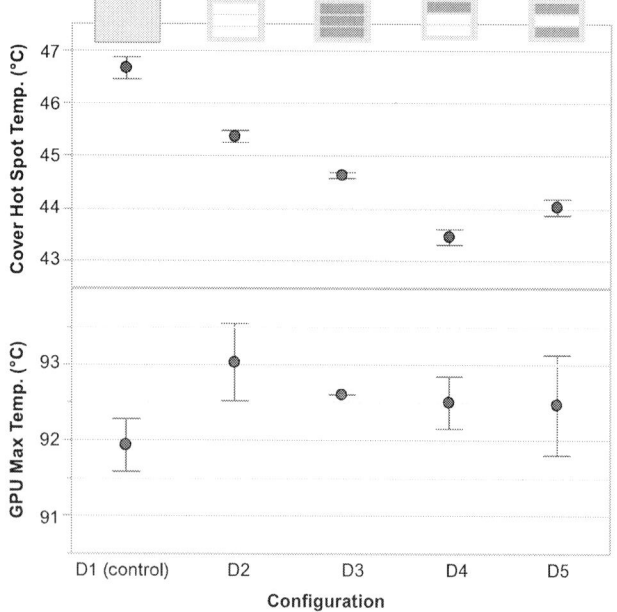

Figure 16: Steady-state graph (means and standard deviations) of back cover hot spot temperature (top) and GPU max temperature (bottom) for all configurations tested in Pixel device, n=3 per configuration.

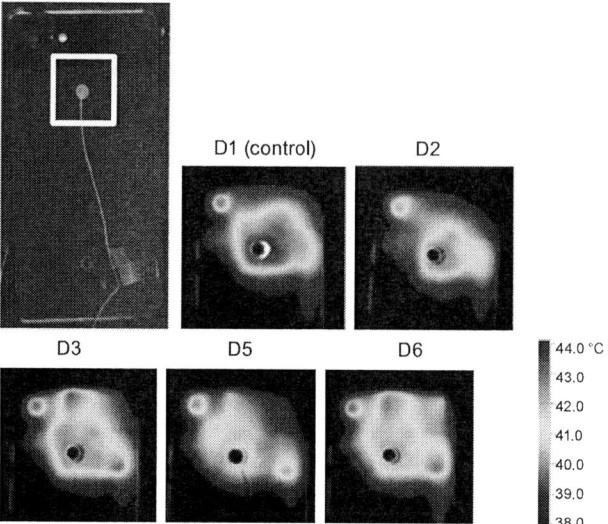

Figure 17: Zoomed in IR images over back cover hot spot for all configurations tested in Pixel device.

Table 6: Pixel Device Results: Back Cover Touch Temperature Study

Configuration	Cover Hot Spot Temp. (°C)		Screen Hot Spot Temp. (°C)		CPU Max Temp. (°C)		GPU Max Temp. (°C)		Slingshot Extreme Benchmark Score	
	Mean	St. Dev.	Mean	St. Dev.	Mean	St. Dev.	Mean	St. Dev.	Mean	St. Dev.
D1 (control)	46.7	0.21	49.7	0.25	84.8	0.17	91.9	0.35	4374.3	1.15
D2 (344 µm)	45.4	0.12	50.5	0.10	86.1	0.51	93.0	0.51	4377.7	1.15
D3 (339 µm)	44.6	0.06	50.1	0.10	85.4	0.65	92.6	0.00	4375.7	1.53
D4 (347 µm)	43.5	0.15	49.9	0.26	85.6	0.17	92.5	0.35	4372.3	2.08
D5 (347 µm)	44.0	0.15	49.9	0.26	85.6	0.51	92.5	0.67	4375.0	1.00

All test configurations produced unique back cover touch temperatures with high precision, and all were distinctly lower than the control (Configuration D1). In agreement with the simulations, Configuration D5 presented the greatest back cover touch temperature reduction at 3.2 °C below the control. Configurations D6, D3, and D2 reduced the back cover touch temperature by 2.7, 2.1, and 1.3 °C, respectively. Screen temperatures increased from the control by < 1 °C for all configurations tested and < 0.5 °C for composite configurations. CPU and GPU temperatures increased from the control by < 1.5 °C for all configurations tested and < 1 °C for composite configurations. The Pixel back cover touch temperature study results validate the directional trend of device surface temperature for the emulated configurations in the simulation study. The directional trend of junction temperatures in simulation was not replicated by the relatively consistent CPU and GPU temperatures in physical device testing. This difference is likely attributed to the complexity of thermal architecture near the SoC in the real Pixel device.

System Performance and User Comfort Study

A continuation study was created to determine the allowable system performance increase when enabled by graphite-insulation composites; Configuration D5 was selected for this study. Out-of-box throttling conditions were restored to the Pixel and all thermal solutions were removed, leaving air only. The back cover touch temperature was measured during steady-state power throttling and recorded for three test runs. Configuration D5 was installed and frequencies were set to match the steady-state cover temperature from the throttled control runs. The appropriate frequencies for testing were determined to be 1996.8 MHz and 596 MHz for the CPU and GPU, respectively. Frequencies, cover hot spot temperature, benchmark score and Frames per Second were measured and compared between the two test scenarios. A smoothed plot of benchmark score, CPU frequency, and GPU frequency vs. run time for all six test runs is displayed in Figure 18. Mean steady-state cover temperature, benchmark score, and Frames per Second are shown in Figure 19. Details are summarized in Table 7.

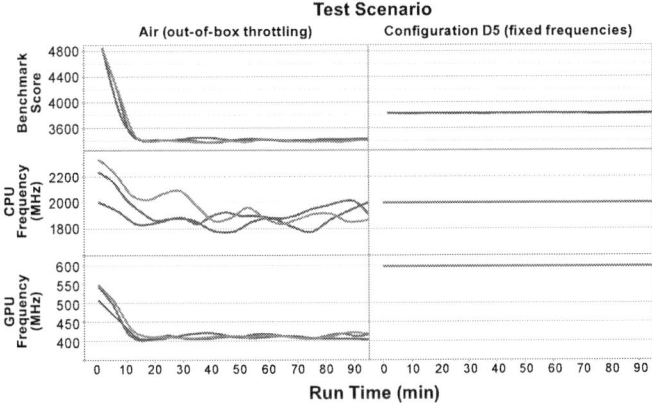

Pixel Device: System Performance and User Comfort Study Benchmark Score and CPU, GPU Frequencies vs. Run Time (by Test Scenario)

Figure 18: Transient graph (smoothed) of benchmark score (top), CPU frequency (middle), and GPU frequency (bottom) for air only, out-of-box throttling (left) and Configuration D5, fixed frequencies (right) in Pixel device, n = 3 per test.

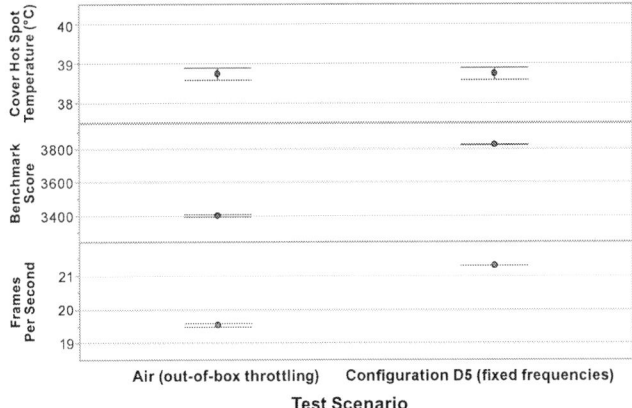

Figure 19: Steady-state graph (means and standard deviations) of back cover hot spot temperature (top), *Slingshot Extreme* benchmark score (middle), and Frames per Second (bottom) for air only, out-of-box throttling and Configuration D5, fixed frequencies in Pixel device, n=3 per configuration.

Table 7: Pixel Device Results: System Performance and User Comfort Study

Test Scenario	Cover Temp (°C)		*Slingshot Extreme* Benchmark Score		Frames per Second	
	Mean	St. Dev.	Mean	St. Dev.	Mean	St. Dev.
Air (out-of-box throttling)	38.7	0.15	3401.0	8.19	19.5	0.06
Configuration D5 (fixed frequencies)	38.7	0.15	3822.7	3.06	21.3	0.00

The mean steady-state cover touch temperature achieved during out-of-box throttling is 38.8 °C in the controlled test environment at 21.7 °C; this temperature is related to UL 60950-1 mobile electronics touch (skin) temperatures at prolonged durations. In this scenario, the mean steady-state benchmark score and Frames per Second are 3401 and 19.5, respectively. When Configuration D5 is placed inside the back cover, the benchmark score is increased to 3823 and Frames per Second increased to 21.6, marking a ~12% increase in system performance, while maintaining the surface temperature limit set for the out-of-box throttling condition.

Summary/Conclusion

Graphite foils with ultra-high spreading capacity and insulation sheets with ultra-low thermal conductivity were combined in a modified Google Pixel 3XL to reduce surface touch (skin) temperatures and increase system performance while minimally impacting the device junction temperature. The experimental results for device surface temperature of five unique thermal configurations were used to validate a comparable simulation study using an axisymmetric thermal conduction model. The resulting surface temperature reductions from insulation-graphite composites exceeded those of air, insulation alone, and graphite alone, when filling the same area and thickness. One insulation-graphite composite configuration was further tested in comparison to an out-of-box condition, and was found to improve system performance in a UL benchmark test by ~12% while maintaining the out-of-box cover surface temperature limits.

The results demonstrated by insulation-graphite composites in Pixel device testing and simulation can be explained by the exceptional thermal properties exhibited by these two materials. Through-plane thermal conductivity for the insulation was measured and calculated using a heat flow method on a TA-FOX 50, a thickness method on an Instron-5565, and a two thickness resistance procedure. In-plane thermal conductivity for high-performance thick graphite was measured and calculated using a thermal diffusivity method on a Bethel TA-33. The heat spreading capacity of the 100 μm high-performance thick graphite was compared to 100 μm synthetic and natural graphite and validated experimentally in a steady-state heat spreading test.

High-performance insulation-graphite composites may have vast utility in the high-powered, thin architectures of mobile electronics. It is important to note that each mobile electronic system may exhibit unique thermal challenges given system power, available space, and/or other constraints. For this reason, the optimal design configuration (area, thickness, orientation) should be determined by virtue of device-specific simulation and testing. The case study presented in this paper demonstrates an art of possibility for enhancing thermal management in mobile electronics; two leading-edge materials, when combined, yield a thermal solution with performance greater than the sum of its parts.

References

1. UL, UL. "60950-1: 2003 Information technology equipment-Safety-Part1: General requirements." (2003).

2. Wagner, Guy, and William Maltz. "Thermal management challenges in the passive cooling of handheld devices." *19th International Workshop on Thermal Investigations of ICs and Systems (THERMINIC)*. IEEE, 2013.

3. Wagner, Guy R. "A study of the maximum theoretical power dissipation of tablets under natural convection conditions." *20th International Workshop on Thermal Investigations of ICs and Systems*. IEEE, 2014.

4. Luo, Zhaoxia, et al. "System thermal analysis for mobile phone." *Applied Thermal Engineering* 28.14-15 (2008): 1889-1895.

5. Bi, C., G. H. Tang, and W. Q. Tao. "Prediction of the gaseous thermal conductivity in aerogels with non-uniform pore-size distribution." *Journal of Non-Crystalline Solids* 358.23 (2012): 3124-3128.

6. Kannuluik, W. G., and E. H. Carman. "The temperature dependence of the thermal conductivity of air." *Australian Journal of Chemistry* 4.3 (1951): 305-314.

7. ASTM C518 – 17 Standard Test Method for Steady-State Thermal Transmission Properties by Means of the Heat Flow Meter Apparatus

8. ASTM F36 – 15 Standard Test Method for Compressibility and Recovery of Gasket Materials

9. LaserComp, Inc.. "Tests of thin samples stacked (using FOX50 instrument)" Application Note AN-TSS © 2008 (October 23).

10. Wagoner, G., Skokova, K.A. and Levan, C.D., "Angstrom's Method for Thermal Property Measurements of Carbon Fibers and Composites", *The American Carbon Society*, CARBON Conference, 1999.

11. Beyerle, R., Smalc, M., Kantharaj, R., Taylor, J., Norley, J., "Thermal Diffusivity Characterization of Thick Graphite Foils", *35th Semi-Therm Symposium*, 2019.

12. *3DMARK® Technical Guide*, Underwriters Laboratories, Fremont, CA, 2020.

13. Xiong, Yin, et al. "Thermal tests and analysis of thin graphite heat spreader for hot spot reduction in handheld devices." *2008 11th Intersociety Conference on Thermal and Thermomechanical Phenomena in Electronic Systems*. IEEE, 2008.

An Analysis of Temperature Variation Effect on Response and Performance of Capacitive Microaccelerometer Inertial Sensors

Jacek Nazdrowicz, Andrzej Napieralski

Lodz University of Technology, Department of Microelectronics and Computer Sciences

Wolczanska St. 221/223

Lodz, Poland

jnazdrowicz@dmcs.pl, napier@dmcs.pl

Abstract

There are many designed and manufactured microdevices including sensors and actuators. The most popular and widely used commercially available examples of these are rotational velocity sensors and linear acceleration sensors. These are commonly referred to as inertial sensors, because their principle of operation is based on the displacement of a solid mass under external force. Because of some limitation, like small size, fabrication methods and fragility, these devices must have as simple of a geometry (structure) as possible. However, even in such simple sensors we can differentiate substructures that behave in some characteristic way. Of course, very important roles are played by the sensing mechanism itself and the geometry, particularly of the substructures included in device, which are strongly determined by phenomena/physics used for sensing physical quantity. In some type of sensors and substructures, deformation is an undesirable effect, such as in capacitive sensors, while in other sensors, such as piezoresistive ones, it is the principle of operation. Obviously, temperature variations can cause changes to geometrical dimensions due to expansions; therefore it influences sensor operation and potentially degrades stability and performance. In this paper, the effects of temperature variations on the output accelerometer performance of the inertial sensors are determined, and thermal deformation of this kind of device is analyzed. The variations of the output capacitances of considered for inertial sensor as well as the variation of resonance frequency and capacitance due to temperature fluctuations, which are simulated, calculated and discussed here.

These were the objectives of this work:

1. Assess deformation (extension) of structure along with temperature growth with reference temperature assumed. This is the basis for further assessment of measurement errors introduced by extension and therefore - capacitances deviations.
2. Assess the influence of temperature variation on particular modes of this device.
3. Assess the influence of temperature variation on capacitance shift, which introduces measurement inaccuracies.

Keywords

MEMS, accelerometer, inertial sensor, capacitances.

Nomenclature

A area, m^2

V volume, m^3

a source length, m

t thickness, m

K sensitivity (m/Hz)

Q quality factor

m mass (kg)

T temperature (K)

c damping coefficient (Ns/m)

X displacement in X direction (m)

k spring constant

v linear velocity (m/s)

a linear acceleration (m/s^2)

F force (N)

E Young's modulus

K sensitivity [$F/(m/s^2)$]

d_0 gap between movable part and substrate or cover (m)

C_1, C_2 capacitances between neighbor static and movable electrodes

ε_0 vacuum dielectric constant

ε relative dielectric constant

β temperature coefficient of Young's modulus

Greek symbols

ρ mass density (kg/m^3)

ω frequency (rad/s)

μ air viscosity (Pa·s)

1. Introduction

The influence of ambient temperature in MEMS accelerometer inertial sensors is very significant in microscale [1,2];. This comes from the fact, that it is fabricated with polysilicon, which is an isotropic material that is very temperature-sensitive material and its physical properties vary with temperature [3]. The ambient temperature generate errors and is a source of deviation of performance. In MEMS devices, performance drift caused by thermal-mechanical coupling is often observed. Calibration and error correction required by temperature variation in MEMS devices is often accomplished with third-order thermal models and external temperature sensors. These can suffer from thermal lag and temperature-induced hysteresis [4]. Temperature variation-based performance shift became very problematic [5]. Multiphysics modeling (including thermal, structural, and electrostatic domains) should be considered and used to reflect behavior of device accurately [6,7]. This is a fundamental approach to optimize such structure [8,9].

The principle of operation of the considered accelerometer is based on capacitance changes due to motion of the inertial mass (combined with electrodes) relative to stationary electrodes. This situation takes place when non-zero acceleration appears along the X-direction and the microaccelerometer senses a change in the electrical capacitance due to the change in separation of adjacent electrodes. As a result, capacitance change is detected as a difference of two capacitances. An appropriate application specific integrated circuit (ASIC) transforms the electrical signals into required output quantity (acceleration).

2. Numerical methods

To assess the behavior of a MEMS accelerometer in a variable temperature environment, Finite Element Analysis was applied. The model of the device was designed in the COMSOL Multiphysics environment to perform multidomain simulations and analysis. Because the considered microaccelerometer is a capacitance-based sensor, structural-electrostatic-thermal analysis was required and performed. The geometrical structure of this device is shown in Fig. 1 and consists of inertial mass, suspension beam serpentines, anchors, static and movable electrodes (as comb structures). Movable electrodes are integral parts of the inertial mass, however one can find geometries with separated electrodes and mass. Static and movable electrodes are detection capacitors. The FEM model includes the following boundary conditions (Fig. 2):

- current temperature homogeneously applied to all device surfaces with reference temperature 293.15K,
- thermal expansion applied to all parts of device,
- body load – test acceleration applied (50m/s²),
- fixed constraint – applied to anchors and stationary electrodes,
- ground – applied to inertial mass with movable electrodes,
- voltage – applied to stationary electrodes,
- symmetry – to limit calculation, applied along symmetry axis.

FEM simulations of this MEMS device were also performed in the Matlab/SIMULINK software package. This model was based on the well-known 2nd order Newton's motion equation:

$$m \frac{d^2x}{dt^2} + c \frac{dx}{dt} + kx = F$$

Young's modulus itself changes along with temperature. The relation between this physical quantity and temperature is as follows:

$$E(T) = E(T_0)(1 + \beta \Delta T)$$

where E(T) is Young's modulus at temperature T, E(T₀) is the initial Young's modulus value. The value of β for polysilicon used for simulations is -80 ppm/K [18].

Thermal Expansion Coefficient, α, has a logarithmic dependency on temperature; for small temperature variation $\Delta T = 10K$, it can be assumed as linear. Fig. 3 presents plots taken from literature and simulations [16] performed by authors (both are similar). We observe, that this dependence is stronger for low temperatures rather than for high ones. Therefore, this effect should not be ignored during thermal analysis, especially in considered range ΔT - 0-100K.

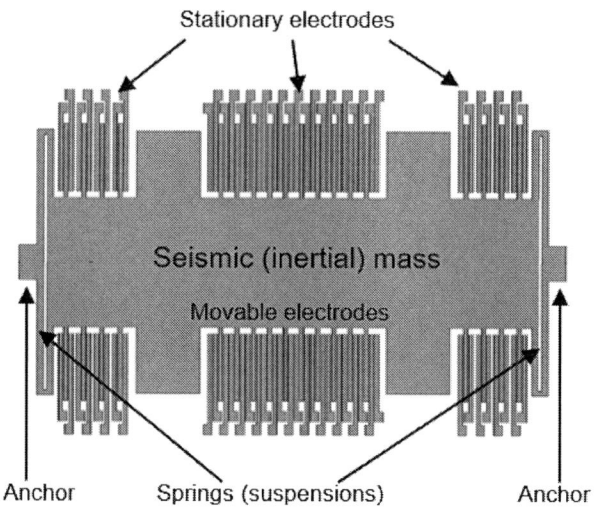

Fig. 1. Structure of MEMS accelerometer.

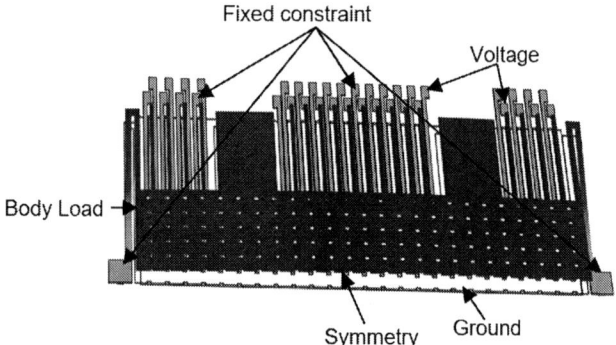

Fig. 2. Boundary conditions applied to MEMS device.

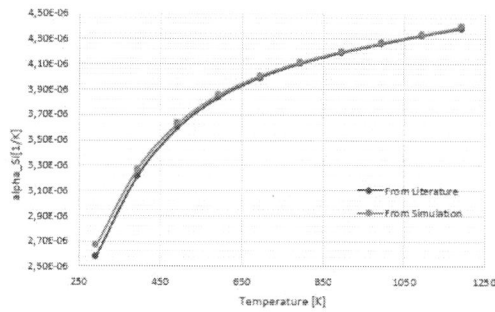

Fig. 3. Thermal Expansion Coefficient dependence on temperature [10,11].

Presented in Fig. 1, the structure of the MEMS accelerometer has 6 sections of comb structures. For movable electrodes (anchored to inertial mass) 0V voltage is applied, for static electrodes: -2.5mV and 2.5V respectively were applied (tests were performed also for higher voltage). This is done to avoid additional electrostatic forces, which can be destructive for precise acceleration measurement. It was assumed here that the temperature of the accelerometer is equal to the ambient temperature, because of the small dimensions of the device and high material thermal conductivity. The accelerometer was simulated for its working temperature range 293.15-393.15K and assumed to be an

92

elastic solid object. Then inertial (linear acceleration) and uniform ambient temperature loads were applied to the model and a modal analysis was performed with thermal-mechanical-electrostatic coupling. Because this structure has an axis of symmetry, symmetry boundary condition is applied for the FEM model, which reduced calculations significantly. The solid mechanics and electrostatics physical domains were linked with the Multiphysics object. The thermal influence on the model was accomplished via the thermal expansion object of Linear Elastic Material Boundary Conditions. To assess displacement and extension of particular parts of the accelerometer, a stationary study was applied. To calculate eigenfrequencies, parametric sweep with stationary and modal analysis were performed.

3. Results

Thermal expansion. As we expected, ambient temperature influences the accelerometer sensor. In Fig. 3, displacements along the X-direction for acceleration a=0 are presented. At a temperature of 293.15K, no meaningful expansions appear. Auxiliary steps performed for different *a* (acceleration) and at constant temperature 293.15K showed that inertial mass moves without any expansion of solid parts. This can be seen with the same displacement for the whole mass (red color). Only suspensions have different displacements.

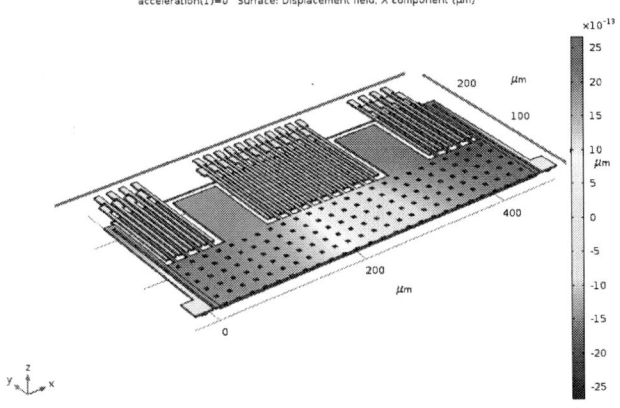

Fig. 3. displacement along X-direction for acceleration a=0.

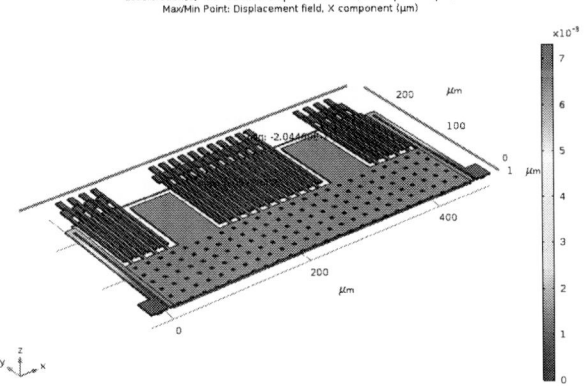

Fig. 4. X-displacement for dT=10K.

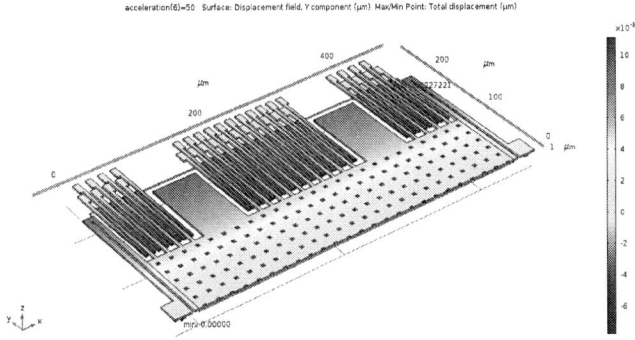

Fig. 5. Y- displacement for dT=10K.

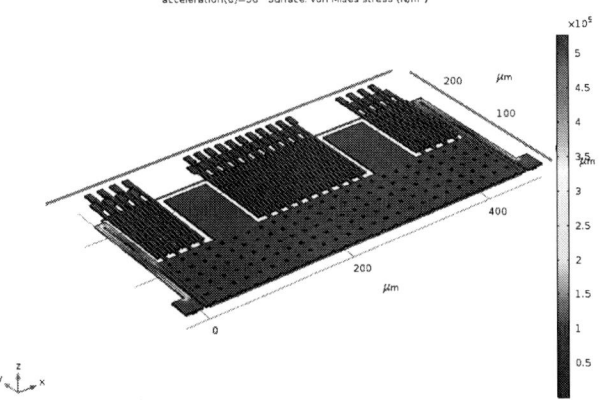

Fig. 6. Von Misses stress distribution.

Fig. 4. shows an example of displacement for dT=10K. Here we see that particular subparts of accelerometer expand along the X and Y directions. Expansion the X-direction causes additional unnecessary stresses and deformations of suspensions on both sides of device. The Y-displacement plotted in Fig. 5 shows that particular electrodes (static and movable) also expand in a variable temperature environment.

Expansion of the inertial mass and particular electrodes naturally causes inaccuracies in measurement, because they cause capacitance changes.

In Fig. 6, von Misses stress distribution is presented. We see that the highest values are in suspensions only. With temperature change, the stress distribution remains unchanged – only the magnitudes grow.

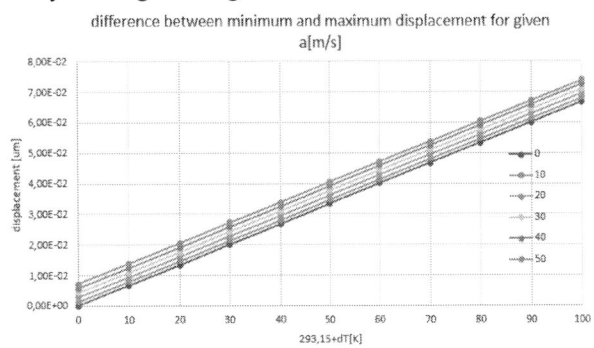

Fig. 7. Difference between minimum and maximum displacement for given acceleration.

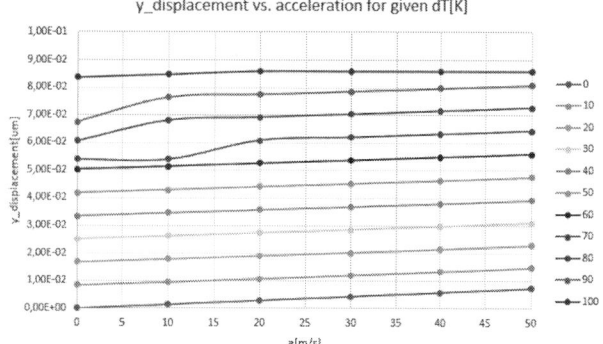

Fig. 8. Displacement in y direction dependence on applied acceleration different temperature.

Fig. 7 and 8 show displacements with different loads applied to the accelerometer. Fig.7 shows the range of displacement (maximum minus minimum) for different accelerations and temperature variations. This displacement difference grows (linearly) along with temperature for each load, therefore electrodes at one end of the device have a different distance between them than those at the other end. This obviously causes inaccuracies in measurement. In Fig. 8 it may be seen that, for each temperature, electrode expansion is almost linear. The temperature and expansion effects have meaningful influence on accuracy, because each time temperature grows by 10K it causes 2-times expansion growth. This shift appears for each acceleration load.

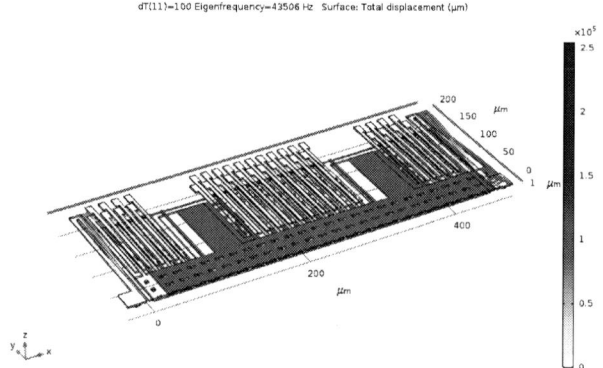

Fig. 9. Mode of vibration for 1st resonance frequency

Variation of eigenfrequency. Fig. 9 shows the mode of vibration for the given accelerometer related to sensing direction. The change of the 1st resonance frequency with temperature is shown in Fig. 10. As we could expect, eigenfrequency decreases linearly as the temperature increases. Results presented in Fig. 10 and 11 are for different spring lengths. These simulations are crucial, because spring length is related to the spring constant which, in turn, is related to the eigenfrequency with the well-known formula: $\omega = \sqrt{\frac{k}{m}}$. Results seen in Fig. 11 confirm nonlinear dependency eigenfrequency on suspension beam length (in whole temperature range).

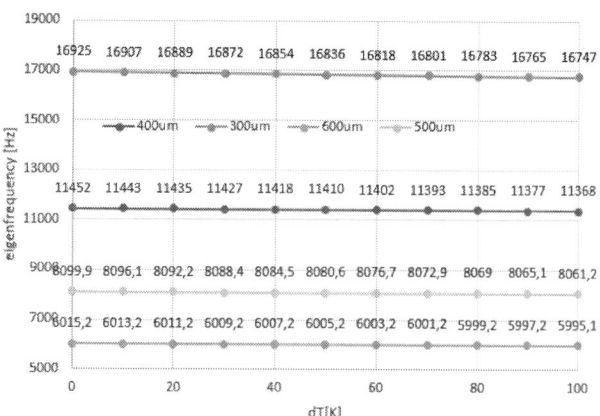

Fig. 10. The change of the 1st resonance frequency along with temperature.

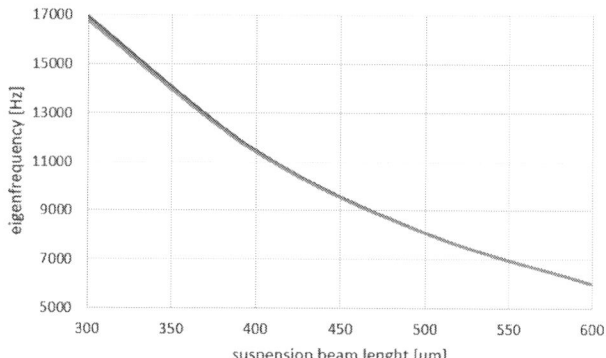

Fig. 11. eigenfrequency dependence on suspension beam length

Capacitance shift
In MEMS accelerometers there are complex structures including many anchored and movable electrodes which form systems of serially or parallelly combined capacitors. The most popular geometry is the comb-like structure (Fig. 4) – applied here. The external acceleration causes displacement and consequently – two different capacitances C_1 and C_2. When acceleration is equal to 0, there is no capacitance output – inertial mass is in equilibrium position. These two differential capacitances are then

$$C_1 = C_2 = C_0 = \frac{\varepsilon_0 \varepsilon A}{d_0}$$

When there is non-zero acceleration applied, displacement of the inertial mass causes changes of capacitances and C_1 and C_2. Capacitance difference $\Delta C = |C_1 - C_2|$ is proportional to displacement. Between ΔC and C_0 there is the relationship

$$\frac{\Delta C}{C_0} = 2\frac{x}{d_0} \Rightarrow \Delta C = 2\frac{C_0 a}{d_0 \omega_0^2}| \qquad K = \frac{\Delta C}{a} = 2\frac{C_0}{d_0 \omega_0^2}$$

Fig. 12 presents the geometry of electrodes commonly applied in accelerometer. Because the accelerometer expands along with temperature variation, it is crucial to analyze how it introduces any unnecessary capacitance shift.

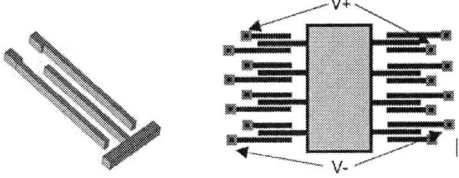

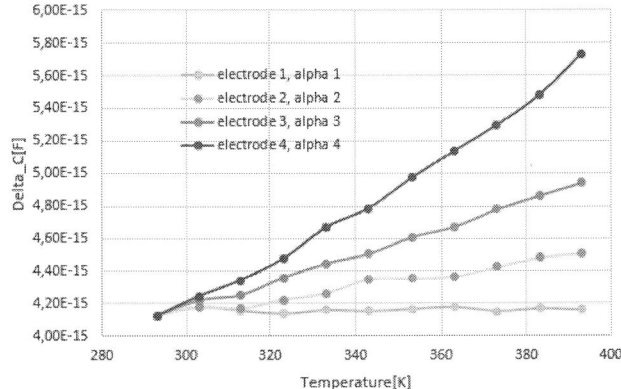

Fig. 12. Capacitance sensing structure and accelerometer.

Fig. 13. Additional capacitance caused by deformation of electrodes thermal expansion ($\alpha_1 < \alpha_2 < \alpha_3 < \alpha_4$).

Results of simulation in the condition of temperature variation deformation are presented in Fig. 13. Capacitance of electrode 1 does not change meaningfully over the entire temperature range. However, for electrodes located away from the middle of comb structure, capacitance temperature dependence increases and, for the edge electrode at 393.15K, the capacitance grew from $6.5*10^{-15}$ to $8.2*10^{-15}$F. Because operational capacitances are 10^{-14}F, these additional ones caused by temperature change should be taken into consideration.

4. Conclusions

The design of MEMS accelerometers is not simple process. Scaling machines to the microworld can be large challenge because these structures are much smaller and much more sensitive to unexpected and undesirable effects. In addition, the device operating environment also has much more influence than in the macroworld. In an idealized situation we would assume operation at a constant temperature without any fluctuations. However, in the real world, inertial sensors must operate over a wide range of temperatures; for applications of commercially used devices (for example smartphones), the influence of temperature cannot be avoided.

Results of simulations show that crucial parameters that the influence on device performance (natural frequency) vary meaningfully as temperature grows. The results show that structure requires individual approach to model and perform simulations of the device in various temperature. Because the accelerometer belongs to the inertial sensor family, it is affected with thermal expansion phenomena. Results show that this produces error because geometrical dimensions of capacitor electrodes vary and give different results in different ambient temperatures.

Analysis results such as those presented here should be an inseparable part of each MEMS accelerometer designing process. First of all, it determines application geometrical structure and suspensions types. Because performance is very sensitive (according to the obtained results) on temperature, it is also crucial to take into consideration all quantities that are dependent on temperature, such as geometry dimensions, damping coefficients, Young's modulus and all derivative quantities like natural frequencies and Q factors.

The presented analysis of the influence of thermal expansion is complex, taking into consideration accelerometer behavior in both mechanical and electrostatic domains under variable temperature. Both domains, mechanical (eigenfrequency) and electrostatic (capacitance), were analyzed separately to simplify calculations to save computational time. In the case of electrostatic analysis with temperature variation, a two-step method was used: in the first step (common for mechanical and electrostatic analysis), simulation of whole device was performed to determine its expansion under temperature variation. The second step included electrode vertex coordinates to determine specific electrode deflection (angle between electrodes and shift caused by expansion). These two parameters became input to the second FEM simulation to perform analysis of single capacitor (sensing element) to obtain capacitance value.

Taking into consideration the mechanical response of the device, the second step was to calculate in COMSOL the eigenfrequency related to inertial mass operational motion directions. This step is crucial, because inertial devices work as a classical accelerometer when $\omega < \omega_n$. Therefore, this step avoids unexpectedly large displacements of the mass that damage the MEMS and allows for assessing the frequency range for operational purposes.

The novelty in the calculation method presented in this paper is the capability to make correction to output capacitance difference caused by acceleration. The big advantage is that it allows designers to create MEMS accelerometers as more accurate sensors by eliminating errors caused by the thermal expansion on the ASIC level. This capacitance difference caused by temperature change can play a very important role, because the total additional capacitance is a sum of particular pair of electrodes additional capacitances. The magnitude of these capacitances can meaningfully influence the final operational value.

References

1. Yang Z., Li X., Simulation and optimization on the squeeze-film damping of a novel high-g accelerometer. Microelectron. J., 37, pp. 383–387, 2006
2. Han J.S., Kwak B.M., Robust optimal design of a vibratory microgyroscope considering fabrication errors. J. Micromech. Microeng., 11, pp. 662–671, 2001
3. Cao H., Li H., Sheng X., Wang S., Yang B., Huang L., A novel temperature compensation method for a MEMS gyroscope oriented on a periphery circuit. Int. J. Adv. Robot. Syst., 10, pp. 1–11, 2013.
4. Prikhodko I.P., Trusov A.A., Shkel A.M., Compensation of drifts in high-Q MEMS gyroscopes using temperature self-sensing. Sens. Actuators A-Phys., 201, pp. 517–524, 2013.

5. Xu L., Yang B., Wang S., Li H., Huang L., Research on Thermal Characteristics and on-chip Temperature-controlling for Silicon Micro-gyroscope. In Proceedings of International Conference on Information and Automation, Shenzhen, China, 6–8 June 2011.

6. Rochus V., Geuzaine C., A primal/dual approach for the accurate evaluation of the electromechanical coupling in MEMS. Finite Elem. Anal. Des., 49, pp. 19–27, 2012.

7. Dai G., Li M., He X., Du L., Shao B., Su W., Thermal drift analysis using a multiphysics model of bulk silicon MEMS capacitive accelerometer. Sens. Actuators A-Phys., 172, pp. 369–378, 2011.

8. Kuramochi N., Toshiba K., Mochiduki K., Tsuchitani M., Application of Robust Design for the Tuning of Resistance-temperature Characteristics in Diodes. In Proceedings of International Symposium on Semiconductor Manufacturing, Santa Clara, CA, USA, 15–17 October 2007.

9. Sadeghian H., Doniavi A., Analysis of quality design techniques for electrostatic actuators. J. Phys.: Conf. Ser., 34, pp. 919–924, 2006.

10. Huang Q.-A., Lee N.K.S., Analytical modeling and optimization for a laterally driven polysilicon thermal actuator, Microsystem Technologies 5 (3), pp133–137, 1999.

11. Paryab N., Jahed H., Khajepour A., Creep and Fatigue Failure in Single- and Double Hot Arm MEMS Thermal Actuators, Journal of Failure Analysis and Prevention 9(2), pp.159-170, 2009.

Measurement of Performance Characterization of Ultra-Thin Vapor Chamber

Professor Wei-Keng Lin, Wen-Hua Zhang, Chien Huang, Ching-Huang Tsai, Kenny Hsaio
T-Global Technology Co.
No.33, Ln. 50, Daren Rd., Taoyuan Dist.,
Taoyuan City 330/Taiwan
wk-lin@tglobal.com.tw

Abstract

The next generation of 5G mobile phone cooling requires many ultra-thin vapor chambers. This paper discusses how to measure the performance characterization of the vapor chamber. All the experiment is processing under the horizontal orientation so that the gravity effect could reduce to minimum. The thermal diffusivity of the two ultra-thin vapor chambers VC-A ($180*80*1mm^3$) and VC-B ($90*90*0.4mm^3$) developed from laboratory are $\alpha_{VC-A}=1.9(cm^2/s)$ and $\alpha_{VC-B}=1.44(cm^2/s)$. Converted to equivalent effective thermal conductivity are $K_A=4833.3$ (W/m.K) and $K_B=3069$ (W/m.K), which is about ten times of that of pure copper. The maximum heat transfer of VC-B is $Q_{VC,max,B}=210(W)$, the axial thermal resistance $R_{th,VC,Z,B}=0.04$ (K/W), and the thermal resistance of vapor chamber $R_{th,VC,B}=0.42$ (K/W). Using the same method to test the thicker vapor chamber of C brand ($90*90*3mm^3$), the $Q_{VC,max,C}=551(W)$, the axial thermal resistance $R_{th,VC,Z,C}=0.08$ (K/W), and the average thermal resistance of the vapor chamber is $R_{th,VC,C}=0.177(K/W)$. The highest thermal conductivity K value vapor chamber with $90X60X0.4mm^3$ designed in ACL currently is vapor chamber VC-D, the equivalent thermal conductivity is 33519 (W/m-k).

Keywords

Thermal diffusivity; ultra-thin vapor chamber; heat pipe

Nomenclature

Symbol	Description
2B	Thickness of the test sample (cm)
A_s	Surface arear of the test sample (cm^2)
α_{1D}	Thermal diffusivity of 1D (cm^2/s)
M/N	Amplitude ratio of T_1 to T_2
ρ	Density of test sample (g/cm^3)
C	Specific of test sample (J/g-k)
L	Measuring distance between T_1 and T_2 (cm)
Δt	The delay time from T_1 to T_2 for sine wave heating cycle (s)
k_{eff}	Effective thermal conductivity of test sample (W/m.k)
$K_{eff,total}$	The effective total thermal conductivity(W/m.k)
$K_{eff,x}$	Effective thermal conductivity in the X direction (W/m.k)
$K_{eff,y}$	Effective thermal conductivity in the Y direction (W/m.k)
M	The amplitude of the temperature change of T_1 for sine wave power.
N	The amplitude of the temperature change of T_2 for sine wave power.
$\dot{m}$	Flow rate (g/s)
T_{out}	Outlet temperature of the water jacket ($^\circ C$)
T_{in}	Inlet temperature of the water jacket ($^\circ C$)
Q_{out}	Heat dissipated from water jacket
$Q_{IN,Actual}$	Aamount of heat actually input the vapor chamber (W)

1. Introduction

As electronic products pay more attention to the trend of efficiency and speed, end users are most afraid of overheating of electronic products, abnormal fan noise, and slow operation. Therefore, under the increasing demand for high-speed computing such as AI and e-sports for various electronic products, it has aroused the interest of all the companies in the future development and dynamics of the thermal industry. Vapor Chamber is like to the principle of Heat Pipe. It is a hollow, closed copper plate that fills a hollow space with liquid and absorbs heat through the metal to evaporate the liquid into a vapor, the heat is then absorbed through the fins and removed, then returns to the liquid state and complete a heat dissipation cycle after cooling. Compared with the heat pipe, the vapor chamber has a larger contact area with the heat source. 5G mobile phones are expected to be announced in 2020. It is currently observed that 5G mobile phones consume will be up to 10W, which is 2.5 times more than that of 4G mobile phones. Therefore, the graphite sheet and heat pipe thermal are not a suitable solution. Thus, Vapor chamber support is required. Under the demand for e-sports and 5G, the ultra-thin vapor chamber solutions will be the mainstream solution.

The main product of the ultra-thin vapor chamber of this paper are jointly developed by the advanced cooling laboratory (ACL) of Tsinghua University in Taiwan. The four most important characterization of the vapor chamber are: the maximum heat transfer Q_{max}, the thermal diffusivity, the axial thermal resistance of vapor chamber and its spread thermal resistance. The spread thermal resistance of vapor chamber measurement is based on the standard set by the Taiwan Thermal Management Association (TTMA) in 2015 [1]. The principle of measurement is the same as the maximum heat flux of the heat pipe [2]. The most important characterization of the vapor chamber is the measurement of thermal conductivity. Due to the diffusion problem, tradition calculation method through Fourier's heat conduction law is not feasibility. This paper proposed to measure the thermal diffusivity of vapor chamber first, then converted this thermal diffusivity value to equivalent effective thermal conductivity. Thermal diffusivity of vapor chamber is measured by Thermal Diffusivity Measurement Instruments (TDMI) developed according to Angstrom theory. Angstrom's theory is about applying a sinusoidal heating cycle to a one-dimensional strip. When heat is transfer to the end of the strip, measuring the length、the time delay between two temperature cycle points and temperature amplitude of these two temperature points, the thermal diffusivity α [3, 4, 5, 6] of the test sample can be obtained. Angstrom's measurement theory is shown in figure 1. Its one-dimensional governing equation can be expressed as equation (1), where $N_{hA}(T-T_a)$ is the heat loss of the surface via thermal convection, or written as equation (2). After the calculation is simplified by the separation variable method, the thermal diffusivity of the object to be tested can be calculated by the heat transfer formula (3). The result is shown in figure 2, in which the horizontal axis represents elapsed time and the vertical axis represents temperature. The blue curve is the temperature at which T_1 changes with the heating cycle; and M is the temperature difference between the highest temperature and the lowest temperature of T_1, that is, the amplitude of the heating cycle. The red curve is the temperature at which T_2 changes with the heating cycle; and N is the amplitude of heating cycle at T_2. According to the Angstrom method, the thermal diffusivity α_{1D} of the one-dimensional test object can be calculated by using the M, N, the delay time Δt and the distance L between T_1 and T_2. In the heat transfer analysis, the thermal diffusivity α_{1D} is the ratio of the thermal conductivity K to the products of density ρ and the specific heat C, expressed by the formula (4).

$$-\left(\frac{\partial T}{\partial t}\right) + \alpha_{1D}\left(\frac{\partial^2 T}{\partial x^2}\right) - \frac{h}{B\rho C_p}(T - T_a) = 0 \quad (1)$$

$$\text{let } N_{hA} = \frac{h}{B\rho C} = \frac{h(2LW)}{B(2LW)\rho C} = \frac{2hAs}{V\rho C} = \frac{h(2As)}{mC}$$

$$-\left(\frac{\partial T}{\partial t}\right) + \alpha_{1D}\left(\frac{\partial^2 T}{\partial x^2}\right) - N_{hA}(T - T_a) = 0 \quad (2)$$

$$\alpha_{1D} = \frac{L^2}{2 \Delta t \left(\ln\frac{M}{N}\right)} \quad (3)$$

$$\alpha_{1D} = \frac{k_{eff}}{\rho C} \quad (4)$$

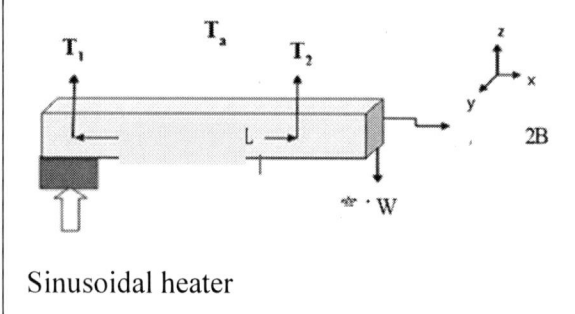

Fig 1. Schematic diagram of temperature trend distribution measured by Angstrom theory

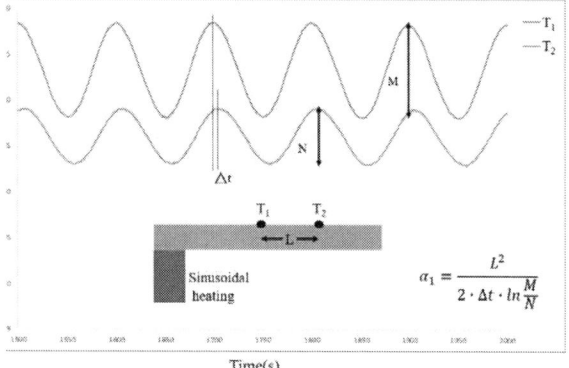

Fig 2. Schematic diagram of one-dimensional heating of Angstrom theory

2. Experimental device

2.1 VC maximum heat transfer measuring instrument

All experiments were conducted with the vapor chamber in the horizontal orientation so that the gravity effect could reduce to minimum.

TTMA defines a vapor chamber is a sort of two-dimension heat pipe with a length and width greater than 30 mm, and a thin vapor chamber refers to a total thickness of the vapor chamber below 1.0 mm, while for the thickness less than 0.5 mm, it is an ultra-thin vapor chamber. The configuration of the standard test of this experiment is shown in figure 3. The condensed water jacket is used as the cooling system. The calculation method of the heat carries away is as shown in equation (5). All experiments were conducted with the vapor chamber in the horizontal orientation so that the gravity effect could reduce to minimum.

$$Q_{out} = \dot{m}C(T_{out} - T_{in}) \quad (5)$$

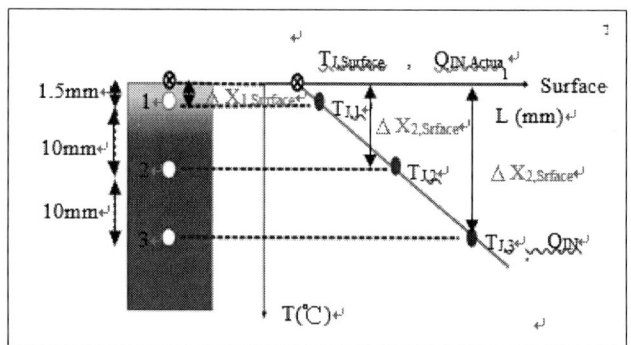

Fig. 3: The configuration of the standard test of heat copper block

The schematic diagram of the experimental configuration with the condensed water jacket as the cooling system is shown in figure 4. The heater, the vapor chamber and the condensate jacket are sequentially arranged from bottom to top. Since the power is supplied to the heating rod and then transmitted to the heated copper block, there is inherent heat loss, and the wattage supplied by the power supply is not necessarily equal to the power dissipation received by the vapor chamber from the heater. Therefore, three holes for placing the thermocouples are arranged on the heater, sequence from top to bottom in order of $T_{J,1}$, $T_{J,2}$, $T_{J,3}$. The actual heat transfer is calculated by the three thermocouples is calculated. figure 4 is a theoretical diagram for calculating the surface temperature and power dissipation of the heater, and defines the heater surface temperature, $T_{J,surface}$, and power dissipation $Q_{IN,Actual}$. The heating area (A_{heater}) is set to $30 \times 30 \text{mm}^2$ and the input power is calculated from the electrical power provided by the DC power supply $Q_{IN} = I \times V$. Since the heat transfer system of the heated copper block complies with Fourier's law, its formula is as shown in the formula (6).

$$Q = -KA_{heater}\frac{\Delta T}{\Delta x} \quad (6)$$

Since the heated copper block is a homogeneous conductor and is a rectangular parallelepiped, K and A_{heater} are constant and a general differential method is used to obtain the formula (7).

$$T_{(X)} = T_1 X + T_2 \qquad (7)$$

Equation (7) shows the temperature curve is linear. It can be seen from figure 4 that the positions of the three temperature points are $\Delta X_{1,Surface}$=1.5mm, $\Delta X_{2,Surface}$=11.5mm and $\Delta X_{3,Surface}$ =21.5mm in order from top to bottom of the heater surface. $T_{J,Surface}$ can be calculated by extrapolation method of equation (8) as shown in equation (9).

$$\frac{T_{J,2}-T_{J,Surface}}{T_{J,1}-T_{J,Surface}} = \frac{11.5}{1.5} \qquad (8)$$

$$T_{J,Surface} = \frac{11.5 T_{J,1} - 1.5 T_{J,2}}{10} \qquad (9)$$

$Q_{IN,Actual}$ of the surface of the copper block can be calculated from Fourier heat conduction law as shown as in equation (10). The K value of the formula (10) represents the thermal conductivity (W/m.K), and A_C represents the cross-sectional area (m^2) of the normal vector with the heat transfer direction.

$$Q_{IN,Actual} = -K A_{heater} \frac{(T_{J,1}-T_{J,Surface})}{\Delta X_{1,Surface}} \qquad (10)$$

Figure 5 is a schematic diagram of the five-point temperature of the nine-square grid issued by TTMA. The five-point temperature is used for comparison at the condensation section of the vapor chamber and the bottom of the condensate jacket. The temperature-measuring points from left to right in the schematic diagram are sequentially T_{UL}, T_{LL}, T_C, T_{UR}, T_{LR} and represent the temperature of upper left of the vapor chamber, the temperature of the lower left, the temperature of the center point, the temperature of the upper right and the temperature of the lower right. Figure 6 is a five-point temperature position diagram of the condensate jacket. The five grooves at the bottom of the condensate jacket with the depth of the groove is 1.3mm and the width is 1.8mm. The lengths of grooves are one 45mm and four 42.5mm. The purpose of measuring the five-point temperature of the nine-square grid is to explore the average temperature of the vapor chamber based on the definition by TTMA. There are two thermal resistances for the vapor chamber, the axial thermal resistance $R_{th,Z}$ and the radial thermal resistance value $R_{th,sp}$ respectively. The axial thermal resistance value $R_{th,Z}$ is calculated as the equation (11). The radial thermal resistance value, also referred to as the diffusion thermal resistance value $R_{th,sp}$. If the radial thermal resistance value is decreased, the surface temperature $T_{J,surface}$ of the heated copper block is lowered. Therefore, the temperature of the vapor chamber can be judged by the magnitude of the axial thermal resistance and the radial thermal resistance. These two thermal resistances can be used as a characterization index of the vapor chamber.

$$R_{th,Z} = \frac{T_{J,Surface}-T_C}{Q_{IN,Actual}} \qquad (11)$$

TTMA additionally defines the thermal resistance of the vapor chamber $R_{th,vc}$ as in (12)

$$R_{th,VC} = \frac{T_{J,Surface}-T_{c,i}}{Q_{IN,Actual}} \qquad (12)$$

Where $T_{C,i}$ is defined as the lowest temperature of the five measuring points of the cooling water jacket. When measuring the experimental data, the heat taken away by the condensate jacket should be calculated and compared with the input power to evaluate the heat loss caused by the experimental

configuration. The water-cooled heat transfer device is mainly composed of a thermostat to constant temperature and pressure water source, condensate jacket, flow meter, power supply control system and temperature measurement system. The heat carried away by this water cooled system is calculated as Eq. (13):

$$Q_{out} = \dot{m} C (T_{out} - T_{in}) \qquad (13)$$

To this end, the definition of heat loss as shown in equation (14), which is used to evaluate the reference value of the measured data and improve the accuracy.

$$Q_{Heat\ loss}(\%) = \frac{Q_{IN,Actual}-Q_{OUT}}{Q_{IN,Actual}} \times 100\% \qquad (14)$$

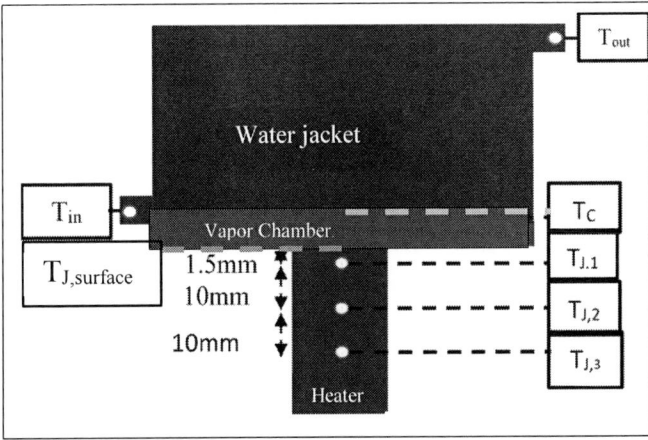

Fig. 4: Schematic diagram for the surface temperature and wattage of the heater

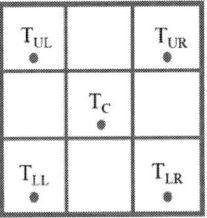

Fig. 5: Schematic diagram of the five-point temperature of nine-square grid

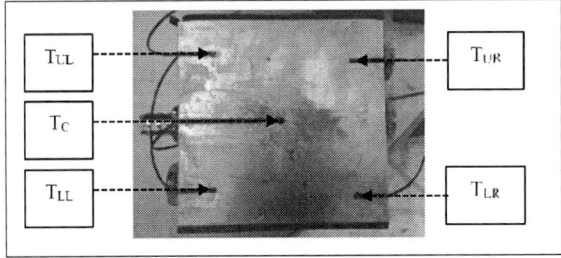

Fig. 6 five-point temperature position diagram of the condensate jacket.

2.2 Performance index of the vapor chamber

The performance indicator of the vapor chamber is to control the water flow or condensation to fix the constant T_C temperature (for example, $T_C = 60$ °C) to obtain the thermal resistance value $R_{th,VC}$ (°C/W) and maximum heat transfer

($Q_{VC,max}$). The maximum heat transfer rate ($Q_{VC, max}$) of the vapor chamber is that the input power when the axial thermal resistance $R_{th,VC,Z}$ is suddenly increased when applying a certain input heat power.

In the case of a completely dried region under the specified T_C, the evaporator temperature versus input power curve has a turning point, the actual input power $Q_{IN,Actual}$ of the vapor chamber at this point called the maximum heat transfer ($Q_{VC,max}$) of the vapor chamber as shown in figure 7. When under the condition of partial (local) drying, there is no apparently turning point of the performance curve as shown in figure 8. At this case, the maximum heat transfer of vapor chamber ($Q_{VC,max}$) is defined the power input at the $T_{J,surface}$ =100°C. The reason for considering $T_{J,surface}$ as the limit of $Q_{VC,max}$ is that all the manufacturers think that the performance indicator of the vapor chamber is no longer meaningful when CPU temperature exceeds 100℃. In another word, all the system manufacturers no longer use this kind of vapor chamber working as the CPU cooler element. Therefore, when at partially dry out condition, TTMA defined the maximum heat transfer is actual input power ($Q_{IN,Actual}$) when at $T_{J,surface}$=100 ℃. Figure 9 shows the maximum heat transfer for the vapor chamber performance Test Instrument (VCPT).

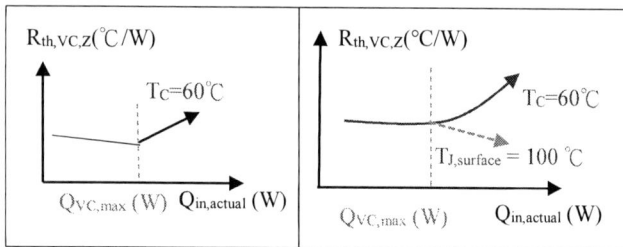

Fig. 7. The curve of axial thermal resistance vs. input power of vapor chamber at completely dry out condition

Fig. 8. The curve of axial thermal resistance vs. input power of vapor chamber at partially dry out condition

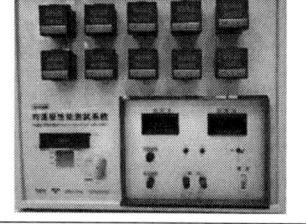

Fig. 9. The maximum heat transfers for the Vapor Chamber Performance Test Instrument (VCPT)

2.3 Thermal Diffusivity Measuring Instrument (TDMI)

Figure 8 shows the Thermal Diffusivity Measuring Instrument [7, 8, 9, 10, 11, 12]. The heat source input is driven with sine wave, which is controlled with an electronic chip to achieve a periodic heating curve. Test sample is placed above the heated copper block in contact, and two temperature detecting points T_1 and T_2 touch the top surface of test sample. The TDMI is mainly composed of three parts: (1) the power supply as shown in Figure 10, (2) the sinusoidal heating power

T.E.C. as shown in Figure 11, and (3) the thermal diffusivity measuring platform as shown in Figure 12.

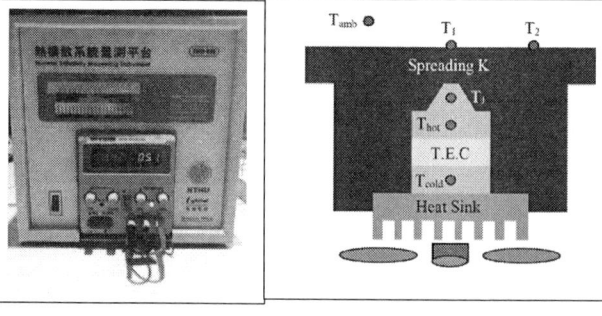

Fig. 10. Power supply of TDMI

Fig. 11 sinusoidal heating power T.E.C.

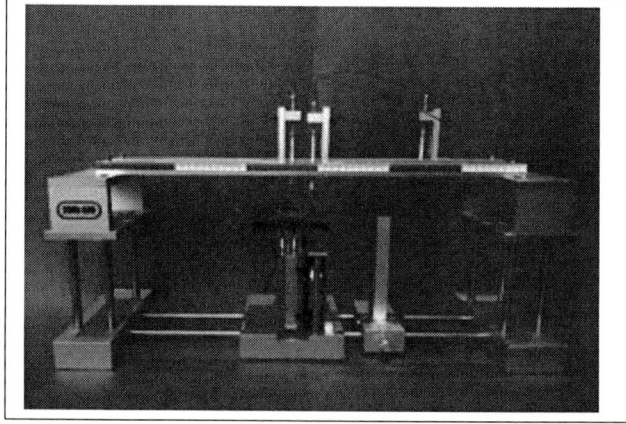

Fig. 12 the thermal diffusivity measuring platform

2.4 Calibration of TDMI

In order to verify the accuracy of the TDMI, calibration experiments must be carried out. The calibration is based on pure material such as copper, tin, lead and aluminum 6061. All the standard diffusivity of these pure material can be found in the literature. After measuring the thermal diffusivity α of these materials by TDMI, and comparing with the standard values, the reliability and accuracy of the instrument can be verified. The data measurement is in the form of full blind test. First, it must be determined that the experimental data is repeatable. Therefore, it is necessary to perform repeatability error analysis on the acquisition of experimental numbers. The repeatability error formula (15) refers to repeating several measurements by the same person during the experiment, and comparing the measured results with the overall average. The relative error is of course as low as possible. Accuracy error refers to the relative error between the test result and the standard value of the golden sample as shown in equation (16). The method compares the test results with the standard values of the test pieces to determine whether the instrument is accurate or not.

$$\varepsilon_{rep} = \frac{|\alpha - \alpha_{ave}|}{\alpha_{ave}} \times 100\% \qquad (15)$$

$$\varepsilon_{std} = \frac{|\alpha - \alpha_{std}|}{\alpha_{std}} \times 100\% \qquad (16)$$

Where: $\alpha_{ave} = \frac{\sum_i^n \alpha_i}{n}$ is the average of several measurements repeated. α_{std} is the standard value of the test sample.

2.5 Definition of thermal diffusion measurement dimension of the test samples

Angstrom's method was originally only applied to the one-dimensional theory test piece, and its shape is a long and thin test piece (generally, the width W<5 cm, length L=10 cm or less) as shown in Fig. 13. However, some of the samples are not suitable for cutting, for example, in a heat pipe, vapor chamber or test piece having a large width (more than 9 cm or more) and has a certain thickness, as shown in Fig. 14, the heat transfer mode may change from one dimension to two dimensions. It is therefore necessary to define a criteria based on the sample size to judge when the one-dimensional mode is applicable as well as the two-dimensional mode. In one-dimensional definition test piece, the width is W, the thickness is T, and the length is L. when in one-dimensional thermal distance x is used as an index as shown in figure 13. The judgment parameter is τ_x=T/W as in equation (16). When in two-dimensional thermal traveling direction x and y are used as an index as shown in figure 14. For the y direction, the width-to-thickness ratio with respect to the y direction is τ_y = T/L as in the equation (17). Theoretically, if the thickness of a pure material sample is fixed and the width is getting larger, the diffusion rate of the sample must change from one dimension to two dimensions. From the heat transfer mode, the thermal resistance in the x direction is $R_{th,x}$, and the thermal resistance in the y direction is $R_{th,y}$. Since the heat transfer direction occurs simultaneously in the X and Y, it is a parallel type, and its total thermal resistance value $R_{th,total}$ is as shown in equation (18). Assuming the measurement distance is L in both the x and y directions, and since it is the same test piece, the cross-sectional area A_C are all the same. Substituting equation (18a), equation (18b), and equation (18c) into equation (18), equation (19) can be obtained. The thermal diffusivity is expressed as thermal conductivity as shown in equation (20), equation (20a), and equation (20b), and then plugging equation 3 into equation (19) to obtain equation (21), that is, the total thermal diffusivity expressed in two-dimensional mode. Assuming that the substance to be tested is a homogeneous substance, the total thermal diffusivity (α_{2D}) in the theoretical two-dimensional mode is twice that of the one-dimensional mode (α_{1D}) as in the equation (22). But in the large-scale vapor chamber of non-homogeneous matter, α_x and α_y may not be the same, so α_x and α_y should be measured simultaneously.

$$\tau_x = \frac{T}{W} \tag{16}$$

$$\tau_y = \frac{T}{L} \tag{17}$$

$$\frac{1}{R_{th,total}} = \frac{1}{R_{th,x}} + \frac{1}{R_{th,y}} \tag{18}$$

$$R_{th,total} = \frac{L}{K_{eff,total}A_C} \tag{18a}$$

$$R_{th,x} = \frac{L}{K_{eff,x}A_C} \tag{18b}$$

$$R_{th,y} = \frac{L}{K_{eff,y}A_C} \tag{18c}$$

$$K_{eff,total} = K_{eff,x} + K_{eff,y} \tag{19}$$

$$\alpha_{total} = \frac{K_{total}}{\rho C} \tag{20}$$

$$\alpha_x = \frac{K_{eff,x}}{\rho C} \tag{20a}$$

$$\alpha_y = \frac{K_{eff,y}}{\rho C} \tag{20b}$$

$$\alpha_{total} = \alpha_x + \alpha_y \tag{21}$$

$$\alpha_{1D,total} = 2\alpha_x \tag{21a}$$

$$\alpha_{1D,total} = \frac{L^2}{2\Delta t(\ln\frac{M}{N})} = \frac{\alpha_{2D,total}}{2} \tag{22}$$

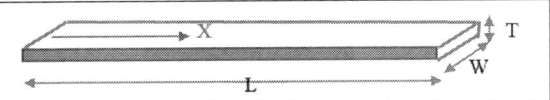

Fig. 13. Schematic of one-dimensional test piece

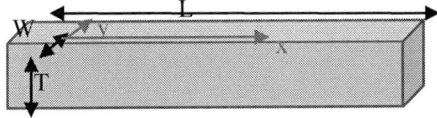

Fig. 14. Schematic of two-dimensional test piece

3. Experiment results and discussion

We have already completed the repeatabilityand the accuracy test on the thermal diffusivity of different materials with different sizes. Due to too many data, we only take out small part of it as shown in following tables.

3.1. The repeatability error verification of TDMI

Table 1, table 2 and table 3 show the repeatability error for Copper, Aluminum and Tin respectively, all repeatability error are less than 5%.

Table 1. Repeatability error Analysis of Thermal Diffusivity for Copper in TDMI Test (L_M=3cm) ($\alpha_{std,cu}$=1.17cm^2/s)

L (mm)	W (mm)	T (mm)	τ=T/W	$\varepsilon_{cu,rep,1}$ %	$\varepsilon_{cu,rep,2}$ %	$\varepsilon_{cu,rep,3}$ %
100	10	0.4	0.04	0%	0.84%	0%
100	20	0.4	0.02	0%	0.91%	0.91%
100	30	0.4	0.01	1.63%	0%	1.63%
100	40	0.4	0.01	3.38%	3.38%	3.38%
100	50	0.4	0.01	2.97%	3.44%	3.44%

Table 2. Repeatability error Analysis of Thermal Diffusivity for Aluminum in TDMI Test (L_M=3cm) ($\alpha_{std,Al}$=0.85cm^2/s)

L (mm)	W (mm)	T (mm)	τ=T/W	$\varepsilon_{rep,1}$ %	$\varepsilon_{rep,2}$ %	$\varepsilon_{rep,3}$ %
100	10	0.4	0.04	0.74%	1.46%	0.71%
100	20	0.4	0.02	0.35%	0.49%	0.13%
100	30	0.4	0.013	1.45%	0.88%	0.57%
100	40	0.4	0.01	2.10%	2.3%	0.2%
100	50	0.4	0.008	0.56%	0.15%	0.41%
100	50	2	0.04	0.75%	0.39%	1.14%

101

Table 3. Repeatability error Analysis of Thermal Diffusivity for Tin in TDMI Test (L_M=3cm) ($\alpha_{std,Sn}$=0.4cm²/s)

L (mm)	W (mm)	T (mm)	τ=T/W	$\varepsilon_{rep,1}$ %	$\varepsilon_{rep,2}$ %	$\varepsilon_{rep,3}$ %
100	10	0.4	0.04	0.74%	1.23%	0.49%
100	20	0.4	0.02	0.84%	2.20%	1.35%
100	30	0.4	0.013	1.26%	1.58%	0.31%
100	40	0.4	0.01	0.33%	0.83%	1.17%
100	50	0.4	0.008	2.48%	1.77%	0.71%

3.2. The accuracy error verification of TDMI

Table 4 shows the accuracy error analysis of copper in TDMI test. The diffusivity of pure copper $\alpha_{std,cu}$ is 1.17 cm²/s. The average value of the experimental amount of 3 times is expressed by $\alpha_{cu,avg}$, The criterion for the judgment is the thickness-to-width ratio of the test sample as an index, the thickness of the test sample T must less than 5 mm or less. When τ<0.02, region is "A" ; when 0.02 <τ<075, regionis "B" ; while 0.075 ≤τ, region is "C". From data analysis of table 4, we obtain following summary:

(I) When $0.022 < \tau < 0.075$ is the " B" region, the 1D mode is adopted, expressed as $\alpha_{i,1D}$

$$\alpha_{i,1D} = \frac{L^2}{2\Delta t (\ln\frac{M}{N})} \qquad (23);$$

Where i is the metal code.

(II) When $\tau \leq 0.022$ is the " A" region, the 2D mode is adopted, expressed as $\alpha_{i,2D}$,

$$\alpha_{i,2D} = \alpha_{i,1D} \times 2 \qquad (24) ;$$

(III) When $\tau \geq 0.075$ is the " C" region, the 0.5D mode is adopted, expressed as $\alpha_{i,0.5D}$,

$$\boldsymbol{\alpha_{i,0.5D} = \alpha_{M,1D} \times 0.5} \qquad \boldsymbol{(25)}$$

And
$$\varepsilon_{i,std,1D} = \left|\frac{\alpha_{i,1D}-\alpha_{std,i}}{\alpha_{std,i}}\right| \times 100\% \qquad (26)$$

$$\varepsilon_{i,std,2D} = \left|\frac{\alpha_{i,1D}\times2-\alpha_{std,i}}{\alpha_{std,i}}\right| \times 100\% \qquad (27)$$

$$\varepsilon_{i,std,0.5D} = \left|\frac{\alpha_{i,1D}\times0.5-\alpha_{std,i}}{\alpha_{std,i}}\right| \times 100\% \qquad (28)$$

The bold font in the table 4 are the accuracy error after partition correction. Table 4 shows that all the accuracy error of copper is controlled within 10% after partition correction as shown by the bold character. Table 5 shows the accuracy error analysis of Aluminum in TDMI test. The bold font in the table 5 are the accuracy error after partition correction. Table 5 shows that all the accuracy error of Aluminum is controlled within 2% after partition correction as shown by the bold character. Table 6 shows the accuracy error analysis of Tin in TDMI test. The bold font in the table 6 are the accuracy error after partition correction. Table 6 shows that all the accuracy error of Tin is controlled within 5% after partition correction as shown by the bold character. Results shows TDMI doing Well, the experimental data is highly reliable.

Table 4. Accuracy error analysis of thermal diffusivity for copper in TDMI Test (L_M=3cm) ($\alpha_{std,cu}$=1.17cm²/s)

No	L (mm)	W	T	$\tau = \frac{T}{W}$	Zone	$\alpha_{cu,avg}$	1-D mode (B) ($\alpha_{cu,1D}$) / $\varepsilon_{cu,std,1D}$	2-D mode (A) ($\alpha_{cu,1D}$ x2) / $\varepsilon_{cu,std,2D}$	0.5D mode (C) ($\alpha_{cu,1D}$ x0.5) / $\varepsilon_{cu,std,0.5D}$
1	100	10	0.4	0.04	B	1.198	**2.4%**	104.78%	48.80%
2	100	20	0.4	0.02	B	1.094	**6.4%**	87%	53.24%
3	100	30	0.4	0.013	A	0.615	47%	**5.12%**	73.71%
4	100	40	0.4	0.01	A	0.60	48%	**2.56%**	74.35%
5	100	50	0.4	0.008	A	0.5826	50%	**0.41%**	75.1%
11	100	10	1	0.1	C	2.396	104%	308.54%	**2.13%**
16	100	10	1.5	0.15	C	2.408	105%	311.6%	**2.9%**
17	100	20	1.5	0.075	C	2.143	83%	266.32%	**8.41%**
22	100	20	2	0.1	C	2.27	99%	288.03%	**2.99%**

Table 5. Accuracy error Analysis of Thermal Diffusivity for Aluminum in TDMI Test (L_M=3cm) ($\alpha_{std,Al}$=0.85cm²/s)

No.	L (mm)	W	T	$\tau = \frac{T}{W}$	Zone	$\alpha_{Al,avg}$	1-D mode (B) ($\alpha_{Al,1D}$) / $\varepsilon_{Al,std,1D}$	2-D mode (A) ($\alpha_{Al,1D}$ x2) / $\varepsilon_{Al,std,2D}$	0.5D mode (C) ($\alpha_{Al,1D}$ x0.5) / $\varepsilon_{Al,std,0.5D}$
1	100	10	0.4	0.04	B	0.85	**0.007%**	100%	50%
2	100	20	0.4	0.02	B	0.84	**1.07%**	97.64%	50.58%
3	100	30	0.4	0.013	A	0.43	49.67%	**1.17%**	74.70%
4	100	40	0.4	0.01	A	0.43	49.30%	**1.17%**	74.70%
5	100	50	0.4	0.008	A	0.42	50.67%	**1.17%**	75.29%
11	100	10	1	0.1	C	1.71	100.9%	302.3%	**0.58%**
16	100	10	1.5	0.15	C	1.70	100.9%	300%	**0%**
17	100	20	1.5	0.075	C	1.70	99.49%	300%	**0%**
22	100	20	2	0.1	C	1.71	101.1%	302.35%	**0.58%**

Table 6. Accuracy error Analysis of Thermal Diffusivity for Tin in TDMI Test (L_M=3cm) ($\alpha_{std,sn}$=0.4cm²/s)

No	L (mm)	W	T	$\tau = \frac{T}{W}$	Zone	$\alpha_{Sn,avg}$	1-D mode (B) ($\alpha_{Sn,1D}$) / $\varepsilon_{Sn,std,1D}$	2-D mode (A) ($\alpha_{Sn,1D}$ x2) / $\varepsilon_{Sn,std,2D}$	0.5D mode (C) ($\alpha_{Sn,1D}$ x0.5) / $\varepsilon_{Sn,std,0.5D}$
1	100	10	0.4	0.04	B	0.41	**1.25%**	105%	48.75%
2	100	20	0.4	0.02	B	0.39	**1.83%**	95%	51.25%
3	100	30	0.4	0.013	A	0.21	47.41%	**5%**	73.75%
4	100	40	0.4	0.01	A	0.2	50.33%	**0%**	75%
5	100	50	0.4	0.008	A	0.19	53.08%	**5%**	76.25%
11	100	10	1	0.1	C	0.81	101.66%	305%	**1.25%**
16	100	10	1.5	0.15	C	0.82	105.41%	310%	**2.5%**
17	100	20	1.5	0.075	C	0.81	103.5%	305%	**1.25%**
22	100	20	2	0.1	C	0.82	105.9%	311.8%	**2.95%**

3.3. Experimental results and analysis of the vapor chamber

The vapor chamber in this experiment has three kinds of VC-A, VC-B and VC-C as shown in Fig. 15. The dimensions are $180\times80\times1$ mm³, $90\times90\times0.45$ mm³ and $90\times90\times3$ mm³ respectively. VC-A and VC-B are ultra-thin vapor chamber and VC-C is a 3 mm-thickness vapor chamber. Table 7 shows the thermal diffusivity results of the vapor chambers VC-Type A, VC-Type B, and VC-Type C measured by TDMI. The effective equivalent thermal conductivity are 4833(W/m.K), 3069 (W/m.K), and 4385 (W/m.K) respectively. Since VC-A does not have a suitable condensate jacket, only the thermal diffusivity is measured. According to the definition of TTMA, the maximum heat transfer of VC-B and VC-C is 210 (W) and 551 (W) (at T_C=60 ℃ ,$T_{J,surface}$=100 ℃); The axial thermal resistance values $R_{th,VC,Z}$ are 0.04 (K/W) and 0.08 (K/W), respectively; the thermal resistance of the vapor chamber $R_{th,VC,B}$ is 0.42 (K/W) and 0.177 (K/W).

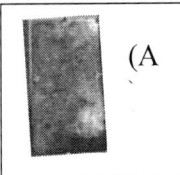 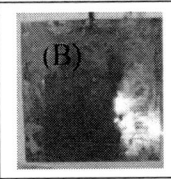 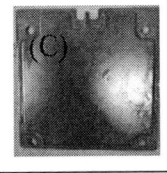

Fig.15. vapor chamber VC-A, VC-B, VC-C entity diagram

Table 7. The thermal diffusivity of the vapor chamber VC-A, VC-B, and VC-C measured by TDMI.

item	VC-A			VC- B			VC- C		
ρ (g/cm³)	6.17			6.49			4.45		
C (J/g.k)	4.123			3.284			4.5		
L×W×T (mm³)	180×80×1			90×90×0.4			90×90×3		
τ=T/W	τ_x=1/80 =0.0125 <0.022	τ_y=1/180 =0.0055 <0.022		τ_x=0.4/90 =0.004 <0.022	τ_y=0.4/90 =0.0044 <0.022		τ_x=3/90 =0.033 >0.022	τ_y=3/90 =0.033> 0.022	
mode	2D mode (A)			2D mode (A)			1D mode (B)		
$\alpha_{1D,total}$ (cm²/s)	α_x= 0.52	α_y= 0.43	$\alpha_{1D,total}$ =α_x+α_y =0.95	α_x= 0.4	α_y= 0.32	$\alpha_{1D,tota}$ =α_x+α_y =0.72	α_x= 0.78	α_y= 2.98	$\alpha_{1D,total}$ =2.19
$\alpha_{2D,total}$ (cm²/s)	= $\alpha_{1D,total}$ *2=1.9			= $\alpha_{1D,total}$ *2=1.44			--		
$K_{VC,TDMI}$ (W/m-k)	K_{VC-A} =1.9*6.17*4.12 3*100=4833			K_{VC-B}= 1.44*6.49*3.284*1 00= 3069			K_{VC-C}=2.19* 4.45*4.5*100= 4385		
Q_{max} (W)				210@T_C=60℃ $T_{j,surface}$=100℃			551@T_C=60℃, $T_{j,surface}$=100℃		
Axial thermal resistance $R_{th,VC,Z}$ (K/W)				0.04			0.08		
Thermal resistance of VC $R_{th,VC}$ (K/W)				0.42			0.177		

3.4. Experimental results and analysis of the pierced vapor chamber

The purpose of this experiment is another way to verify of the accuracy of TDMI, table 8 showed these the results of experiment. Let the vapor chamber A and B are pierced, expelled all the water vapor out of chamber so that the vapor chamber becomes an empty copper plate, and measured the thermal diffusivity by TDMI is 1.09 (cm²/s) and 1.12 (cm²/s) respectively, converted to equivalent thermal conductivity of 370.18 (W/m.k) and 379.8 (W/m.k), that are the same as the thermal conductivity of pure copper. Hence TDMI for the measurement of the k value of the vapor chamber has a certain accuracy and reliability.

Table 8. Thermal diffusivity of the pierced vapor chamber VC-D and VC-E test by TDMI

item	VC-A			VC- B			Pure copper
ρ(g/cm³)	8.933			8.933			8.933
C(J/g-℃)	0.383			0.383			0.383
L×W×T (mm³)	180×80×1			90×90×0.4			
τ=T/W	τ_x=1/80 =0.0125 <0.022	τ_y=1/180 =0.0055 <0.022		τ_x=0.4/90 =0.004 <0.022	τ_x=0.4/90 =0.0044 <0.022		
mode	2D mode (A)			2D mode (A)			
$\alpha_{1D,total}$ (cm²/s)	α_x= 0.260	α_y= 0.280	$\alpha_{1D,total}$ =α_x+α_y =0.541	α_x= 0.2931	α_y= 0.2619	$\alpha_{1D,total}$ =α_x+α_y =0.555	α =1.17
$\alpha_{2D,total}$ (cm²/s)	= $\alpha_{1D,total}$ *2=1.09			= $\alpha_{1D,total}$ *2=1.12			--
$K_{VC,TDMI}$ (W/m-k)	K_{VC-A} =1.082*8.933*0.383* 100=370.18			K_{VC-B} =1.11*8.933*0.383* 100= 379.8			K_{Cu}=380
Error (%)	2.58%			0.05%			

Two other of ultra-thin vapor chamber VC-D and VC-E were designed and made by ACL as shown in figure 16 and figure 17. The characterization of these two types of vapor chamber were shown in table 9. In Table 9 shows the equivalent effective thermal conductivity of vapor chamber VC-D is 33519.56 (W/m-k), while for VC-E is 22837(W/m-k) and that is the highest thermal conductivity k value vapor chamber designed in ACL at present.

 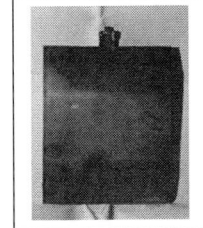

Fig. 16. VC-D customer made entity diagram thickness 0.4mm made by ACL

Fig. 17. VC-E 60x90x0.4 mm³ entity diagram made by ACL

Table 9. The thermal diffusivity of the vapor chamber VC-D, and VC-E measured by TDMI.

item	VC-D			VC-E			
ρ (g/cm^3)	5.387			5.76			
C (J/g.k)	3.604			3.54			
L×W×T (mm^3)	27*60*0.4			60×90×0.4			
τ=T/W	τ_x=0.4/27 =0.0148 <0.022	τ_y=0.4/60 =0.0066 <0.022		τ_x=0.4/60 =0.0066 <0.022	τ_y=0.4/90 =0.0044 <0.022		
mode	2D mode (A)			2D mode (A)			
$\alpha_{1D,total}$ (cm^2/s)	α_x= 4.30	α_y= 4.33	$\alpha_{1D,total}$ =αx+αy=8.6	α_x= 2.76	α_y= 2.84	$\alpha_{1D,total}$ α_x+α_y=5.6	
$\alpha_{2D,total}$ (cm^2/s)	=$\alpha_{1D,total}$ *2=17.265			=$\alpha_{1D,total}$ *2=11.2			
$K_{VC,TDMI}$ (W/m-k)	K_{VC-D} =33519.56			K_{VC-E}= 22837			

4. Conclusions

4.1 After the vapor chamber is pierced, the vapor chamber are turned into empty copper plates, the thermal diffusivity of the measured empty copper plate are converted into thermal conductivity of 372.9 (W/m.k) and 383 (W/m.k) respectively, and that is similar to the thermal conductivity of the pure copper.

4.2 Since there is no standard diffusivity value of vapor chamber, the test results can only be used as a reference. In this study, three vapor chambers were tested. The thermal conductivity of VC-A was 4833.3 (W/m.k), and VC-B was about 3069 (W/m.k). VC-C is about 4385 (W/m.k).

4.3 The axial thermal resistance values $R_{th,VC,Z}$ of VC-B (T=0.4mm) and VC-C (T=3mm) are 0.04 (K/W) and 0.08 (K/W), respectively. It is also proved that the thicker the thickness of vapor chamber, the higher the axial thermal resistance value of vapor chamber.

4.4 Thermal resistance $R_{th,VC}$ for VC-B (T=0.4mm) and VC-C (T=3mm) are 0.42 (K/W) and 0.177 (K/W) respectively. It is also implied that the thicker the vapor chamber, the smaller the thermal resistance of the diffusion, but the bigger the axial thermal resistance. When the thickness of the vapor chamber is thinner, the diffusion is more difficult, but the axial thermal resistance is relatively smaller.

Acknowledgments

This project is supported by National Science Council, Taiwan, MOST 108-2622-E-007-016-CC2. Acknowledge NSC for their support for this project. Professor Wei-Keng Lin also relies on the support of T-Global Technology Co., Ltd and the experiment can be completed.

References

[1] http://www.thermal.org.tw/Events/news-more.asp? BtfiZsf=

[2] http://www.thermal.org.tw/Events/news-more.asp? BtfiZst=

[3] J.E. Parrott and A.D. Stuckes, "Thermal Conductivity of Solids", Pion Limited, London, 1975.

[4] V. Mirkovich, "Comparative method and choice of standards for thermal conductivity determinations", Journal of the American Ceramic Society, vol. 48, pp. 387-391, 1965.

[5] D. Hughes and F. Sawin, "Thermal conductivity of dielectric solids at high pressure", Physical Review, vol. 161, p. 861, 1967.

[6] A. Tomokiyo and T. Okada, "Determination of thermal diffusivity by the temperature wave method", Japanese Journal of Applied Physics, vol. 7, p. 128, 1968.

[7] Chen-I Chao, Wei-Keng Lin,* Shao-Wen Chen, & Han-Chou Yao, "Feasibility of the ÅNGSTRÖM Method in Performing the Measurement of Thermal Conductivity in Vapor Chambers", Heat Transfer Research 47(7), 617–632, 1064-2285/16/ \$35.00 © 2016 by Begell House, Inc. (2016) , (MOST 104-ET-E-007-005-ET)

[8] Wei-Keng Lin, Wen-Hua Zhang, Pei-Hsun Wu, "Thermal Diffusivity Measurement of Vapor Chamber ", Proceedings of the 3rd World Congress on Momentum, Heat and Mass Transfer (MHMT'18), 2018, ICMFHT18, Budapest, Hungary – April 12 – 14,2018.04.02

[9] Wei-Keng Lin; Wen-Hua Zhang; Ching-Huang Tsai; Shao-Wen Chen," The development measuring skill for the thermal conductivity of heat pipe, graphite sheet and vapor chamber", ICCMST 2019, Japan, May 22-25

[10] Wei -Keng Lin, "Design of Thermal Diffusivity Measurement for Material", IMAPS 140 S. Santa Cruz Ave, Los Gatos, CA 95030 – USA 2018.11.08,

[11] Wei-Keng Lin, "Measurement of Performance Parameters of Ultra-Thin Vapor Chamber under Microgravity", IMPACT-Iaac 2019 Conference, Oct. 23th-25th at Taipei Nangang Exhibition Center, Taiwan.

[12] Wen-Hua Zhang, Wei-Keng Lin, Ching-Huang Tsai, Pei-Hsun Wu and Shih-Kuo Wu, "A Thermo-electric Apparatus for Thermal Diffusivity and Thermal Conductivity Measurements", Energies 2019, Volume 12, Issue 22, 4238

Numerical Investigation of Coolants for Chip-embedded Two-Phase Cooling

Pritish R. Parida*, Timothy Chainer

IBM T. J. Watson Research Center, Yorktown Heights, NY
*prparida@us.ibm.com

Abstract

Inter-chip cooling, where a liquid coolant is passed between the layers of stacked chips, is an enabling technology for realizing significant computational performance improvements through three-dimensional (3D) integration of microelectronic components. The development of this cooling technology requires high fidelity thermal models to evaluate the device and system performance under different operating environments. In the present work, a Eulerian multiphase model developed for predicting two-phase flow and heat transfer behavior in parallel micro-channels and micro-pin fields was used to compare the thermal performance of four different refrigerants – R1234ze, R245fa, R134a and R600a, for chip-embedded two-phase cooling. Results shows that medium- and high-pressure refrigerants such as R1234ze R600a and R134a with a small density ratio, low viscosity, large latent heat of vaporization and low surface tension helps achieve low device temperatures while reducing the pressure drop across the device resulting in smaller variation in saturation as well as device temperature profile.

Keywords

Two-phase cooling, micro pin-fins, thermal modeling, flow boiling.

Nomenclature

3D	-	Three Dimensional
A	-	Area concentration, [1/m]
Cp	-	Specific heat capacity, [J/kg-K]
d	-	Diameter, [m]
d_h	-	Hydraulic diameter, [m]
F	-	Force terms, [-]
f	-	Wall area fraction function, [-]
G	-	Mass flux, [kg/s-m^2]
g	-	Gravitational Constant, [m/s^2]
H	-	Latent heat, [J/kg]
h	-	Specific Enthalpy
I	-	Identity matrix, [-]
MFR	-	Mass flow rate, [kg/hr]
$\dot{m}$	-	Interphase mass transfer rate
p	-	Pressure, [kPa]
Q	-	Intensity of interphase heat exchange
q	-	Input/wall heat flux, [W/m^2] or [W/cm^2]
R	-	Interphase momentum exchange force
S	-	Source term
T	-	Temperature, [K]
t	-	Time, [s]
v	-	Velocity, [m/s]

Greek Symbols

α	-	Volume Fraction, [-]
λ	-	Bulk viscosity
μ	-	Shear viscosity
ρ	-	Density, [kg/m^3]
σ	-	Surface Tension, [N/m]
τ	-	Stress-strain tensor

Subscripts

crit	-	Critical
i	-	Interfacial
l	-	Liquid
lV	-	Liquid-vapor interphase
sat	-	Saturation
V	-	Vapor

1. Introduction

Heterogeneous integration of microelectronic chips in the form of three dimensional (3D) stacks of is an enabling technology providing a path for increasing computational performance [1-3]. However, traditional single chip cooling solutions which utilize heat sinks or cold-plates thermally coupled to the backside of a die-package cannot satisfy the cooling requirements of high-power 3D chip stacks [4, 5]. Inter-chip cooling, where a coolant is passed between the layers of stacked chips, is a solution to this thermal management challenge [6-8]. The use of water (most commonly used coolant) for interlayer cooling is challenging as it requires robust isolation from power and signal interconnects to protect against electrical shorts and high frequency signal transmission losses [8-10]. Alternately, chip-to-chip interconnect compatible dielectric coolants such as R1234ze, R245fa, R134a, etc. can be used. However, they cannot be operated in single-phase cooling mode due to their poor heat transfer characteristics. This necessitates their operation in two-phase cooling mode where the dielectric fluid is converted from liquid to vapor phase as it flows through the embedded micron-scale cavities [9-12]. The cooling is achieved by utilizing the latent heat of vaporization of the dielectric coolant, whereby large amounts of heat can be removed with low coolant flow rates and small changes in coolant and chip temperatures across the coolant cavity/channel [13-18].

Interlayer liquid cooling adds a new set of constraints to chip stack integration including microchannel dimensions for coolant flow, provisions for coolant supply and return, coolant operating pressure, coolant properties, etc. [9]. An evaluation of trade-offs between channel dimensions, coolant properties and pressure drop to determine the range of channel dimensions, interconnect/pin-fin spacing for different heat flux criteria while accounting for process fabrication, electrical layout, mechanical integrity and computational performance., is required [11]. For example, fabrication process and requirements for electrical connection between the stacked chips, limits the microchannels height [9]. Additionally, the development of this cooling technology requires high fidelity thermal models to evaluate the device and system performance under different operating environments.

Over the last couple of decades, significant progress has been made on the development of detailed physics simulation of two-phase flows through mini-/micro-channels. These simulations cover a wide range of exemplary cases ranging from compact and simple geometries such as single bubble simulation in a single uniform cross-section channel [19-23] to complex flow domains such as radially expanding micro-channels populated with micro-pin fins [24]. For example, Gorle et al. [19, 20] used a volume-of-fluid (VOF) method [25] and integrated it with Lee model [26] to capture the mass and energy transfer between the liquid and vapor phases in a uniform cross-section micro-channel. Parida et al. [27, 28] developed a Eulerian multiphase model for simulating two-phase flow through micro-channels populated with pin-fins.

In a prior work, this Eulerian multiphase model was applied to the design of embedded radial expanding channels for cooling of a high-performance microprocessor chip [24]. The cooling structures, including flow directing walls and orifices as well as a pin field, were designed with constraints compatible with future 3D structures including through silicon vias (TSV's). The detailed validation study against experimental data showed that the junction temperatures can be predicted with good accuracy in various flow regimes starting from sub-cooled boiling regime to annular flow regime [24, 28]. In the present work, this validated Eulerian multiphase conjugate thermal model has been utilized to compare the performance of different dielectric coolants, namely R1234ze, R245fa, R134a and R600a, for chip-embedded two-phase cooling. Section 2 describes the criteria for refrigerant selection and comparison of different properties of the refrigerants being considered. Sections 3 and 4 summarizes previously reported model formulation and validation results, respectively. Section 5 provides a detailed comparison of thermal performance of different refrigerants considered.

2. Refrigerants

Four single component refrigerants, namely R1234ze, R245fa, R134a and R600a, were considered for this study. These refrigerants were chosen based on the followings thermodynamic and transport properties [14]:

a. Manageable saturation pressure at coolant operating temperatures from the point of view of handling and mechanical loading of the chip package,

b. Liquid-vapor density ratio less than 200 and low fluid viscosity for lower two-phase pressure drop,

c. High liquid and vapor phase thermal conductivities and latent heat of vaporization for good heat transfer,

d. Low flammability level,

e. Low ozone depletion (ODP)

f. Low global warming potential (GWP).

R245fa is considered a low-pressure refrigerant with a saturation pressure (P_{sat}) of 178 kPa (absolute) at a saturation temperature (T_{sat}) of 30 °C while R134a is considered a high-pressure refrigerant with a P_{sat} = 770 kPa at T_{sat} = 30 °C. R1234ze and R600a are both considered medium-pressure refrigerants with saturation pressures of 578 kPa and 405 kPa, respectively at a saturation temperature of 30 °C. Figure 1 shows the saturation temperature variation with absolute pressure for these four refrigerants generated using REFPROP [29]. The low-pressure refrigerant (R245fa) is most sensitive to the changes in pressure with an average rate of 0.1283 K/kPa while the high-pressure refrigerant (R134a) is the least sensitive with an average rate of 0.0380 K/kPa. Table 1 summarizes the saturation temperature sensitivity to pressure. The higher the sensitivity with pressure, the higher the expected variation in saturation temperature along the flow direction within the micro-channel.

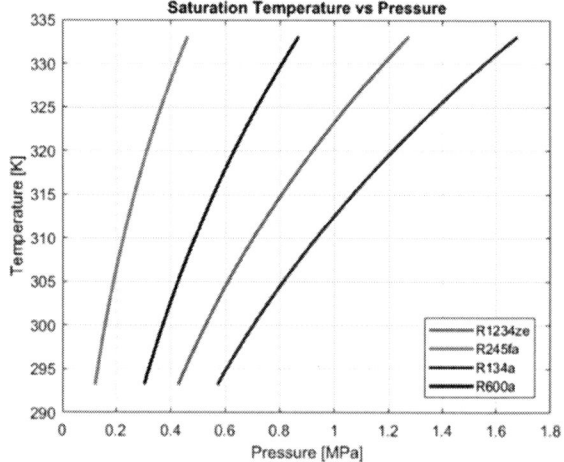

Fig. 1. Variation of saturation temperature with pressure.

Table 1. Sensitivity of saturation temperature with pressure

Coolant	Average Rate [K/kPa]
R1234ze	0.0498
R245fa	0.1283
R134a	0.0380
R600a	0.0742

Figure 2 shows the pressure-enthalpy diagram of the four refrigerants generated using refrigerant property data from REFPROP [29]. The horizontal axis shows the coolant enthalpy relative to the coolant's liquid state enthalpy at 1 atmospheric pressure (=101.325 kPa). R1234ze, R245fa and R134a have an identical relationship between pressure and enthalpy while R660a's enthalpy shows a stronger sensitivity to pressure with a roughly 2x wider P-h diagram. A wider P-h diagram means higher latent heat of vaporization and is preferred from the perspective of coolant's capacity to remove heat at a given mass flow rate.

Table 2 summarizes the physical properties of the four coolants studied. A third order polynomial fit used in the simulations for capturing the saturation temperature changes along the flow direction, is also included in the table. R600a's density in the liquid phase is ~50%, and specific heat capacity and latent heat of vaporization is roughly double compared to the other three coolants. R245fa is 2-2.5x more viscous in the liquid-phase, has a 2-3x higher density ratio and ~1.5x higher surface tension compared to the other three coolants.

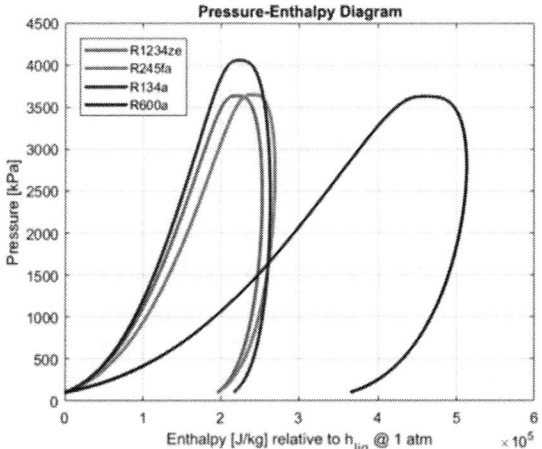

Fig. 2. Pressure-enthalpy diagram for the four coolants.

Table 2. Physical properties of the four coolants

Property	Units	R1234ze	R245fa	R134a	R600a
ρ_L	[kg/m3]	1146.4	1309.8	1187.5	544.31
$C_{p,L}$	[J/kg-K]	1402.9	1345.8	1446.5	2463.3
μ_L	[μPa-s]	188	348.85	183.1	143.4
k_L	[W/m-K]	0.0725	0.0848	0.079	0.0875
ρ_V	[kg/m3]	30.523	12.177	37.535	10.48
$C_{p,V}$	[J/kg-K]	998.57	993.29	1065.5	1835.3
μ_V	[μPa-s]	12.46	10.78	11.91	7.631
k_V	[W/m-K]	0.014	0.0137	0.0143	0.0174
h_{LV}	[J/kg]	163060	184020	173100	323330
σ	[N/m]	0.0082	0.123	0.0074	0.0095
ρ_L/ρ_V	[-]	37.6	107.6	31.6	51.9
T_{sat} [K] = a0 + a1*p + a2*p² + a3*p³ where p is pressure in [MPa]					
a0		244.76	261.88	253	252
a1		161.67	320.21	88.87	172.7
a2		-137.23	-585.80	-36.45	-135.4
a3		55.84	500.33	7.115	50.85

3. Model Formulation

An Eulerian multiphase model [25] available in ANSYS® Fluent® – a commercial 3D numerical solver, was used as the starting framework for the present study. This model treats each phase (liquid and vapor) as interpenetrating continua and uses the concept of phasic volume fraction to keep track of different phases within a computational cell. In this model, a set of continuity, momentum and energy equations is solved for each phase. Coupling between liquid and vapor phases is captured through the common pressure field and through momentum and energy exchange sub-models such as interphase drag, lift, continuum surface force, heat and mass transfer models across the fluid-solid interface, etc.

Governing Equations:

The following section describes the set of conservation equations governing the Eulerian multiphase flow simulation:

Continuity Equation for the liquid (l) phase [25]:

$$\frac{\partial}{\partial t}(\alpha_l \rho_l) + \nabla \cdot (\alpha_l \rho_l \vec{v}_l) = \dot{m}_{Vl} - \dot{m}_{lV} + S_l \quad (1)$$

where, α_l, ρ_l, v_l is the volume fraction, density and velocity, respectively, of the liquid phase, $\dot{m}_{Vl}$ is the mass transfer into the liquid (l) phase from the vapor (V) phase and $\dot{m}_{lV}$ is the mass transfer from the liquid phase to the vapor phase. S_l is the mass source term for the phase l.

Momentum Equation for the liquid (l) phase [25]:

$$\frac{\partial}{\partial t}(\alpha_l \rho_l \vec{v}_l) + \nabla \cdot (\alpha_l \rho_l \vec{v}_l \vec{v}_l) = \begin{aligned} &-\alpha_l \nabla p + \nabla \cdot \overline{\overline{\tau}}_l + \alpha_l \rho_l \vec{g} \\ &+ \vec{R}_{Vl} + \dot{m}_{Vl} \vec{v}_{Vl} - \dot{m}_{lV} \vec{v}_{lV} \\ &\vec{F}_l + \vec{F}_{lift,l} + \vec{F}_{wl,l} + \vec{F}_{vm,l} + \vec{F}_{td,l} \end{aligned} \quad (2)$$

where, $\overline{\overline{\tau}}_l$ is the liquid phase stress-strain tensor expressed as:

$$\overline{\overline{\tau}}_l = \alpha_l \mu_l (\nabla \vec{v}_l + \nabla \vec{v}_l^T) + \alpha_l \left(\lambda_l - \frac{2}{3}\mu_l\right)\nabla \cdot \vec{v}_l \overline{\overline{I}} \quad (3)$$

μ_l and λ_l are the shear and bulk viscosity of phase l, $\vec{v}_{Vl}$ is the interphase velocity. $\vec{F}_l$ is a body force term which can be used to model an external body force or a volumetric force such as continuum surface (tension) force. $\vec{F}_{lift,l}$ is the lift force, $\vec{F}_{wl,l}$ is the wall lubrication force, $\vec{F}_{vm,l}$ is the virtual mass force and $\vec{F}_{td,l}$ is the turbulent dispersion force, $\vec{R}_{Vl}$ is an interphase momentum exchange (drag) force between the liquid and vapor phases and p is the pressure shared by both liquid and vapor phases.

Energy Equation for the liquid (l) phase [25]:

$$\frac{\partial}{\partial t}(\alpha_l \rho_l h_l) + \nabla \cdot (\alpha_l \rho_l \vec{v}_l h_l) = \begin{aligned} &\alpha_l \frac{\partial p}{\partial t} + \overline{\overline{\tau}}_l : \nabla \vec{v}_l - \nabla \cdot \vec{q}_l \\ &+ S_l + Q_{Vl} + \dot{m}_{Vl} h_{Vl} - \dot{m}_{lV} h_{lV} \end{aligned} \quad (4)$$

where, h_l is the specific enthalpy of the liquid phase, $\vec{q}_l$ is the heat flux, S_l is a source term, Q_{Vl} (= $- Q_{lV}$) is the intensity of heat exchange between the liquid and vapor phases, and h_{Vl} (= h_{lV}) is the interphase enthalpy (latent heat). Almost all the interphase interaction terms require sub-models or constants to define the interfacial area concentration to predict the magnitude of interphase mass, momentum and energy transfer rates across phase interfaces. The interfacial area concentration is defined as the interfacial area between two phases per unit mixture volume and for two phase boiling flows, it is expressed as:

$$A_i = \frac{6(1-\alpha_V)\min(\alpha_V, \alpha_{V,crit})}{d_V(1-\min(\alpha_V, \alpha_{V,crit}))} \quad (5)$$

where, $\alpha_{V,crit} = 0.25$ and d_V is the vapor (or dispersed phase) bubble diameter.

A complimentary set of continuity, momentum and energy equations were solved for the vapor (V) phase. Additionally, k-ε mixture turbulence model equations were solved to capture the turbulent interaction between the liquid and vapor phases. At the fluid-solid interfaces (walls), a mechanistic wall boiling model was used to capture the energy transfer mechanism directly from the wall to both the liquid and vapor phases [24]. In the solid domain, only the thermal energy equation was solved. Further details on the conservation equations, sub-models used and on the wall heat transfer models can be found in Fluent theory guide [25].

4. Model Validation

This section summarizes the prior validation study of the Eulerian multiphase conjugate model performed for: (i) Uniform cross-section parallel micro-channels [27] and (ii). Micro cavity populated with micro-pin fins [28].

Figure 3 shows the unit structure of the uniform cross-section micro-channel geometry that was modeled. The fluidic channel is 10 mm long with a cross section of 100 μm x 100 μm and a 100 μm long and 50 μm wide inlet restrictor to suppress any coolant backflow and mitigate flow instabilities. A conformal hex-dominant mesh with inflation layers near interfaces/walls was utilized. A grid independency study was also conducted which showed that a mesh size of ~1.5 million elements with ~1 million elements in the fluid domain was enough. In this case, the model was validated against data reported by Szczukiewicz et al. [30, 31] for flow boiling of R245fa in uniform cross-section micro-channels. Table 3 lists the boundary conditions. The experimental study had 67 such repeating unit structures with a total width of 1 cm. The total device level mass flow rate was 4.6 kg/hr. Figure 4 highlights the excellent agreement between the average heater surface temperature prediction by the model and the corresponding experimental data.

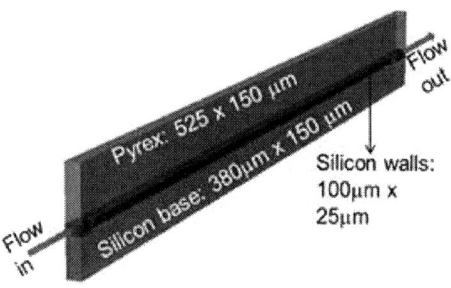

Fig. 3. Isometric view of the computational domains for the uniform cross-section microchannel validation study [20].

Figure 5 shows an isometric view of the micro-cavity populated with micro-pin fins. The model is 200 μm wide unit structure and consists of one fluid and two solid (silicon and glass) domains. The fluid domain consists of the inlet plenum, an orifice and a micro-cavity with single row of pin fins. Although, the silicon domain extends the entire length of the

fluid domain, a uniform heat flux boundary condition was applied only under the pin-fin region. Tables 4 lists all the boundary conditions. Like the uniform cross-section micro-channel case, conformal hex-dominant mesh with inflation layers near interfaces/walls was utilized.

Table 3. Boundary conditions for uniform cross-section micro-channel validation study.

Fluid domain	R245fa
Inlet	G=1895 kg/s-m², 37.7 [°C]
Outlet Pressure	213 kPa (T_{sat} = 35.2 [°C])
Solid domain	Silicon / Pyrex
Bottom Surface	Uniform heat flux: 47 W/cm²
Top Surface	Adiabatic

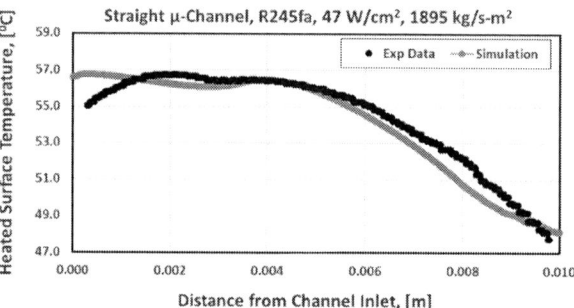

Fig. 4. Average heated surface temperature prediction and comparison against data [27].

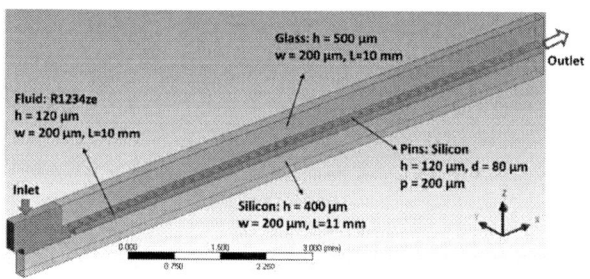

Fig. 5. Isometric view of the computational domains for the micro pin-fin validation study [28].

Table 4. Boundary conditions for micro pin-fin validation study.

Fluid domain	R1234ze
Inlet	G=482 kg/s-m², 27.2 [°C]
Outlet Pressure	544.1 kPa (T_{sat} = 28 [°C])
Solid domain	Silicon / Pyrex
Bottom Surface	Uniform heat flux: 38.9 W/cm²
Top Surface	Adiabatic

Figure 6 shows the width-average junction temperature profile prediction and its comparison with the experimental data. Experimental device tested had 50 such repeating structures with a total width of 1 cm and total device level mass flow rate of 2.08 kg/hr. For this case as well, the temperature prediction was observed to be in very good agreement with the experimental data having maximum errors within ± 2 °C.

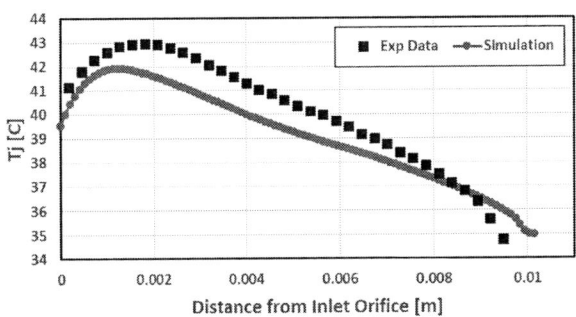

Fig. 6. Width-average junction temperature profile prediction and its comparison against data for the micro pin-fin case [28].

5. Coolant Performance Comparison

The above described micro-cavity populated with micro-pin fins case (Figures 5 and 6) was used for the coolant performance comparison. Depending upon the coolant, reference pressure in the model was adjusted to ensure a saturation temperature of about 28 °C at the outlet. From a cooling system perspective, such an outlet condition is easily achievable in a pumped two-phase cooling loop by selecting a suitable condenser performance level as well as by selecting a suitable diameter tubing between the device and the condenser [32]. At the inlet, the coolant was fed into the channels at 27.2 °C. Such an inlet temperature can be ensured by utilizing either a sub-cooler or a pre-heater upstream of the device. Table 5 summarizes the different simulation cases. Case 1 is the same as the micro-pin fin validation case and, was used as the reference for the comparative study. For cases 1 to 4, the mass-flux was kept the same. In cases 1 to 3, the expected exit vapor quality was 35 – 40% depending upon the latent heat. In case 4, the expected vapor quality was ~20% due to much higher latent heat of vaporization for R600a. So, to match the exit vapor quality with cases 1 to 3 another case (namely, case 5) with roughly half the mass flux was also simulated for the coolant R600a. Note that the inlet sub-cooling in each case depends upon the pressure drop across the micro-cavity as well as upon the relationship between the coolant's saturation temperature and pressure.

Table 5. Simulation cases for the comparative study

Case #	Coolant	Inlet (G, T_{in})	Outlet (T_{sat} = 28 [°C])
1	R1234ze	482 kg/s-m², 27.2 °C	P = 544.10 kPa
2	R245fa	482 kg/s-m², 27.2 °C	P = 165.44 kPa
3	R134a	482 kg/s-m², 27.2 °C	P = 727.28 kPa
4	R600a	482 kg/s-m², 27.2 °C	P = 382.46 kPa
5	R600a	244 kg/s-m², 27.2 °C	P = 382.46 kPa

Figure 7 shows a comparison of the width average heated surface temperature. R134a shows the best thermal performance with a 2 °C lower peak temperature and smaller temperature variation from inlet to outlet, compared to the reference R1234ze case. On the other hand, the low-pressure

coolant R245fa shows the worst thermal performance with 7 °C higher peak temperature as well as a larger temperature variation from inlet to outlet, compared to the reference case. R600a at the lower mass flow rate shows a temperature profile that is very similar to the reference case.

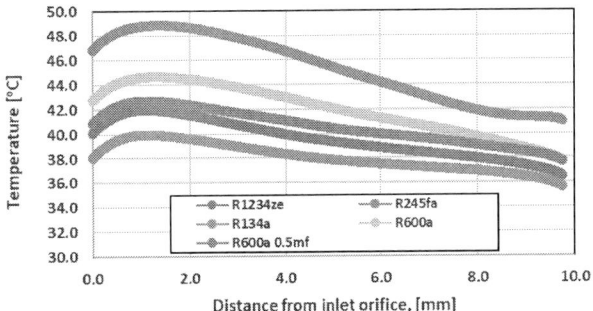

Fig. 7. Comparison of width-average junction temperature profile prediction for the five different micro pin-fin cases.

Figure 8 shows the silicon temperature at the heated surface as well as at the silicon-coolant interface for all the five cases. The blue colored vapor volumes in Figure 8 are regions (or computational cells) where the coolant is mostly vapor with a vapor volume fraction greater than > 0.9. Figure 9 shows the vapor volume fractions (i) at the symmetry plane (south) and (ii) at cavity mid-plane (60 μm from channel base). Note that the orifice is equidistant from the north and south symmetry planes while the pins are centered 90 μm from the south symmetry plane. This creates an asymmetric flow behavior within the cavity with vapor favoring the north side of the cavity. In all the cases, near the channel entrance region including the orifice and a few pins downstream of the orifice, the flow is in sub-cooled boiling regime. In this region, large gradients in temperature profile exist which, is an artifact of the complex heat transfer behavior involving thermally developing flow, sub-cooled boiling, flow recirculation immediately downstream of the inlet flow restrictor (orifice) and peripheral heat spreading. The peripheral heat spreading results in an effective wall heat flux that is slightly lower than that in the bulk of the channel. This combined with the extremely high heat transfer rates associated with hydro-dynamically developing flow in the channel entrance region causes a positive slope in the temperature gradient along the flow direction. Secondly, the flow recirculation regions immediately downstream of the inlet restrictor, are relatively lower pressure regions and because, the local fluid saturation temperature is a strong function of the local absolute fluid pressure, the fluid in the recirculation regions has a higher susceptibility to boil. As a result, these locations act as potential nucleation sites for boiling. In case of the micro-pin fins, the trailing edge of the pins are also low-pressure regions that act as preferential nucleation sites. The vapor bubbles that form in these preferential nucleation sites do not aggregate there due to velocity and density gradients between the vapor bubble and bulk liquid. Further, the low surface tension of the coolant also helps in re-wetting the wall once a bubble moves away from the wall.

109

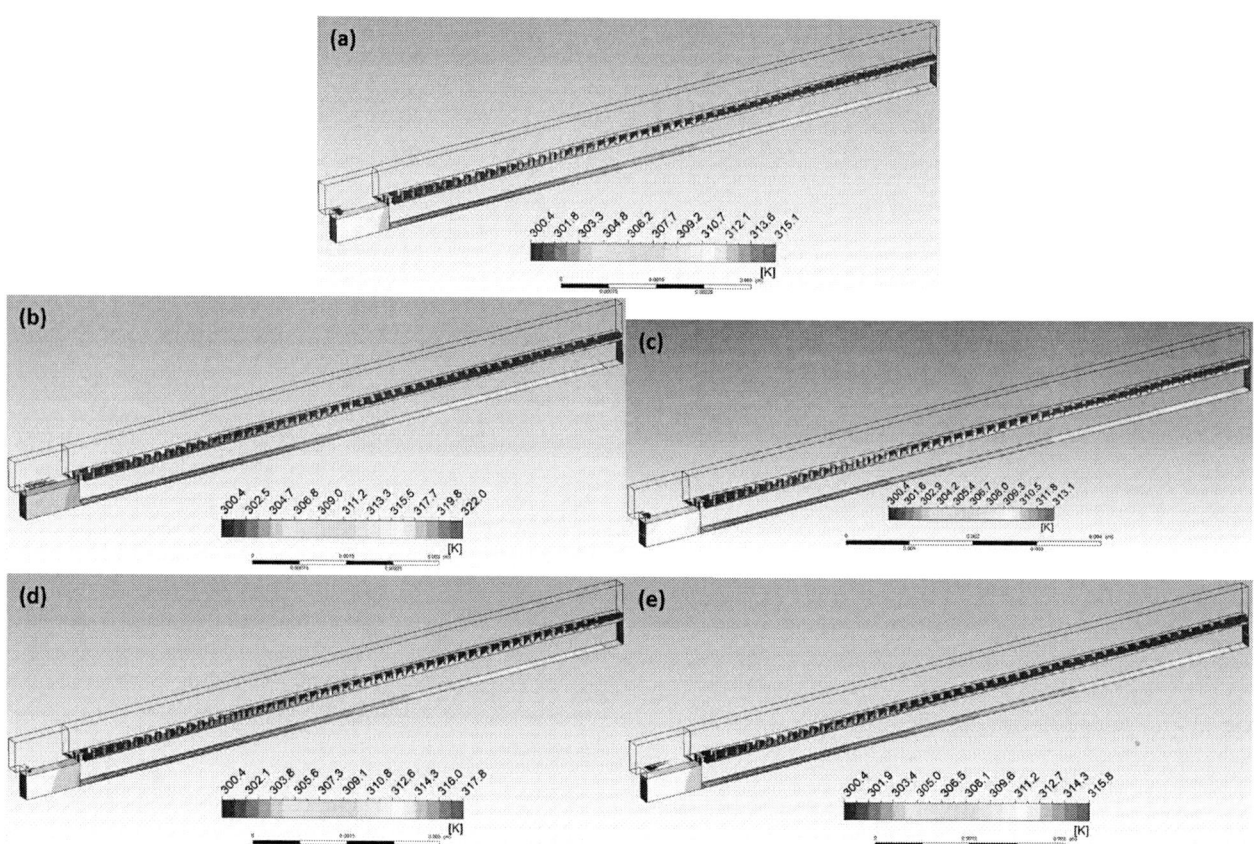

Fig. 8. Silicon temperature and vapor volumes for an applied heat flux of 38.9 W/cm² cooled by (a) R1234ze, (b) R245fa, (c) R134a and (d) R600a at a mass flux of 482 kg/m²-s and (e) R600a at a mass flux of 244 kg/m²-s.

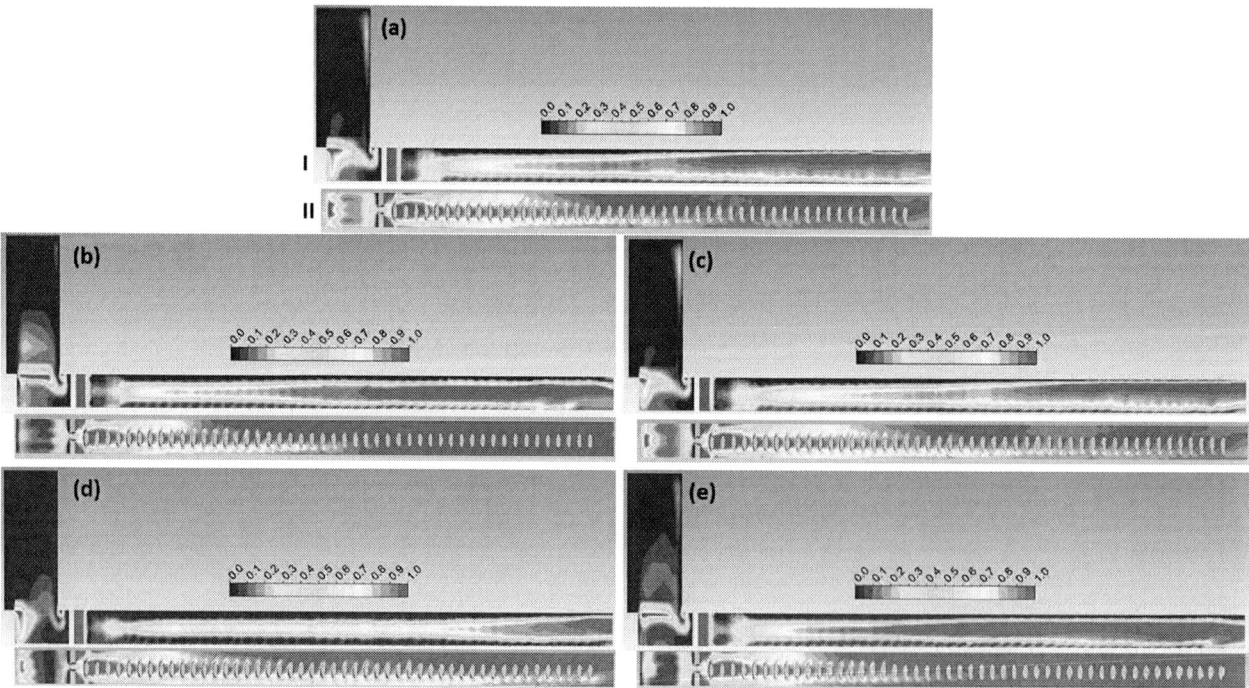

Fig. 9. Vapor volume fractions (I) at the south symmetry plane and (II) at the cavity mid-plane for an applied heat flux of 38.9 W/cm² cooled by (a) R1234ze, (b) R245fa, (c) R134a and (d) R600a at a mass flux of 482 kg/m²-s and (e) R600a at a mass flux of 244 kg/m²-s.

In the inlet plenum, a couple of recirculation regions can be observed for all the cases. At the first location, which is far away from the inlet orifice and is in contact with silicon, vapors are formed due to cavitation and grow due to heat transfer from silicon. Since the state of the coolant in the inlet plenum is sub-cooled, the vapors that form at the first location shrink in size or disappear before reaching the inlet orifice. At the second location, which is just before the orifice and is in contact with glass, the vapors are formed due to cavitation only and are smaller in size as compared to that at the first location. Ideally, such sharp features upstream of the inlet orifices should be avoided as flow boiling in upstream sections can cause flow instabilities and can result in a pressure drop across the orifice that is relatively higher than that expected for a single-phase flow. This increase is due to reduced flow cross-section in the presence of vapors in the inlet plenum. In all the five cases, the amount of vapor generated is small enough to not impact the pressure drop significantly or cause any unfavorable flow instabilities. The amount of vapor generated is different in each case due to different degrees of coolant sub-cooling, flow velocity and physical properties. Case 2 (R245fa) and Case 4 (R600a) show the least amount of vapor in the inlet plenum due to relatively high sub-cooling (> 4 °C). Low coolant flow velocity as in case 5, also helps in reducing the amount of vapor in the inlet plenum.

As we move away from the inlet further into the channel, the fluid undergoes sub-cooled boiling since the local pressure is higher than the saturation pressure. Further downstream into the channel, local pressure inside the channel drops until it reaches local saturation conditions and then it follows the coolant saturation curve. In this region, the fluid is highly susceptible to boiling and thus, the rear half of the channel is filled with large amounts of vapor suggesting annular flow regime in the rear half of the channel. In Figure 9, the vapor volume fractions on the symmetry plane highlight the thinning of liquid film from inlet to outlet. Note that in Case 4 (Fig. 9(d)), the liquid film is relatively thick compared to other cases due to relatively high latent heat of vaporization resulting in relatively less amount of vapor in the cavity. In case 2, due to a relatively high density-ratio, the cavity gets filled with vapors at a relatively shorter distance from the inlet. Large amounts of vapor in the cavity result in a steeper pressure gradient in the rear half of the channel which in turn results in a steeper drop in the local saturation temperature. This, combined with lateral heat spreading and extremely high heat transfer rates associated with annular two-phase flow, result in a sharp drop in the heater surface temperature. In Case 1, the drop in local saturation temperature from inlet to outlet is ~2 °C while the drop in the junction temperature is ~6 °C. This steeper drop in junction temperature is due to increasing heat transfer rates in the annular flow regime with increasing vapor quality. In Case 2, the change in junction temperature is nearly 8.5 °C with a change in saturation temperature of 6.3 °C from inlet to outlet.

Table 6 summarizes the key performance predictions for all the five cases. The mass flow rate reported is at the device

level. R245fa, which has a high density-ratio, surface tension and liquid phase viscosity, shows the highest pressure drop and highest junction temperature. A high density-ratio means that the vapors occupy a larger portion of the channel resulting in an increased flow impedance. A high surface tension results in slower bubble departure from and re-wetting of the boiling surface resulting in increased wall super-heat and larger bubbles near the walls which reduces the effective cross-section for flow of the coolant in liquid phase. A high liquid phase viscosity in a reducing cross-section due to presence of vapors further worsens the flow impedance for flow of the coolant in liquid phase. On the other hand, R134a, which has a low density-ratio, surface tension and liquid phase viscosity, shows a low pressure drop and lowest junction temperature. In case 5, R600a, having the lowest liquid-phase viscosity, a medium density-ratio, a medium surface tension and high latent heat of vaporization, shows the lowest pressure drop and a junction temperature profile that is comparable to the reference R1234ze case. Compared to R1234ze and R134a, R600a has slightly higher surface tension which negatively impacts its heat transfer performance at the interface.

Table 6. Performance comparison summary

Cases	#1	#2	#3	#4	#5
Coolant	R1234ze	R245fa	R134a	R600a	R600a
MFR [kg/hr]	2.08	2.08	2.08	2.08	1.05
Tj max [°C]	42	48.9	40	44.7	42.7
Tsub_inlet [°C]	2.8	7.1	2.1	4.1	2.6
Δp total [kPa]	34.5	41.5	28.8	37.9	21.1
Δp orifice [kPa]	4.2	4.7	3.9	7.6	2.5
Vapor Quality [%]	39.5	35	37.2	19.7	39.6
Pumping Power [W]	0.017	0.018	0.014	0.040	0.011

The pressure drop across the orifice was observed to be about 20% of the total pressure drop in case 4 and about 12% for all other cases. In case 4, due to a relatively lower density, the flow velocity is nearly double compared to all other cases resulting in a relatively higher pressure drop across the orifice. Comparing cases 4 and 5, we can see that the lower flow rate is better from both temperature (~2 °C lower) and pumping power (4x lower) points of view, which is strikingly different from what one would observe for single-phase flows.

6. Conclusions

A Eulerian multiphase model, developed and validated for simulating two-phase flow through uniform cross-section micro-channels and micro-cavity populated with micro pin-fins, was used for comparing the performance of different dielectric coolants, namely R1234ze, R245fa, R134a and R600a, for chip-embedded two-phase cooling. Results shows that medium- and high-pressure refrigerants such as R1234ze R600a and R134a with a small density ratio, low viscosity, large latent heat of vaporization and low surface tension helps achieve low device temperatures while reducing the pressure drop across the device resulting in smaller variation in saturation as well as device temperature profile.

Some of the key properties to consider while selecting a coolant are listed below:

(i) A coolant with large latent heat of vaporization is preferred as it can be flowed at lower mass flow rates resulting in lower pressure drop and pumping power.

(ii) A coolant with relatively low density-ratio occupies a relatively smaller portion of the cavity in vapor-phase resulting in lower pressure drop.

(iii) A coolant with low surface tension is results in helps with quicker bubble departure from and re-wetting of the boiling surface resulting in reduced wall super-heat and smaller bubbles near the walls that are easier to disperse due to velocity and density gradients between the vapor bubble and bulk liquid.

(iv) Low pressure/ recirculation zones such as those downstream of an orifice or in the trailing zone of a pin-fin, are preferential sites for nucleation and helps in increasing the wetted surface area for heat transfer.

(v) Presence of sharp interfaces in the inlet plenum as well as heat spreading into the inlet and outlet plenums negatively impact both the heat transfer and pressure drop characteristics across the whole computational domain.

(vi) An increase in the heat transfer rate with vapor quality in the annular flow regime and a decrease in fluid saturation temperature along the length of the cavity due to reduction in local fluid pressure results in the decrease in junction temperature from inlet to outlet. Due to this effect, low-pressure coolant such as R245fa show a large temperature variation from inlet to outlet while high-pressure refrigerant such as R134a show a small temperature variation.

References

[1]. Haron, N. Z., and Hamdioui, S., "Why is CMOS scaling coming to an END?" Proc. Design and Test Workshop, 2008. IDT 2008. 3rd International, pp. 98-103.

[2]. DARPA BAA 12-50, "Intrachip/interchip Enhanced Cooling (ICECool) Fundamentals" June 2012.

[3]. DARPA BAA 13-21, "Intrachip/interchip Enhanced Cooling (ICECool) Applications", Feb 2013.

[4]. G. Loh, Y. Xie. 3D Stacked Microprocessor: Are We There Yet? IEEE Micro. May-June 2010.

[5]. E. G. Colgan, R. J. Polastre, J. Knickerbocker, J. Wakil, J. Gambino, K. Tallman, "Measurement of Back End on Line Thermal Resistance for 3D Chip Stacks", Proceedings 29th IEEE Semi-Therm Symposium, March 2013.

[6]. T. Brunschwiler, B. Michel., H. Rothuizen, U. Kloter, B. Wunderle, H. Oppermann, and H. Reichl, "Interlayer Cooling Potential in Vertically Integrated Packages", Microsyst. Technol., 15(1):57–74, 2009.

[7]. T. Brunschwiler, S. Paredes, U. Drechsler, B. Michel, Y. Cesar, W.and Leblebici, B. Wunderle, and H. Reichl, "Heat-Removal Performance Scaling of Interlayer Cooled Chip Stacks", 12th IEEE ITherm, pp. 1–12, 2010.

[8]. Brunschwiler T., "Interlayer Thermal Management of High Performance Microprocessor Chip Stacks", Cuvillier Verlag Gttingen, Berlin, 2012.

[9]. Schultz, M., Yang, F., Colgan, E., Polastre, R., Dang, B.,

Tsang, C., Gaynes, M., Parida, P. R., Knickerbocker, J. and Chainer, T., "Embedded Two-Phase Cooling of Large 3D Compatible Chips with Radial Channels", Journal of Electronic Packaging, vol. 138(2), 2016.

[10]. F. Yang, M. Schultz, P. Parida, E. Colgan, R. Polastre, B. Dang, C. Tsang, M. Gaynes, J. Knickerbocker, T. Chainer, "Local Measurements of Flow Boiling Heat Transfer on Hot Spots in 3D Compatible Radial Microchannels", Proceeding of ASME InterPACK/ICNMM, July 6-9, CA, 2015.

[11]. T. J. Chainer, M. D. Schultz, P. R. Parida, M. A. Gaynes, "Improving Data Center Energy Efficiency with Advanced Thermal Management", IEEE Transactions on Components, Packaging and Manufacturing Technology, vol.7, issue 8, pp. 1228 – 1239, 2017.

[12]. Brunschwiler T. et al., "Embedded Two Phase Cooling for 3D Silicon Integration", Proc. GOMACTech 2015.

[13]. Kandlikar, S. G., 2014, "Review and Projections of Integrated Cooling Systems for Three-Dimensional Integrated Circuits," J. Electron. Package, 136(2), p. 11.

[14]. Kandlikar, S. G., 2012, "History, Advances, and Challenges in Liquid Flow and Flow Boiling Heat Transfer in Microchannels: A Critical Review," J. Heat Transf.-Trans. ASME, 134(3).

[15]. Thome, J. R., 2004, "Boiling in microchannels: a review of experiment and theory," International Journal of Heat and Fluid Flow, 25(2), pp. 128-139.

[16]. S. Krishnamurthy and Y. Peles. Flow boiling of water in circular staggered micro-pin fin heat sink. Int. J. Heat and Mass Transfer, 51:1349–1364, 2008.

[17]. A. Reeser, A. Bar-Cohen, and G. Hestroni. High vapor quality two phase heat transfer in staggered and inline micro pin fin arrays. In Back-side and Interlayer Two-Phase Cooling, 14th IEEE ITHERM Conference, pages 213–221, 2014.

[18]. C. L. Ong, S. Paredes, A. Sridhar, B. Michel, and T. Brunschwiler. Radial hierarchical microfluidic evaporative cooling for 3-d integrated microprocessors. In Proc. 4th European Conference on Microfluidics, Limerick, 2014.

[19]. Gorle, C., P. Parida, F. Houshmand, M. Asheghi, K. Goodson, "Volume-of-fluid simulation for predicting two-phase cooling in a microchannel", Proc. 67th Annual Meeting of the APS DFD, San Francisco, 2014.

[20]. C. Gorle, P. R. Parida, H. Lee, F. Houshmand, M. Asheghi, K. Goodson, "Validation Study for VOF Simulations of Boiling in a Microchannel", Proceeding of ASME InterPACK / ICNMM, July 6-9, San Francisco, CA, 2015.

[21]. Fang, C., David, M., Rogacs, A., and Goodson, K., "Volume of Fluid Simulation of Boiling Two-Phase Flow in a Vapor-Venting Microchannel", Frontiers in Heat and Mass Transfer, 1, 013002 (2010).

[22]. Anjos, G., "A 3D ALE Finite Element Method for Two-Phase Flows with Phase Change", Thesis No. 5426, EPFL, 2012.

[23]. Cioncolini, A., and Thome, J., "Algebraic Turbulence Modeling in Adiabatic and Evaporating Annular Two-Phase Flow," Int. J Heat Fluid Flow, 32, 805–817, 2011.

[24]. Parida, P. R., Sridhar, A., Vega, A., Schultz, M., Gaynes, M., Ozsun, O., McVicker, G., Brunschwiler, T., Buyuktosunoglu, A., Chainer, T., "Thermal Model for Embedded Two-Phase Liquid Cooled Microprocessor", Proceedings of 16th IEEE ITherm Conference, Orlando, FL, May 30 – June 2, 2017.

[25]. Ansys, Inc., "ANSYS Fluent Theory Guide", Release 15.0, Chapter 17, pp. 511-600, Nov 2013.

[26]. Lee, W., "A pressure iteration scheme for two-phase flow modeling". In Multiphase transport fundamentals, reactor

safety, applications, T. Veziroglu, ed. Hemisphere Publishing, Washington DC. 1980.

[27]. P. R. Parida, H. Tsuei, T. J. Chainer, "Eulerian Multiphase Conjugate Model for Chip-Embedded Micro-Channel Flow Boiling", Proceeding of ASME InterPACK / ICNMM, July 6-9, San Francisco, CA, 2015.

[28]. Parida, P. R. and Chainer, T., "Eulerian Multiphase Conjugate Model Development and Validation for Flow Boiling in Micro-Pin Field", Proceedings of 15th IEEE ITherm Conference 2016, Las Vegas, NV, May 31-June 3.

[29]. Lemmon, E.W., Huber, M.L., McLinden, M.O, NIST Standard Reference Database 23: Reference Fluid Thermodynamic and Transport Properties-REFPROP, Version 9.1, National Institute of Standards and Technology, Standard Reference Data Program, Gaithersburg, 2013.

[30]. S. Szczukiewicz, "Thermal and Visual Operational Characteristics of Multi-Microchannel Evaporators using Refrigerants", EPFL Theses No. 5594, 2012.

[31]. S. Szczukiewicz, N. Borhani, J. R. Thome, "Two-Phase Flow Boiling in a Single Layer of Future High Performance 3D Stacked Computer Chips", IEEE ITHERM Conference 2012.

[32]. P. R. Parida, M. Schultz, T. J. Chainer, "Sim2Cool: A Two-Phase Cooling System Simulator and Design Tool", Proceeding of IEEE ITherm, May 29 - June 1, San Diego, CA, 2018.

Validated Model Calibration for Simulation Aided Thermal Design

R. Cioban[1,2*], Sz. Szőke[3], Z. Kórádi[3], D. Zaharie-B.[1], C. Leordean[1]

1. Robert Bosch Srl, Engineering Center Cluj, 1 Robert Bosch, Jucu 407352, Romania
2. Babes-Bolyai University, Faculty of Physics, 1 Kogalniceanu Mihail, Cluj-Napoca 400084, Romania
3. Robert Bosch Kft, Engineering Center Budapest, 104. Gyömrői út, Budapest 1103, Hungary

raul.cioban@ro.bosch.com

Abstract

The search for better, more accurate thermal models is an ongoing challenge in many industries, including automotive. In most automotive electronic control units (ECU) the heat-generating electronic components are soldered to a circuit board and the quality of the associated component models play a major role in getting the right answers from the ECU thermal model. Our goal is to improve the temperature predictions of finite element (FE) analysis by improving the accuracy of the component models. We describe the implementation details and obtained results of a method that calibrates individual thermal component models to experimental datasets. In particular, we used multiple experimental test setups in which different heat paths were active in the studied component, and the resulting measurement results were taken as reference for the FE model optimization process. The focus of the proposed paper is the FE component model optimization process, but also touches on experimental aspects, like spread and noise in the data. Results presented are for a MOSFET in LFPAK88 (SOT1235 [7]) package, which employs soldered clip connections instead of the more traditional wire bonding, and as consequence presents at least three significant heat paths within the package: through the heat-slug (Drain), through the pins (Source), and through the thin plastic mold. Our optimized FE model describes all these heat paths with a high degree of accuracy, leading to excellent results in the model validation experiments performed on a PCB-mounted test vehicle.

Keywords

Thermal model, optimization, verification, LFPAK.

1. Introduction

Thermal management is essential in the automotive electronics industry. Poor thermal design can lead to delays in the development process or may negatively affect reliability of the product. Simulation models are widely used from early stages of development. By creating numerical representations of systems, the thermal performance of designs can be checked and optimized for different applications. In order to have good system representations, models have to be accurate down to component level. Component calibration based on thermal impedance measurements is the preferred method of achieving this goal [1]. This is done either using a manual iterative approach based on structure function representation of the heat path [2, 3] or by an automated optimization process based on direct thermal impedance calibration [4]. Using calibrated components allows the optimization of other design elements such as footprint design, component placement, thermal interface material (TIM), etc.

FE thermal models of electronic components with detailed internal geometry are generally not available from suppliers, or, when available, may not always meet the user's accuracy requirements for a particular application. In our view, experimental validation of available or generated models is necessary to ensure correct simulation results. Validation of every component model in every application would be an unrealistic burden on the ECU development process. Instead, we calibrate and verify our component models against test setups with sufficient diversity to guarantee reasonable accuracy in very different cooling scenarios of different ECUs. As a result, such validated models can be used without restriction, greatly simplifying the work of simulation engineers.

The scope of this paper is mainly to present the challenges of thin package MOSFETs' measurements and FE model calibration. Second, it is to validate the calibrated model using test boards, and to show the usefulness in studying the impact of different footprints and PCB layout details on overall cooling performance. Components in LFPAK-style package were chosen as test specimens for the current study due to ever-stronger miniaturization trends.

2. Experimental and numerical methodology

A thermal system may be fully described in several ways, for instance by its response to a 1W power step excitation, also called thermal impedance (Z_{th}). Datasheets of power MOSFETs - and of other electronic components - typically include a Z_{th} curve, which describes thermal behaviour when the component is mounted on an ideal heatsink, and $R_{th\_junction-to-case}$ values for typical and/or worst-case parts. However, this data is of limited practical use when SMD components are soldered on a PCB and cooled through thermal vias, or when cooled through their plastic mold. The same component will have very different internal heat flow distributions when mounted on a good heatsink, or on a PCB, or in a through-mold cooling configuration, and every such configuration will present a different Z_{th} curve, even for the (early) time ranges where one would expect the package properties to dominate the step response. With regard to FE model calibration, we can use this observation to our advantage: even though one single Z_{th} curve is not enough to optimize a detailed 3D model, by measuring several Z_{th} curves in very different cooling conditions, we can gather enough reference data to tune all parts of the FE model. For instance, a good representation of the heat path through the pins is particularly important in case of packages with copper clip connections on top of the chips, like the LFPAK88. In our presented approach, we chose the following three conductively cooled test setups for a complete thermal characterization of the studied power MOSFET package [5]:

1. "slug-down" or bottom cooling (drain tab on water-cooled cold plate, pins in the air)
2. "through-pin" cooling (pins soldered to a copper block, which in turn is attached to a cold plate, drain and package body in the air)
3. "through-mold" or topside cooling (plastic mold on cold plate facing down, drain tab facing up towards air)

Radiative heat transfer was not separately investigated on this particular device under test (DUT), but we found it to be relatively straightforward to determine the emissivity coefficient of the mold compounds with the help of an infrared camera. Cooling effect of the ambient air by convection has limited to no impact on the component Z_{th} compared to conduction in most of our measurements, therefore no special care was taken to limit the convective cooling.

Thermal impedance measurements have been carried out using commercial T3Ster™ equipment. We have made efforts to get a high signal-to-noise ratio (SNR), and maximize the amplitude of the excitation, but we found practical limits to increasing the power step. This resulted in experimental data with limited accuracy, and had to be critically considered during the model optimization process.

After careful analysis of our measurement results, we concluded that there should not be *one* particular measurement considered as reference for the simulation model tuning, instead we should define a target Z_{th} *range*, with minimum and maximum $Z_{th}(t)$ values at each time point. The target range must cover all measured Z_{th} curves, so the sample-to-sample variations are reflected in the optimized model(s). In addition, an arbitrary tolerance factor can be taken into account, to have at least a minimum defined spread between the minimum and maximum target lines. This tolerance factor may be useful when a low number of samples are measured, or the measurement lines intersect in a narrow region. It may even include a-priori knowledge about production tolerances:

$Z_{th\_target\_avg}$ (t) = average (Z_{th_DUT1} (t),Z_{th_DUT2} (t), … Z_{th_DUTn} (t));

$Z_{th\_target\_min}$(t)=min(Z_{th_DUT1}(t), Z_{th_DUT2}(t),   … Z_{th_DUTn}(t), $Z_{th\_target\_avg}$(t)*(1-p));

$Z_{th\_target\_max}$(t)=max(Z_{th_DUT1} (t), Z_{th_DUT2} (t), … Z_{th_DUTn} (t), $Z_{th\_target\_avg}$ (t) *(1+p))

Where n= number of DUTs measured, p= tolerance factor and t= time (logarithmically spaced).

The optimization process used for calibration is a manual, iterative approach. The thermal impedance results (Z_{th}) of the measurements and simulations are converted into structure functions (SF) [6]. The structure function is a representation of the cumulated R_{th} and C_{th} of the entire heat path, and the corresponding R_{th} and C_{th} values of each layer from an essentially 1-D heat path can be identified. Based on this, a comparison is made between the measurement and simulation results, and the material properties of the first structure showing a mismatch in the heat path are adjusted, until a good agreement is achieved between the two results. Every structure

is taken in the order it appears in the heat path, and its corresponding material property adjusted [3]. This method is used to successively calibrate all three heat paths of the component, first starting with the main path though the slug.

Although the structure function is considered a good indicator for the required changes in the material properties, the final quality of the model was checked and assessed by using direct comparison of the simulated and target Z_{th} curves, in order to avoid any errors caused by numerical transformations involved in obtaining the SF.

This approach was chosen over an automated method using optimization software based on direct comparison of Z_{th} curves, in order to have a better understanding of the component construction and the sensitivity of its heat transfer through each structure in the component.

For the FE model simulation, two commercial software packages were used: an FEA solver Ansys Mechanical and a CFD solver Mentor Graphics FloEFD. This approach was chosen in order to establish if the derived models are valid and interchangeable between different software packages with different solvers. For calculation of structure functions on both measured and simulated data, the T3Ster Master software was used.

The first step in creating the FE simulation model is creating a representative geometry. In order to correlate the measurement results to the simulation model, a good representation of the geometry is required. In this regard, analytical measurement techniques (cross-sectioning, optical microscopy, chemical de-capsulation) were used to access and describe the inner structure of the component.

3. Results and discussions

The component chosen for characterization is a BUK7S1R0-40H type power MOSFET in LFPAK88 (SOT1235) package (Figure 1), made by Nexperia. This component was chosen as the current study object due to its novel package design concept, high thermal performance, and the foreseeable importance of auxiliary heat paths when mounted on PCB. The main structures in the heat paths of the LFPAK88 package [7] are the silicon die, solder die attach on both sides of the silicon chip, copper slug at the bottom and copper clip at the top. In comparison to the typical aluminium bonding wire, used in the D2Pak packages, this new technology comes with an improved thermal path through the source pins due to the copper clip, becoming an important aspect in the component modelling process.

Figure 1*: LFPAK88 package [7]*

3.1. Thermal impedance measurements

Three different cooling test setups were used during Z_{th} recordings in order to activate three different heat paths in the

(DUT): the "main" heat path through the Drain tab, and the two "secondary" heat paths through pins and mold.

The slug-down test setup is shown in Figure 2. The DUT is pushed to the cold plate by a steel, hollow needle. Force is controlled by adjusting a combination of spring and screw, and is measured by a weight cell. In all our experiments, the pushdown force was around 10(+/-2)N. Between the DUT and cold plate we used a TIM paste having a thermal conductivity of 1 W/mK.

Figure 2. Slug-down cooling test setup, with one of the water-cooled cold plates used in the experiments.

The water-cooled heatsink provides a low-resistance heat path to the 25 °C water. $R_{th\_junction-to-water}$ was measured in the range of 1.7 K/W, and the R_{th_JC} of the DUT was found to be around 0.3 K/W (TDIM method, [8]). In order to get acceptable SNR on T3Ster equipment, one needs to have at least 40 mV useful signal (change in voltage), which translates to about 20 K of temperature swing when using the body diode of the MOSFET as temperature sensing element. Heating up the DUT, by using the same body diode as heating element (voltage drop of ~0.7V) would require more than 30 A of heating current to achieve the target temperature swing in our test setups.

$$\Delta T = R_{th} * P_{diss} = R_{th} * U_{ds} * I_d$$

This amount of current would be possible to generate from a measurement equipment point of view (using the T3Ster Booster with an external power supply), but is impractical to connect to small SMD-style DUTs due to cabling difficulties (Figure 3). To carry the large heating current thick copper cables are required, and soldering them to the DUT without reflowing the solder on its back is increasingly difficult with thicker cables.

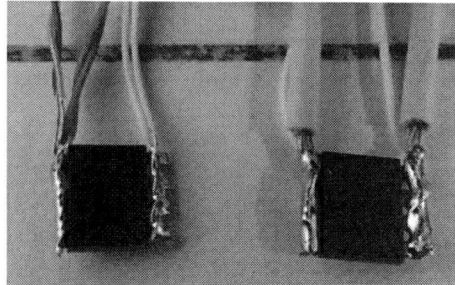

Figure 3. Different cable thicknesses on a thin LFPAK88 sample

Thick cables may also exert force on the DUTs during measurements, in addition to the pushdown needle, hindering repeatability. In order not to alter the original thermal

behaviour of the DUTs, we have chosen reasonably thin cables (0.25 mm and 1.00 mm diameter) in our experiments, capable of safely carrying a reduced heating current of 5 A to 20 A for the short duration of our measurements, achieving less than ideal temperature swings. Power step amplitude was between 3W - 14W in our slug-down experiments.

The SNR of the area of interest on the recorded Z_{th} curves is further reduced by the following effect: in the slug-down test setup, the contribution of the DUT package itself is only 20% of the total R_{th} value. As Z_{th} has to be recorded from steady-state to steady-state, the heating power must be chosen to match the total R_{th} of the test setup. The resulting SNR may be good on the long timescales, but low temperature swing is achieved within the DUT package, and therefore low SNR on the time range dominated by the DUT itself (early values, < 10 ms).

A combination of limited heating current and very low R_{th_JC} of the DUT lead to poor SNR, and random variations in the recorded Z_{th} curves can be seen. Random variations due to poor SNR can be partially mitigated by using a repeat-and-average technique, however the time devoted to measuring 1 DUT is increased 4, 16, 64 fold for an improvement of 1, 2, 3 bits, respectively. We used a maximum of 16 repetitions in our investigations. Sample results are shown in Figure 4.

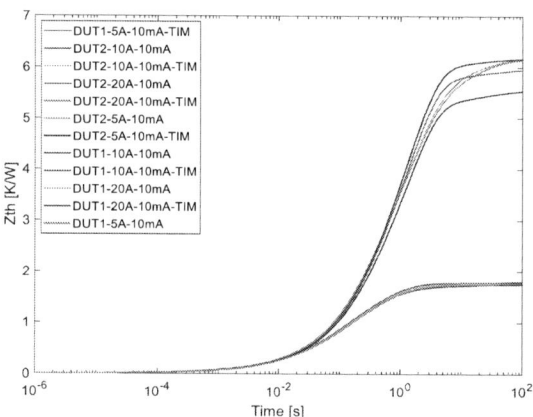

Figure 4. Sample Z_{th} results of slug-down cooling setup, with and without TIM, using different heating currents. Upper curves are without TIM, lower curves with TIM.

Variability in the results was observed also on a sample-to-sample basis, and this could not be reduced by tweaking our test setups, so we have to presume that manufacturing variations (e.g. thickness of die-attach solder) are the root cause of a significant spread in Z_{th} curves for timescales t < 100 µs (Figure 5). At later times (t = 1ms ... 10ms, where the package dominates the response) the relative difference between the average and the individual samples is ~ 5%.

Initial transient correction was applied to the measured datasets using square-root approximation between 40 us and 180 us. Same initial correction settings were used on simulation results during the optimization process. A number of n=6 samples BUK7S1R0-40H from the same batch were measured in slug-down configuration, a value of p= 0.1 was used in the evaluation.

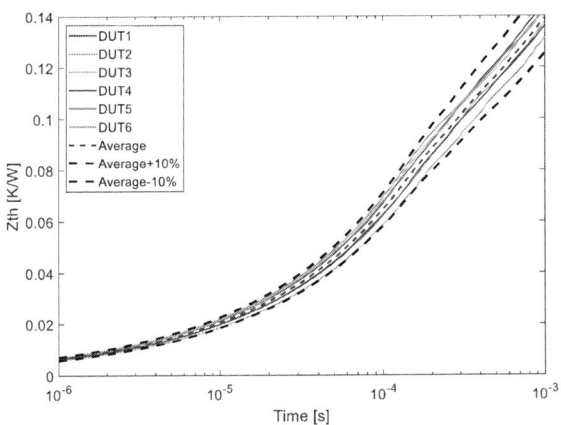

Figure 5. *Close-up of the measurement results shows sample-to-sample variations in the early thermal responses of measured DUTs, slug-down cooling configuration. Dashed lines denote the minimum and maximum target curves calculated from the sample average, and the arbitrary defined p=10% tolerance factor (n=6).*

The through-mold cooling test setup is very similar to the previous one, but the DUT is placed with the mold compound facing down, toward the heatsink, with addition of a TIM layer, while the metallic Drain tab is facing up, into the air. The same pushdown needle and force setting was used as in the slug-down measurements.

In this setup the thickness of the cables used for electrical contacting the DUT are of crucial importance. The LFPAK88 package is so thin, that cables larger than ~ 1 mm diameter would cause the DUT to stand off the large, flat surface of the cold plate. As a workaround, either very thin cables, or a raised surface (e.g. a suitably wide copper block) should be used for thermal contact. The latter is visible in Figure 6. $R_{th\_junction-to-water}$ was measured in the range of 15-20 K/W, and the R_{th_J-Top} of the DUT was found to be around 4 K/W, using TDIM method [9]. Power step amplitude was about 0.7 W when using 1 A heating current. We measured n=5 samples in through-mold cooling configuration, and used p= 0.05.

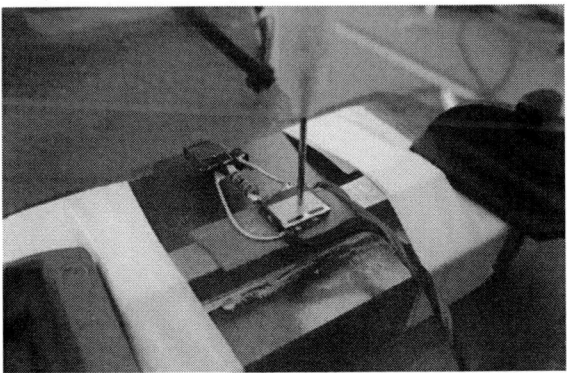

Figure 6. *Details of the through-mold cooling test setup.*

For through-pin cooling tests, we used two different setups. Our first setup consisted of a DUT individually soldered to one end of a large copper block (~ 50 x 15 x 5 mm), which in turn was attached on the other end to a water-cooled heat exchanger (Figure 7a). To minimize unintended heat transfer from DUT to ambient air, a heat isolating foam was used around DUT and partially around the copper block. Only the Source pins were used for heat transfer through the copper block, as the Gate terminal was needed for biasing and other electrical tests. Thinnest available cables (0.1 mm diameter) were used to electrically contact the Drain tab.

The second setup consisted of five DUTs soldered to a flat copper block (~100 x 50 x 5 mm), as shown in Figure 7b. The copper block was attached to a water-cooled cold plate using a TIM. By using this setup the measurement and sample preparation time was much reduced compared to the first one.

As with the previous test setup, a relatively large total $R_{th\_junction-to-water}$ of the test setup (15 K/W) and the thin cables limited the total power dissipation, with negative impact on the quality of measured data in the early parts of Z_{th}. Still, the relevant part of the package (Source pins and copper clip) could be adjusted in the FE model to fit the data. Measurement results are shown in Figure 7c.

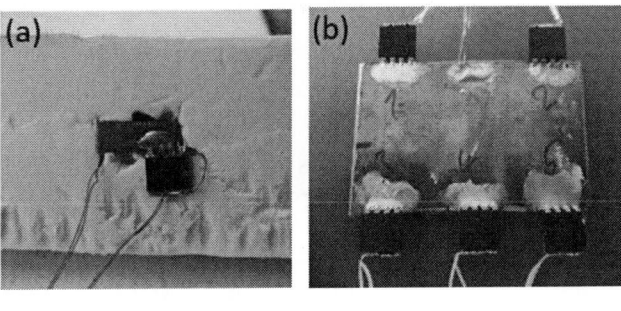

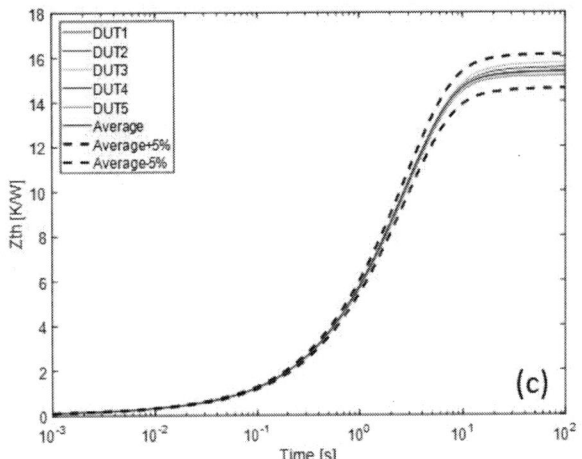

Figure 7. *(a) Pin cooling measurement first setup; (b) Pin cooling measurement second setup; (c) Measured Z_{th} curves of pin-cooling test setup (n=5, p=5%).*

3.2. Geometry creation

A key factor in performing a good calibration is an accurate geometry model. Poor geometrical accuracy would create difficulties in identifying and recreating all the features present in the structure functions of the measurements. Creating an accurate geometrical model requires first to measure all the component dimensions through a series of analytical methods. For measuring the outer dimensions of the component, optical microscopy is the easiest and fastest method. This approach was used to measure the size of the component package. For investigating the inner structure of the component and determining the dimensions of each structure, the cross-sectioning method is required. By embedding the component in resin, then grinding it until the required plane is reached, the areas of interest can be assessed using a metallographic microscope (Figure 8a). Using this technique, several planes were defined and measured and the geometrical dimensions (*width, length, thickness*) of the die, die attach, copper slug and copper clip were determined. Based on the gathered data, a geometrical model was adjusted using a CAD software SpaceClaim (Figure 8b, c).

3.3. Optimization

For setting up the simulation model, a design of experiments approach was used to find appropriate mesh settings and boundary conditions. In order to achieve a 1D heat flow, a set temperature boundary condition of 0^0C, the same temperature as the ambient temperature, has been chosen for the bottom of the heatsink. This simplification of modelling with 0^0C is possible due to the current approach of using temperature independent materials. Unlike the measurement, which was done during cool down, the simulation was run in a heat up behaviour for simplicity. A logarithmic time stepping is required in order to have fine time stepping up to the 10 ms time frame where the heat travels through the component package, and coarser time stepping when the heat reaches the measurement setup. This time stepping was implemented using APDL code. The initial material properties were chosen based on standard literature material properties for silicon, solder and copper. Density and specific heat are coupled pre-factors in the time domain of the differential equation describing the time dependent heat propagation for a three-dimensional system. Considering this,

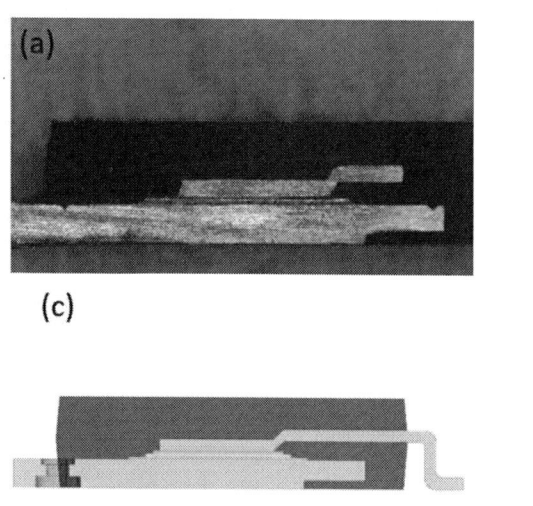
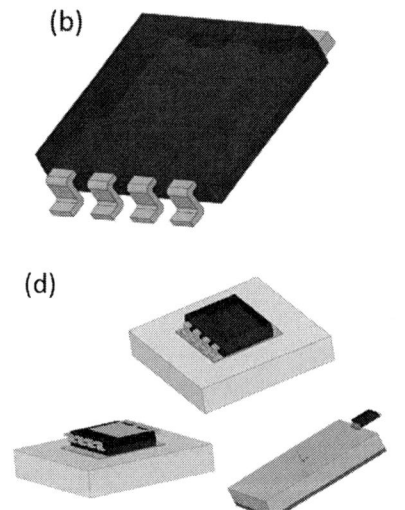
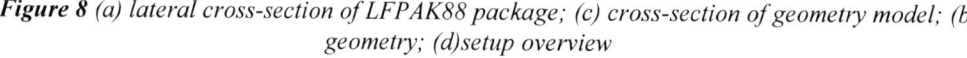

Figure 8 *(a) lateral cross-section of LFPAK88 package; (c) cross-section of geometry model; (b) geometry; (d)setup overview*

In order to have an accurate representation of the power load surface area, a separation between the active and passive areas was done. This was done based on coverage area of the top solder (clip solder).

For representing the measurement setups of the three configurations, estimations were made for the TIM dimensions and heatsink. Compared to the surface of the actual heatsink the model representation in plane is small, but considered sufficient due to the out of plane main heat path. For the pin cooling measurement, the copper block used in the setup was measured and accurately represented in the geometry model. For the mold cooling setup the same approach as for the slug down cooling was used with a TIM layer and a simplified representation of the heatsink (Figure 8d).

only density was chosen as a varying parameter for the sake of simplicity in the calibration process. The results are extracted and evaluated for the average temperature of the active surface area of the silicon die. These considerations were used in both software packages for the component calibration.

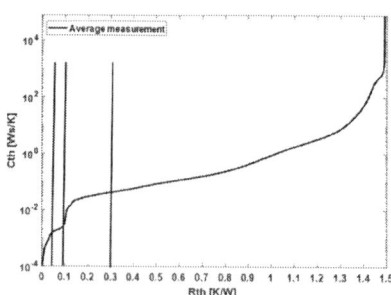

Figure 9. *SF of slug down average measurement with R_{th} of component structures marked*

For **calibration of the main heat path**, the slug-down cooling setup was used. Assuming a simplified 1D-heat flow, this would reduce the problem to identifying and fitting three layers in the structure functions: silicon die, solder and bottom copper. By analysing the structure functions corresponding to the measurement Z_{th}, three distinct regions of the slug down heat path inside the component can be identified (Figure 9). The first region up to 0.05 K/W may be assigned to the heat flowing through the die. The initial increase in thermal capacitance is a representation of the heat flow through the silicon as the source of the load - at least this was the usual conclusion in case of simpler, D2Pak-style packages, where there was no copper clip on top of the Silicon die. The second characteristic between 0.05 and 0.09 K/W representing the heat flow though the solder beneath the silicon, and it is identified as the subsequent increase of the partial R_{th} and partial C_{th}. The third and final distinguishable characteristic is between 0.09 and 0.3 K/W and it describes the heat flow through the thin copper slug. This region is characteristic by the sharp increase of the cumulative thermal capacitance. The variations found in the measurement results for the initial region describing the heat flow through the component, is attributed on one side to noise in the measurement data and secondly to small differences between the geometry of the components like variation in the copper slug and die attach volumes or voiding of the die attach. The thermal resistance beyond 0.3 K/W corresponds to heat passing through the thermal interface material and cold-plate heatsink. The variation found in the measurement data for this region can be attributed to the cumulated effect of the initial variation in part, but also to the manual application of the thermal interface material that led to differences in thickness between the different measurement setups.

The calibration of the model was done by a similar analysis of the structure functions calculated from the simulated Z_{th} responses, in which the corresponding R_{th} and C_{th} of each layer was identified and compared to the measurement results. Based on these findings, subsequent adjustments of the thermal conductivity (TC) and density of each structure in the heat path were done starting from the silicon die and moving outwards to the heatsink. After each material property change, a new simulation was run and a new comparison was made.

During the calibration process, it was noticed that even in ideal heatsink conditions, the 1D heat flow assumption in this case is an oversimplification of reality. Due to the structural configuration of this type of component, the properties of the top (clip) solder and to an extent the bottom (slug) solder was found to have a significant impact on the initial part of the SF (<0.05 K/W), where normally heat is assumed to pass only through the silicon die. For the top (clip) solder, this is due to the large surface in direct contact between the Silicon active area and clip solder. Heat actually goes simultaneously from the active area (top of the chip) into the bulk of Silicon and into the top solder. The chosen initial transient correction settings have a strong impact on the SF values and shape corresponding to the aforementioned structures. In order to account for this influence, adjustments had to be considered for all three layers during the silicon material calibration. It

was deemed necessary to decouple the top and bottom solder material properties from each other, so that geometrical variations or solder voiding can be accounted for in the simulation model through changed material properties.

The region above 0.3 K/W R_{th}, representing the measurement setup, proved difficult to fit mainly due to uncertainties in the heatsinks' internal dimensions and structure (water channels' shape and location), and their influence on the 3D heat spreading. Although an important aspect to take into account, the accurate representation of the measurement setup was considered beyond the scope of this investigation. Using this approach, a good agreement was achieved between the measurement results and simulation model in the component region of the structure function up to 0.3 K/W R_{th} (Figure 10a, b). Although the structure functions were used as a guide for the calibration, the final assessment of the fit was done based on Z_{th} comparison. A sufficiently good fit was considered reached when the calibrated Z_{th} was brought within the limits of the maximum and minimum target Z_{th} curves.

The next step in the modelling process is calibrating the material properties of the **secondary heat paths** through the pins and the mold. The measurement setup of the pin cooling was the first setup built with similar boundary conditions to the slug down cooling. Using the same assumption of a 1D heat path would reduce the problem to identifying and calibrating the material properties of the clip solder and copper clip. As the influence of the clip solder was established and accounted for in the previous calibration step, no further calibration of this material was found necessary. Additional changes to the solder clip is also not recommended as it would require a parallel change in silicon properties. Starting from the material properties derived in the previous step for the silicon, solder and copper slug, only the copper clip was initially considered for calibration, identified as the sharp increase in thermal capacitance after 0.09 K/W. However, it was noticed during the investigations that in the pin cooling setup, also the slug heat capacity had a significant contribution to the thermal capacitance of the same region as the copper clip. This is attributed to the shorter heat path towards the copper slug compared to the end of the pins. Thus, in the microsecond time frame, the heat is dissipated in both directions, until it saturates the copper slug. Considering the copper slug properties fixed as to keep the agreement found in the previous calibration, the thermal conductivity of the clip was lowered for a good fit to the pin cooling measurement. Another increase in thermal capacitance was found starting from 0.5 K/W that could not be accounted for by the material properties of the clip. It was identified as a contribution of the mold's heat capacity. To achieve a good agreement between the data sets, an adjustment of the mold density and thermal conductivity was required. As a result of these adjustments, a good fit was obtained for the simulation results (Figure 10c,d).

The final step in the component calibration process involved the comparison to the top-side cooling setup measurement. Using the same setup as for the slug down cooling, the component – now with updated material properties – was rotated and positioned with the mold in contact with the TIM. Due to the pin cooling calibration

119

having a representation of the mold as well, no additional calibration was necessary and the model rather confirmed the calibration results from the previous steps (Figure 10e, f).

The previously described steps were implemented in ANSYS Mechanical, while the exact steps using FloEFD were different, as different persons were performing the two optimizations. Even so, comparing the *results* of the model tuning process between the two FE software, no significant differences could be seen. The small shift in C_{th} between the two solvers in the beginning of the structure function can be attributed to different mesh settings between the models.

While the silicon and clip solder thermal conductivities are relatively low, the calibrated value for the chip solder is higher than expected. These values fit the measured data, but as was found during the calibration, the solder properties have an impact also on the initial part of the thermal response, where we would theoretically expect only the silicon contribution, so different combinations of these three material properties might give similar results in Z_{th} and SF. Similarly, in the pin calibration setup, the bottom (clip) and top (chip) solder was noticed to have a cumulated contribution to the thermal response of the system.

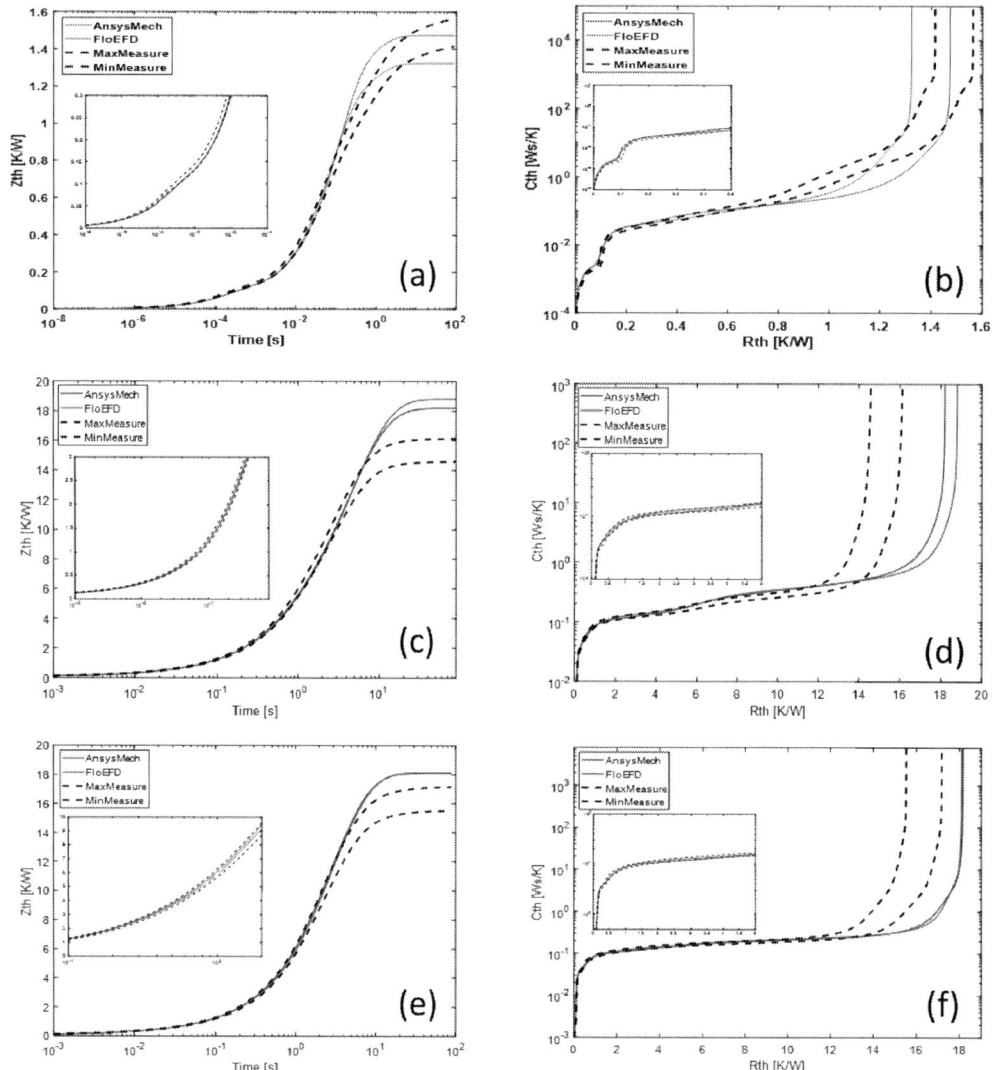

Figure 10 *(a) SF of the slug down calibration (b) Z_{th} of the slug down calibration (c) SF of the pin cooling calibration (d) Z_{th} of the pin cooling calibration (e) SF of the top side cooling calibration (f) Z_{th} of the top side calibration p=5%*

The resulting properties (Table 1) are different from the generally expected values for the corresponding materials. Although the smaller differences in the copper slug properties can be attributed to geometrical tolerances, the conductivity of the silicon and solder have obvious deviations from what one may consider physically plausible values, and it's difficult to attribute just to geometrical influences.

Due to the observed overlap of the effects of different material layers, it is expected for devices with dual cooling paths (e.g. LFPAK-style packages) to end up with optimized models with very different material properties, but having the same overall response. The same effect may prevent extraction of real material properties from Z_{th} measurements, even if using structure functions for visualisation and analysis.

Material/Prop	Density (Kg/m³)	TC (W/mK)	Specific heat(J/KgK)
Silicon	2300	40	1000
Solder chip	8400	100	210
Solder clip	3200	20	210
Copper slug	10000	450	280
Copper clip	8300	310	280
Mold	4000	1	820

Table 1 Calibrated material properties for a transient correction of 40-180 µs

Another important aspect to consider is the transient correction, which was done at a very early stage of the thermal impedance curve in order to remain in the thermal response region of the silicon even though the signal was still noisy. In order to investigate the importance of the transient correction of the measurement data on the resulting material properties of the model, a different correction setting was applied on the measurement data in the 450-900 µs region where a better signal to noise ratio was present. Using the same methodology as before a new calibration was done (Figure 11) and a new set of material properties was derived (Table2).

Material/Prop	Density (Kg/m³)	TC (W/mK)	Specific heat(J/KgK)
Silicon	2300	145	1000
Solder chip	8400	80	210
Solder clip	2800	80	210
Copper slug	9000	310	280
Copper clip	8300	310	280
Mold	4000	1	820

Table 2 Calibrated material properties for a transient correction of 450-900 µs

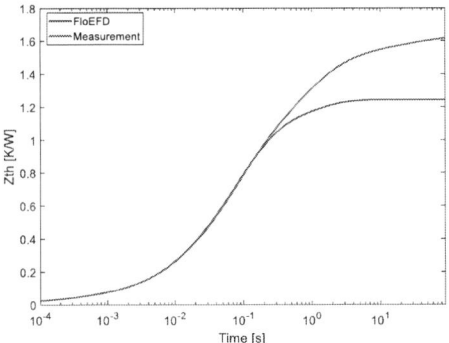

Figure 11 *Slug down comparison of new calibration for a transient correction of 450-900 µs*

Comparing the two results, a significant difference can be found between the derived properties, highlighting the importance of the reference data chosen (or the different correction settings of the data), but also the difficulties of modelling thin packages compared to previous experience on larger D2Pak packages [5].

3.4. Model validation

Validating a model in general, means comparing its response to a reference response, for instance to a measurement, to analytical calculation, or to a response of a more accurate (more detailed) model. In our case, the latter two are not available, and we have already made our best to generate the most accurate measurement data in the first place, to which the models were tuned to fit almost perfectly. There would be little point in repeating measurements in the same test setups. We decided to look at new test setups where several heat paths are active simultaneously, and compare measurement with model prediction. Because the DUT under investigation is an SMD component, it is reasonable to put it in a test vehicle consisting of a PCB. We could choose a test setup where the PCB is attached to a heatsink or a setup where the PCB is just hanging in the air. For sake of simplicity, we chose the second option.

Initially we designed a test card very similar to what is used for measuring R_{th_JA} of electronic components: almost the whole PCB surface was devoted to cooling the most important cooling port of the DUT, which is the Drain tab, or slug (see Figure 12a. After performing measurements on these test vehicles, we realized that the large R_{th} of the PCB-to-air interface and the heat spreading within the PCB dominates the thermal response. The thermal resistance of the component was ~50x lower than the external resistances, so it was impossible to get a good SNR, or to see the effect of the package in the total response: even a package with halved thermal conductivities would not make a discernible difference in the response.

Our second attempt in designing a test vehicle followed this reasoning: if there is no way to get precise verification of the package model in PCB based experiments, at least we should be able to prove that by further simplification of the model (e.g. disabling one of the secondary heat paths) the response would be way off. This task is possible to implement in a layout that cools both the Drain tab and the Source pins in an equal way. Such a layout is shown in Figure 12b.

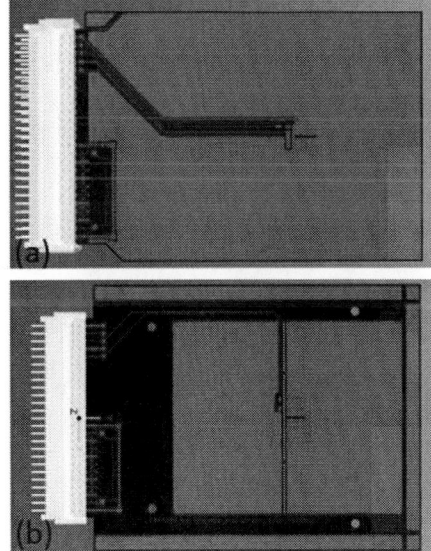

Figure 12 *Test cards used for model validation, top layer copper of the layout. (a) Initial layout, large cooling area assigned only to Drain; (b) Improved layout, half of the PCB area assigned for cooling the Drain pad, other half area assigned to cool the Source pins*

This second, two-sided PCB features large dedicated cooling areas for the copper slug (drain) and source pins of the component. In the final PCB layout, the FR4 between the two areas was also removed (with "slits") to lower the thermal interaction, and physically separate the two areas. This way the setup can exercise the two most significant heat paths of the component (through the slug and through the pins) and the setup simulations are expected to show significant differences between the calibrated and non-calibrated models.

The DUT mounted on test board with the measurement setup (JEDEC standard still air chamber, acc. to JESD51-2A) was implemented in the CFD software environment (Figure 13a), and a steady-state conjugate heat transfer simulation was run to simulate the airflow and heat transfer on the surface of the board and the component. These boundary conditions were mapped to a transient thermal simulation with no convection (featuring only radiation and heat conduction in solids), where the component, the footprint (thermal vias) and the board (explicit copper layers and FR4) were modelled to higher details (Figure 13b). This mapping was performed by transferring the heat transfer coefficients (HTCs) on the test PCB and component surfaces.

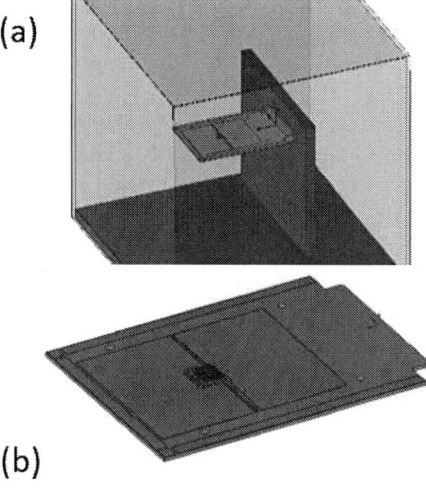

Figure 13. *(a) CAD model of the test PCB in the JEDEC still air chamber (used for airflow simulation);(b) Detailed FE model of the test PCB with the mounted LFPAK88 component (including explicit vias in the footprint)*

A mismatch between the two structure functions is observed when comparing the measurement with the simulation. Such a mismatch found in the middle (PCB) and final regions of the structure function is to be expected due to a few factors. First, the distribution of the copper thickness throughout the PCB is considered uniform, whereas cross-sections of the board at different positions indicate a variation of the copper thickness. The component footprint can also contain geometrical inaccuracies, like different via hole diameters, via plating thickness or even unwanted solder filling some of the vias. Second, standard database material properties and not measured values are used for the board copper and FR4. Standard, literature material properties are a good starting point for a rough estimation, but to get an accurate representation of the model, the range of the

properties must be narrowed down. Finally, the HTCs derived from a steady-state simulation with an ideal test setup might also differ from the measurement. Considering these effects, the heat conductivity of the PCB copper layers and vias, the density of the FR4 in the footprint area, and the HTC values on the PCB surface were slightly adjusted to get a more accurate fit between measurement and simulations, and this optimized parameter set for the PCB test board was used in all subsequent simulations.

Both measurement and simulated data evaluation showed that $R_{th,JA}$ values of the PCB test setups placed in still-air chamber are very large (>15 K/W) compared to the $R_{th,JC}$ value of the components (<0,5 K/W). This means that the overall thermal performance is dominated by the heat spreading in the PCB, and the PCB-to-air bottleneck, and also the measurement features poor signal-to-noise ratio in the early time range (<100 ms) of the Z_{th} because of the limited allowable power dissipation, and small temperature swings within the LFPAK88 package. However, if we manually "detune" our component model (by changing material parameters or changing the geometry), the relative contribution of the two main heat paths within the component will also change, and the different utilization of the large drain and source copper areas of the PCB will show a visible effect on the simulated results. This can be clearly seen on the thermal impedance and structure function responses of the different simulation models (Figure 14a, b). The component model with the correct geometry and materials fits the measurement within 5% tolerance range, while the component model with the same geometry but not calibrated (standard library) materials shows a too optimistic result, in this particular case.

Also, if we simplify the LFPAK88 component model by removing elements of the copper clip, as if no information would be available about the internal structure, (Figure 14a, b) result, showing significant mismatch not only in the early component transient, but also in the later parts. Steady-state values become higher by >20%, which is a good proof that modelling the heat transfer through the pins is a necessity for a universally applicable LFPAK88 thermal model.

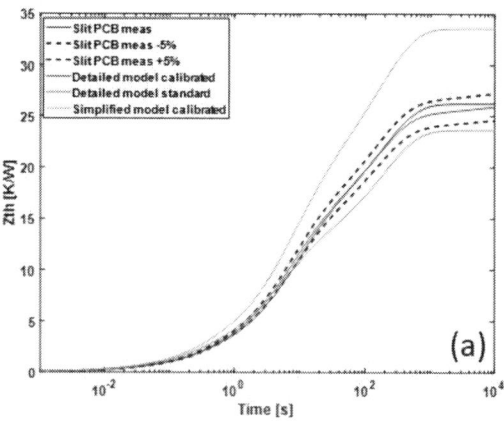

Figure 14. *(a) Comparison of the measured and different simulated thermal impedance curves;*

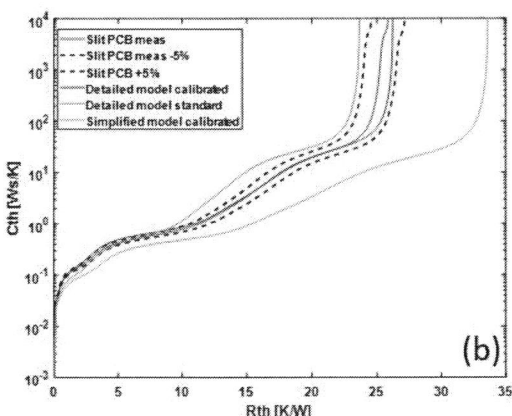

Figure 14. *(b) Comparison of the measured and different simulated structure function curves.*

Importance of the heat flowing through the pins is also shown by the steady-state temperature distribution on the surface of PCB and component. One of the PCBs has been painted matte black, put in a standard size (1 cubic foot) still-air chamber, and photographed with an infrared camera through an opening in the top of the chamber; results are shown in figure 15a. Note the separation cut (slit) between the Drain and Source copper areas acts as a good thermal decoupling element, and the temperature rise of the PCB near the Source pins is solely due to the heat flowing out of those pins, there is no cross heating from the Drain tab. In this particular experiment, heating power was regulated to 1.5 W and the ambient air was at 25 °C. A corresponding simulation result in FloEFD environment with the same power dissipation is depicted in figure 15b, which shows the outstanding match of the simulation models (component and PCB) to measurement.

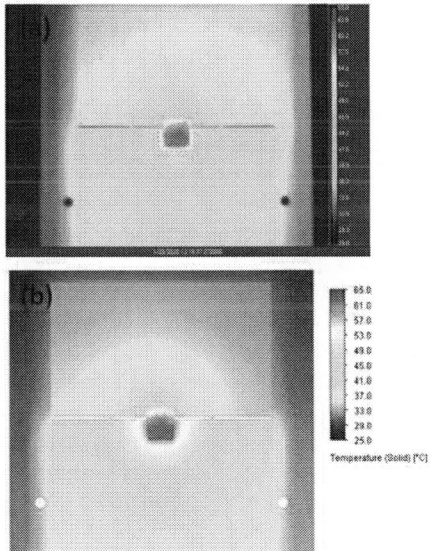

Figure 15: *(a) Measured (b) simulated temperature distributions at steady-state over the test-PCB*

4. Conclusions

The component was measured and its model was calibrated using the comparison of the structure function based on an iterative approach, adjusting the different material properties of the layers. The resulting model exhibits an extremely good match to the measured reference datasets.

The tuned model was checked against an independent measurement for validation, using a PCB test board. FE simulations showed a reasonable match with measured data, and showed that the revised PCB test card design (where equal copper cooling areas are allocated to both Drain and Source) is a suitable test vehicle to prove that no further simplifications are possible, and therefore all existing details are required in our tuned model.

Obtaining a reliable simulation model with very good agreement to measurement opens the way to assess and compare thermal performance of different footprint variants. This information is relevant when developing new products and ensures that the optimal design is reached in terms of thermal performance and manufacturability.

Acknowledgments

The authors wish to thank János Frank for working out the details of test setups, Dániel Balog for the test-PCBs, and the entire ThID team at Bosch Engineering Center Budapest for performing the measurements. The research activity was supported in-house by the CoC Reliability ZUV-15, ZUV-20 and RE-07 projects.

References

1. T. Shinoda, "Three thermal simulation & test innovations for electronics equipment design", Mentor Graphics Engineering Edge issue 7, vol. 1, p. 56-61, 2018;
2. Z. Qiu, J. Zhang, P. Ning, X. Wen, "Transient thermal FE-model calibration based on thermal structure functions for power modules", 2017 IEEE Transportation Electrification Conference and Expo, Asia-Pacific;
3. R. Bornoff, A. Vass-Varnai, "A detailed IC package numerical model of calibration methodology", 29th IEEE Semiconductor Thermal Measurement and Management Symposium;
4. Sanchit Tandon, E Liu, Thomas Zahner, Sebastian Besold, Wolfgang Kalb, Gordon Elger, "Transient thermal simulation of high power LED and its challenges", 2017 18th International Conference on Thermal, Mechanical and Multi-Physics Simulation and Experiments in Microelectronics and Microsystems (EuroSimE)
5. Sz. Szőke, Z. Kórádi, "Component Model Calibration Using Transient Thermal Test Methods and Multiple Measurement Setups", 25th International Workshop Thermal Investigations of ICs and System (Therminic 2019)
6. V. Székely "A new evaluation method of thermal transient measurement results", Microelectronics Journal issue 3, vol. 28, p.277-292
7. https://www.nexperia.com/packages/SOT1235.html
8. "Transient dual interface test method or the measurement of the thermal resistance junction to case of semiconductor devices with heat flow through a single path", JEDEC Standard JESD51-14, Nov. 2010.

Effects of Solder Voiding on the Reliability and Thermal Characteristics of Quad Flatpack No-lead (QFN) Components

Ross Wilcoxon, Dave Hillman and Tim Pearson
Collins Aerospace
400 Collins Road NE
Cedar Rapids, IA
ross.wilcoxon@collins.com

Abstract

Test vehicles were assembled with Quad Flatpack No-lead (QFN) components and subjected to thermal cycle conditioning. Components were inspected to measure voiding within the component thermal pads for different solder alloys and different configurations of microvias in the test boards. The resulting distributions of voiding showed that lead-free solders had substantially more voiding and that the presence of microvias increased the voids in components with tin-lead solder. Thermal modeling was used to assess whether the presence of these voids would adversely affect the thermal characteristics of the QFN. This modeling indicates that voiding in the range measured in this study would have little impact on the component's thermal resistance.

Keywords

Quad flatpack no-lead (QFN), Solder voiding, thermal resistance

Acronyms

CTE	Coefficient of thermal expansion
FEM	Finite element model (or modeling)
I/O	Input / output
PCB	Printed circuit board
QFN	Quad flatpack no-lead
R	Junction-to-board thermal resistance
REL	Family of lead-free solders
SAC305	Sn/Ag3%/Cu0.5% solder
SnPb	Tin-lead solder

1. Introduction/Background

Quad Flatpack No lead (QFN) components are soldered directly to a circuit board without compliant leads. They typically include a large solder pad, which is directly under the die, that provides a mechanism for holding the part onto the circuit board as well as the primary thermal path and electrical ground for the component. This center pad, referred to in this paper as the thermal pad, is surrounded by one or more rows of input/output (I/O) pads (interconnect pads) that provide electrical connections between the circuit board and the die. Figure 1 shows conceptual views of a representative QFN. Figure 1a) shows the thermal and I/O pads on the bottom of the component while Figure 1b) shows a cross-sectional image of a QFN attached to a circuit board. For reference, the printed circuit board (PCB) is shown with both microvias, which provide a vertical interconnect

between outer layers of copper traces, and thru vias that extend completely through the PCB.

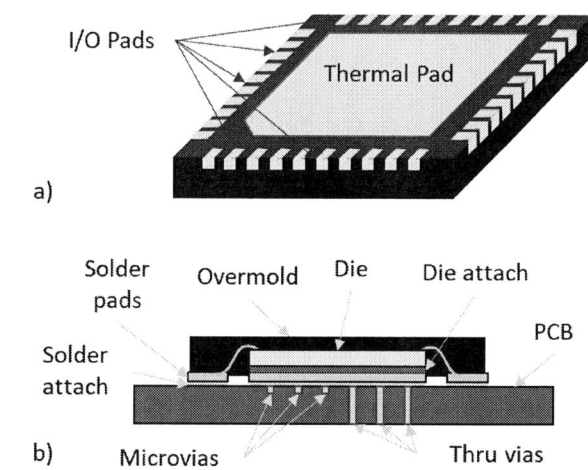

Figure 1. QFN package a) view from bottom of component, b) cross section of component assembled to printed circuit board

The increasing availability of X-ray inspection/analysis technology applied to the printed circuit assembly process has led to a greater recognition of the presence of voids in solder joints. Voids can be important in solder joints because they can adversely affect their mechanical reliability. Voids in QFN thermal pads can also affect the component's thermal resistance.

A number of studies have investigated QFN solder joint reliability and thermal characteristics; many of these have also addressed voiding. In an early study of QFN design issues, Tee, et al used thermal cycle testing as well as simulations to determine that reliability increased with smaller components, thinner die higher coefficient of thermal expansion (CTE) molding compound and the presence of solder fillets [1]. Modeling and testing of larger QFNs with multiple rows of I/O showed that smaller die and larger interconnect pads likewise improved reliability [2]. Anselm and Ghaffarian used thermal shock conditioning to evaluate three different QFN component sizes; only the largest (10mm) component exhibited substantial failures [3]. That study observed voids in the thermal pad of the QFNs that they tested, but stated that the solder joints met IPC specifications. Yun, et al investigated the impact of conformal coating on QFN parts and concluded that thermal cycle reliability was actually somewhat improved by the presence of one of the conformal coats use in their study [4]. Similar to previous studies, that work also showed that smaller die and higher CTE molding compound also improved reliability, particularly in parts without conformal coating. Pearson, et al thermal cycle tested four sizes of QFNs assembled to test boards with four different solder alloys and concluded that finer pitch parts tended to exhibit higher reliability and that there was no substantial reliability difference among the four solder alloys (eutectic tin lead and three lead-free solders) [5].

Lall et al, investigated the impact of voids in the interconnect pads on component reliability [6]. In that study, components were assembled to test boards and then inspected with X-ray to characterize the voids. Typical voiding was measured and used to define void geometries in finite element models (FEM). This model showed that thermal strains in the interconnect pad solder substantially increased in the regions near the void. Hillman, et al [7] used X-ray inspection to quantify voiding in the thermal pads components described in [5] and concluded that the presence of voids in the thermal pad did not adversely affect the interconnect pad solder joint integrity.

Many manufacturers of QFN components provide PCB design guidelines for ensuring that they meet reliability and thermal performance specifications. Reference [8] is an example of these guidelines, which invariably describe the importance of including thermal vias within the PCB in the areas to which QFNs are assembled. These vias provide a reduced thermal resistance path from the QFN thermal pad to copper layers in the circuit board from which dissipated power cans spread to a larger PCB area. Geiger, et al presented a very thorough study on how printed circuit board design can affect QFN voiding [9]. They found that thru vias in the thermal pad led to reduced voiding, due to the fact that the thru vias provide an exit path for materials that outgas during solder assembly. They also investigated factors that led to solder protruding into unfilled vias, which presumably would improve thermal characteristics. Gadepalli et al assessed how different solder paste patterns used in assembling QFNs to circuit boards influenced solder voiding [10].

Many publications have described analyses for determining the thermal resistance of QFN components. Codecasa et al describe a calculator for determining compact model parameters for components including QFNs [11]. They show that, for example, a larger die reduces the QFN thermal resistance. Arzhanaov et al describe a method for generating models to determine the thermal resistance of QFNs mounted to circuit boards [12]. This work showed that the number of thermal vias and copper layers in the circuit board substantially influence the package thermal resistance. Simulations showed that voiding in the thermal pad had a relatively small impact on the QFN's thermal resistance; increasing the voiding to as high as 80% only increased the overall thermal resistance by ~8%. The study also included thermal testing of components, which were purposefully assembled to have substantial voiding. The measured thermal resistance of those components showed good agreement with the values predicted by FEM.

This paper extends the reliability and voiding results described in References [5] and [7] by including data on the relationships between the presence of microvias in the PCB under the component and voiding in the thermal pad solder. These results are used to guide simulations of a component to assess the impact of voids on component thermal resistance.

2. Reliability Testing

Test boards for thermal cycle evaluations were designed to be populated with four different QFN packages (10mm size with 88 I/O on a 0.4mm pitch, 10mm with 72 I/O on a 0.5mm pitch, 7mm with 48 I/O on a 0.5mm pitch and 6mm with 48 I/O on a 0.4mm pitch). The FR-4 test board was 2.01±0.2mm thick and included 10 internal copper layers that mimicked ground and signal planes. The QFNs were daisy-chain test components that allowed the continuity of the solder joints to be continuously monitored during thermal cycle testing. Eighteen samples of each of the four component configurations could be assembled to a single test board. The test boards included multiple configurations of microvias in the interconnect pads and thermal pad. For the purposes of this paper, three configurations are important – two of which are shown in Figure 2. The '5 Vias' configuration included one via in the center of the thermal pad and a set of four vias between each corner of the thermal pad and its center. The 'Array' was a 10x10 array of vias across the area of the thermal pad. The third microvia configuration was 'None', in which no microvias were present. The plated microvias had a diameter of 0.127mm (5 mil).

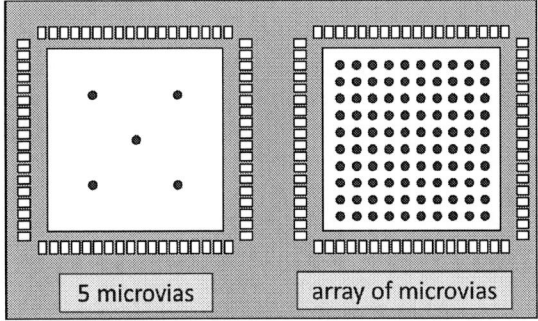

Figure 2. Microvia configurations used in test board

Test boards were assembled with four different solder alloys: eutectic tin-lead (SnPb), tin/silver/copper (Sn/Ag3%/Cu0.5%, i.e., SAC305), Sn/Ag Sn/Ag3% Bi 2-3% Cu 0.5-0.7% trace Sb Ni + proprietary grain refiners (REL22) and Sn/Ag 0.5-0.7% Bi 1.8-2.2% Cu 0.6-0.8% (REL61). In total, more than 2900 QFN components were included in the study; this included 140 to 180 samples for each component/solder alloy combination.

After assembly, the test boards were inspected with automated X-ray equipment to measure the voiding in the thermal pad of each QFN component. Figure 3 shows an example of a typical X-ray inspection of a QFN

Figure 3: Example of voiding in QFN thermal pad

After the test boards had been inspected to determine the amount of voiding in each component, they were placed into a thermal chamber for thermal cycle conditioning of -55°C to +125°C with a minimum 10 minute dwell at each temperature extreme and a maximum temperature ramp of 10°C/minute. The continuity of the components was continuously monitored throughout thermal cycle conditioning by an event detector in accordance with the IPC-9701 specification. Test boards were subjected to 3000 total thermal cycles.

Further information on the test vehicles, assembly procedures and testing approach is available in Reference [7].

3. Thermal Analysis

A simplified finite element model of the 72 I/O, 10mm QFN package style was created in ANSYS Workbench. Due to symmetry in the QFN package, the model included a quarter of component with insulated boundary conditions along the centers of the component. Figure 4 shows the QFN model in which the two front edges are the axes of symmetry and the two rear edges show the I/O pads. The silicon die was 5mm x 5mm x 0.35mm.

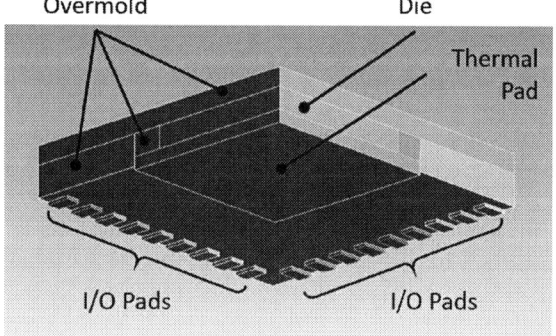

Figure 4. Quarter Model of QFN Package (view from bottom)

The thermal and I/O pads were attached to a quarter model of a 3cm x 3cm x 1.5mm PCB. The PCB included a 10x10 array of 0.3mm vias that each contained two cylindrical bodies of the same diameter. The cylinder nearest the QFN was 0.25mm tall; the other cylinder was 1.25mm long to fill the rest of the volume within the PCB. The small cylinder was used to represent microvias; both cylinders combined were used to simulate thru vias.

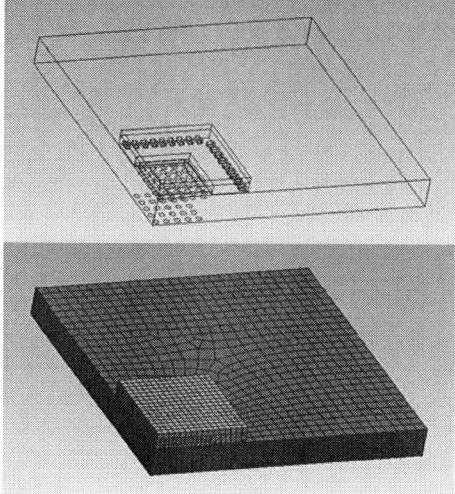

Figure 5. Quarter model of QFN on PCB model: top – line drawing showing features, bottom – with mesh

Table 1 shows the QFN material properties used in the model (these values are shown to two significant digits of values reported in the indicated references). The thermal conductivity of the PCB assumed two copper signal layers and two copper ground planes, but no thermal vias other than the 10x10 array. The nominal die attach between the silicon die and copper was assumed to be a silver filled epoxy with thermal conductivity of 6W/mK and 30 micron bond line thickness, leading to an interface conductance of 2e5 W/m²K. This parameter was varied by ±50% to assess the impact of die attach on overall thermal resistance.

The effect of the solder thermal conductivity was accounted for by assuming that the solder was 0.1mm (4 mil) thick, leading to a conductance for SnPb solder of 5e5 W/mK. Visual inspection of cross-sectioned QFNs used in reliability testing showed that voids tended to extend through the entire solder thickness from the PCB to the QFN thermal pad. Therefore, t effective thermal conductivity of a solder pad with voids would be equal to the solder conductivity multiplied by (1-v), where v is the amount of voiding (between 0 and 100%). The model accounted for solder by setting the thermal conductance between the QFN and the PCB and vias as 5e5*(1-v) W/mK for voiding, v, ranging from 0-90%.

Table 1. Material Properties used in Thermal Model

Material	Conductivity (W/mK)
Silicon	150 [13]
Overmold Epoxy	1 [13]
Copper	390
PCB	15 (x-y), 0.5 (z)
SnPb Solder	50 [14]
SAC Solder	~60 [14]

The vias used in the model were quite large at 0.3 mm (11.8 mil) in diameter. In comparison, the microvias in the test board were 0.127mm (5 mil) in diameter. The larger vias were used in the model to avoid the extremely fine mesh that would be needed to simulate the small vias. The thermal conductivity of the vias was then adjusted by multiplying the estimated effective thermal conductivity of the copper plated vias (200 W/m/K) by the ratio of areas of the actual to simulated vias, leading to a nominal via thermal conductivity of 35W/mK.

Two thermal boundary conditions were included in the model. Since a quarter model was used, 0.25W of heat flux was applied to the top surface of the die to simulate 1W total power dissipation. A convection coefficient of 10kW/m^{2}K with ambient temperature of 0°C was applied to the two edges of the PCB, which were 1 cm from the outer edges of the QFN. The size of the PCB and convection coefficient applied to its outer edges were selected with a goal of creating a model that accounted for the effects of thermal spreading without being dominated by the thermal resistance associated with natural convection from the PCB or introducing errors that can result from fixed boundary temperatures.

4. Results

Solder Voiding and Reliability

Figure 6 compiles the measured cycles to failure for each component and its void percentage. Results are segregated by solder alloy type. Components with lead-free solder tended to have substantially more voiding than those with SnPb solder. This figure indicates that there is little, if any, correlation between voiding in the thermal pad and the thermal cycle reliability of QFN components. Reference [7] more specifically assessed the test data to confirm that component reliability was not changed by voiding as high as 40-50%. Reference [5] reports failure distribution statistics for each component and solder alloy combination and also shows metallurgical cross sections that provide further insight into the failure mechanisms of the test population.

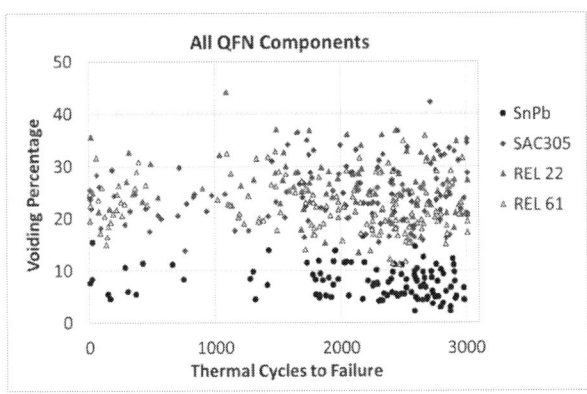

Figure 6: Void results vs. thermal cycles: all samples (from [7])

The data analyzed in References [5] and [7] did not address the specific impact of microvias in the PCB regions under the thermal pads. The test vehicle included different configuration of microvias in the thermal pad, with the specific configuration depending on the component style.

Figure 7 plots the probability distributions of voids for the different solder alloys. Test data are shown with symbols while lines indicate the normal distributions corresponding to the mean and standard deviation for each data set. Data for two circuit board microvia conditions are shown: solid lines and symbols correspond to 'no microvias' while dashed lines/open symbols correspond to results for the 'full array' of microvias as depicted in Figure 2. The plot shows two general trends. First, the lead-free solder alloys (SAC305, REL22 and REL61) had substantially greater voiding than the tin-lead (SnPb) solder. The presence of an array of microvias tended to increase the voiding observed in the QFNs with SnPb solder, but microvias did not appear to affect voiding in the components with lead-free solders. The components with 5 microvias had similar void distributions as those with no microvias.

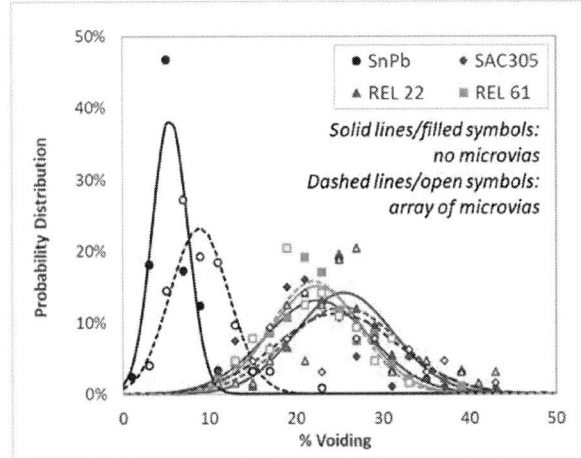

Figure 7: Voiding distributions for 10mm/0.5mm pitch QFN

Table 2 summarizes the voiding statistics for the four solder alloys and the different microvia configurations. For the SnPb solder, the voiding was greater for the full array of microvias but 5-microvia configuration has the same voiding results as with no microvias. All of the lead-free solder alloys led to substantially more voiding with the REL 22 solder having slightly more voiding than the other lead-free solders. The amount of voiding in the lead-free solders does not appear to be influenced by the presence of microvias.

Table 2: Void statistics for 10mm/0.5mm pitch/72 I/O QFN

Solder Alloy	Thermal Pad Microvias	# of samples	minimum	maximum	mean	median	standard deviation
SnPb	None	61	1.3%	10%	5.4%	5.2%	2.1%
	5 Vias	63	1.5%	12%	5.3%	4.9%	2.1%
	Array	62	2.2%	22%	8.9%	8.1%	3.4%
SAC305	None	47	12%	39%	23%	22%	6.1%
	5 Vias	32	13%	33%	22%	22%	5.0%
	Array	32	13%	42%	25%	25%	7.1%
REL22	None	46	16%	43%	26%	25%	5.6%
	5 Vias	32	17%	37%	26%	27%	4.2%
	Array	32	11%	43%	25%	25%	6.6%
REL61	None	47	10%	39%	22%	22%	5.3%
	5 Vias	32	10%	32%	23%	22%	5.2%
	Array	32	12%	33%	22%	22%	5.1%

A number of factors could potentially have led to the voiding characteristics illustrated in Figure 7 and Table 2. The substantially higher voiding in the lead-free solders could be due to inherent differences in the solder alloys and/or the higher processing temperatures used for those solder alloys. Previous research has shown that voiding in QFNs increases when they are processed at higher temperatures [9]. The small increase in voiding seen in the SnPb components with an array of vias (~3%) may simply be masked in the lead-free components by their substantially larger average void levels.

Effect of Voids on Thermal Reliability

Figure 8 shows an example of a simulation result for assessing the QFN thermal resistance. This also illustrates the edges to which the 10kW/m²K convection coefficient boundary condition was applied. Since the power input to the die was 1W (0.25W to the quarter model) and the edges were essentially at 0°C, the maximum temperature on the die corresponds to a thermal resistance, R. This resistance provides a relative metric for comparing the thermal characteristics of different configurations of the package, but it is not a standard resistance parameter such as $\theta_{junction-to-ambient}$ or $\theta_{junction-to-board}$.

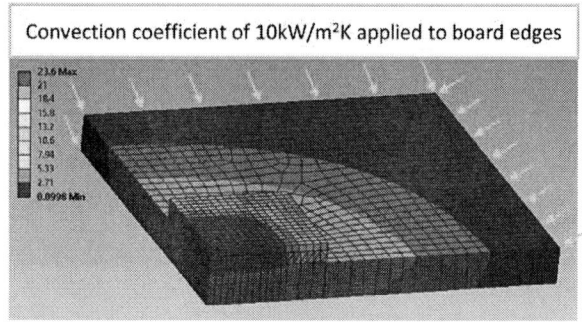

Figure 8. Example of thermal modeling results

Figure 9 shows thermal resistance values of the QFN package with different levels of voiding, different via configurations and different die attach conditions. For the 'no vias' configuration, the 10x10 array of cylinders in the model that filled the vias were modeled as PCB material. For the 'microvias' configuration, only the short 0.25mm tall cylinders were modeled as Via material (z-direction conductivity – 35 W/mK) while the longer cylinders were PCB material. For 'thru vias', all cylinders were modeled with Via material.

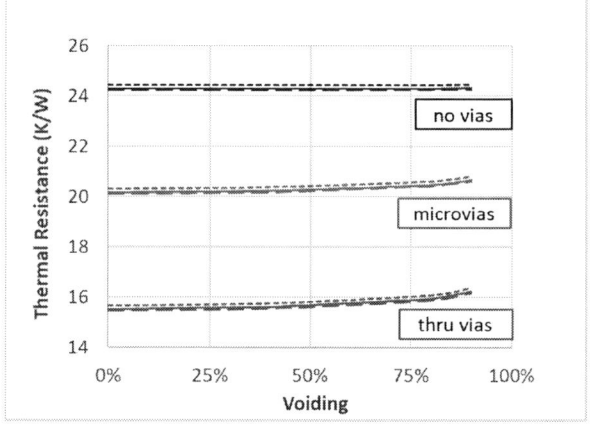

Figure 9. Effect of voids, vias and die attach on thermal resistance

Thermal conductance of the die attach between the silicon die and copper pad was also varied in the simulations shown in Figure 9. Dashed lines in the figure indicate die attach materials with thermal conductance 50% larger and smaller than the nominal value. The results in Figure 9 indicate that the via configuration substantially influences the thermal resistance of the QFN while voiding (up to ~80%) and die attach (in the range included in the simulation at least) both had little influence on the package thermal performance.

Figure 10 shows thermal resistance values as a function of voiding for the full array and a somewhat crude representation of the '5 Vias' configurations, in which only two of the vias in the array were assigned to have Via material properties. These two vias were located in the vicinity of the center via and one corner via (recognizing that to be fully accurate, the "center" via should have only been a quarter of a cylinder rather than a full cylinder). These simulations were done with only the nominal die attach material thermal conductance. The results indicate that the 5 Via configurations had higher thermal resistance than their Array counterparts; but even a few vias do provide some improvement. Again, the effects of voiding in the thermal pad were negligible.

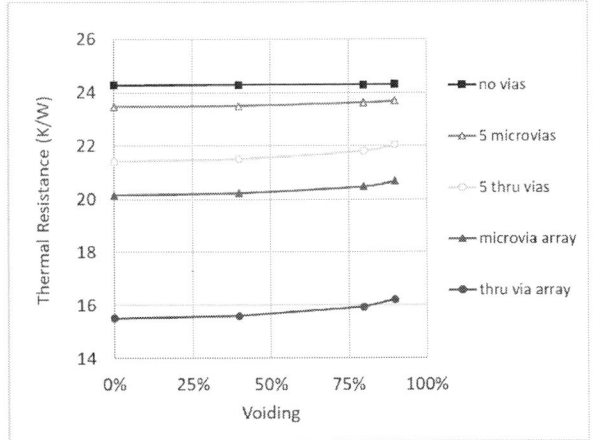

Figure 10. Thermal resistance for different vias and voiding

Additional simulations were conducted to assess the sensitivity of thermal resistance results to the material properties used for the PCB and vias. Figure 11 shows the thermal resistance values for the different via configurations and with PCB in-plane thermal conductivities of 15±50% W/mK. Results shown in this plot are for the case of no voiding. This plot indicates the significant level that the QFN thermal resistance is influenced by the PCB conductivity.

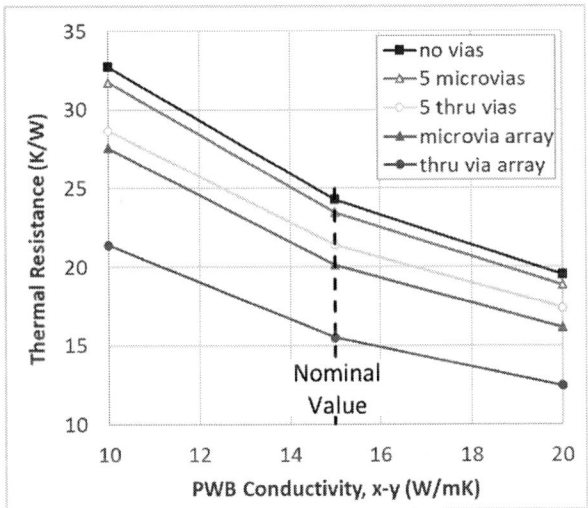

Figure 11. Effect of PCB x-y thermal conductivity on thermal resistance

The modeling approach of using oversized vias to reduce meshing complexity did introduce uncertainty to the analysis because the effective thermal conductivity of the via actually varies with via diameter. Figure 12 shows thermal resistance values for QFNs with full arrays of microvias and thru vias, with via diameters of 0.6mm (11.8 mil) for a range of Via material thermal conductivities. The nominal value used in this analysis (35 W/mK) is near the 'knee' of the curve in

which lower conductivity begins to affect results. The maximum resistance (at via conductivity of 0.5 W/mK) corresponds to the 'None' condition for no vias.

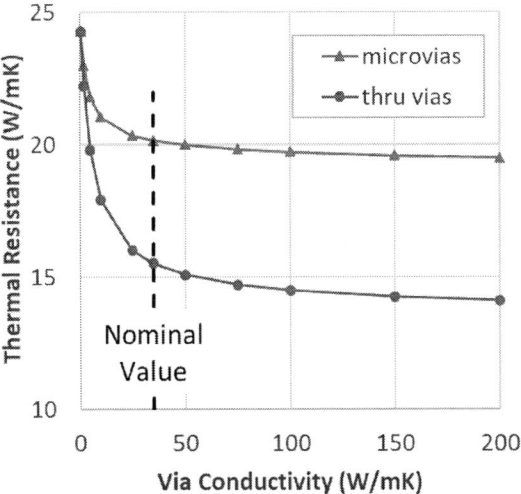

Figure 12. Effect of via thermal conductivity on thermal resistance

5. Discussion

The small impact of voiding on the QFN thermal resistance agrees with previously reported results [12]. Figure 13 of that study showed junction temperature vs. power dissipation for three levels of voiding, which was artificially introduced by applying solder mask over part of the exposed thermal pad prior to reflow. The slopes of those lines correlate to the overall junction-to-ambient thermal resistance of the components. The component with 10% voiding had a thermal resistance of ~56K/W; this increased by ~2% in the component with 70% voiding. This result also validates the recommendations in IPC specifications for circuit board assembly, which indicate that reasonable levels of thermal pad voiding (~50% or less) have negligible impact on the QFN thermal resistance [15].

The presence of microvias in the circuit board thermal pad increased the voiding in the QFN thermal solder pad. Within the range of voiding observed in the testing, voiding had no significant impact to either the thermal resistance to the circuit board or in the QFN solder joint reliability. However, the presence of vias under the thermal pad did improve the overall thermal conductivity into the board and therefore led to an overall reduction in thermal resistance.

6. Conclusions

X-ray inspection of QFN components assembled to thermal cycle test vehicles was used to determine the distribution of voiding within the thermal pad of QFNs with different solder alloys. This showed that components assembled with tin-lead had approximately 6% voiding while those with lead-free solders had average voiding in the range of 20-25%. The presence of an array of microvias in the QFN thermal pad increased the voiding, particularly in the components with tin-lead solder. Subsequent analysis of thermal cycle testing results (reported in references [5] and

[7]) showed that the voids did not degrade the solder joint reliability.

Finite element modeling of the QFN component on a circuit board indicated that, even at the maximum levels identified in the test vehicle (>40%), voiding in the thermal solder pad attachment would not adversely affect component thermal resistance. While microvias may increase voiding, the benefits through their improvement in conduction into the circuit board far outweigh the negligible increase in thermal resistance caused by that voiding. In components with lead-free solders, which had substantially more voiding than the components with tin-lead solder, the minimal thermal impact of voiding would be even further reduced by the fact that the lead-free solder thermal conductivity is typically ~20% higher than tin-lead.

The model used in this analysis was relatively simplistic. However the assessments to evaluate its sensitivity to the assumed die attach characteristics, via thermal conductivity and PCB planar thermal conductivity indicate that the fundamental conclusions are reasonably robust. For the range of conditions evaluated in this study, die attach conductivity and voiding have little impact on overall junction to board thermal resistance while the via configuration and PCB conductivity can have significant effects.

Combining the simulation and thermal cycle testing results provides clear evidence that, while vias may increase solder voiding, the resulting voiding does not degrade the QFN reliability or thermal resistance. Since vias do improve overall thermal resistance, designers requiring good thermal performance should not be reluctant to incorporate as many vias as are reasonable from a manufacturing perspective.

References

1. Tee, T.Y. et al., "Comprehensive board-level solder joint reliability and testing of QFN and PowerQFN packages", *Microelectronics Reliability*, Vol. 43, No. 8, pp. 1329-1338 (2003) DOI:10.1016/S0026-2714(03)00184-7

2. Li, L., "Reliability Modeling and Testing of Advanced QFN Packages", *Proc. of the IEEE 63rd ECTC* (2013) DOI: 10.1109/ECTC.2013.6575653

3. Anselm, M. and Ghaffarian, R.,"QFN Reliability, Thermal Shock, Lead-free vs. SnPb, Microstructure", *Proc. of the 16th IEEE ITherm* (2017) DOI:10.1109/ITHERM.2017.7992505

4. Yun, C. et al, "Reliability Assessment of QFN Components for Aerospace Applications", *Proc. of the IEEE 66th ECTC* (2016) DOI:10.1109/ECTC.2016.186

5. Pearson, T. et al., "Solder Joint Integrity Evaluation of Bottom Terminated Component (BTC) Subjected to Thermal Cycling", *Proc. of the SMTAI Conf.* (2019)

6. Lall, P. et al., "Effect of Thermal Cycling on Reliability of QFN Packages", *Proc. of the 17th IEEE ITherm* (2018) DOI:10.1109/ITHERM.2018.8419496

7. Hillman, D. et al., "Bottom Terminated Component (BTC) Void Concerns: Real and Imagined", *Proc. of the SMTAI Conf.* (2019)

8. https://d3uzseaevmutz1.cloudfront.net/pubs/appNote/AN315REV1.pdf (accessed Dec. 30, 2019)

9. Geiger, D. et al., "The Impact of Via and Pad Design on QFN Assembly", *SMT Magazine*, Vol. 31, No. 11, pp. 16-32 (2016)

10. Gadepalli, H. et al., "Voiding and Thermal Resistance Modeling and Characterization for a QFN Assembly", *Proc. of the ASME IMECE* (2010) DOI:10.1115/IMECE2010-38391

11. Codecasa, L. et al., "TRAC: A Thermal Resistance Advanced Calculator for Electronic Packages", *Energies*, Vol. 12, No. 6 (2019) DOI:10.3390/en12061050

12. Arzhanov, B. et al., "Thermal Evaluation of Printed Circuit Board Design Options and Voids in Solder Interface by a Simulation Tool", Int. Journal of Mechanical and Mechatronics Engineering, Vol. 10, No. 3, pp. 494-505 (2016) DIO:10.5281/zenodo.1111897

13. Bendaou, O. et al, "Thermal characterization of a QFN electronic Package accompanied by a reliability study based on a response surface approach", *Proc. of the 4th IEEE CIST* (2016) DOI:10.1109/CIST.2016.7804965

14. Wilson, J., "Thermal Conductivity of Solders", *Electronics Cooling Magazine*, August 2006, https://www.electronics-cooling.com/2006/08/thermal-conductivity-of-solders/

15. "Design and Assembly Process Implementation for Bottom Termination Components", IPC-7093, March 2011

Thermal Characterization of a Virtual Reality Headset during Transient and Resting Operation

Rachel McAfee, Cole Haxton, Matthew Harrison, Joshua Gess
Oregon State University
204 Rogers Hall
Corvallis, OR 97331
mcafeera@oregonstate.edu

Abstract

Virtual Reality (VR) is a powerful tool for maintenance process development, engineering design, pedagogy, and combat training. The evolution of the gaming industry has driven the demand for comfortable and reliable VR performance. By using the headset's user datasheet defined MicroController Unit's (MCU) operational temperature, the thermal resistance of the headset used in this study was found to have an external resistance, R_{ja}, of 29.1 K/W, but 28.6% of this heat load is transmitted to the user. Relying on the user's body, specifically the forehead, as a heat sink results in uncomfortable perspiration during usage. Collected data show that after 2 hours of operation, the temperature increases 2.8°C on average when the headset is removed from the user and placed at rest while still operational. The maximum temperature increase is 5.6°C at the top of the VR headset, the surface nearest the internal MCU. This temperature spike proves that the headset needs more effective convective surface area in order to maintain a steady and comfortable operational temperature during "headset on" and "headset off" usage as the user's body was not available as a heat sink in the latter mode. A copper sheet has been added inside the headset to thermally connect all of the external surfaces on the device, effectively increasing the optimal convective heat transfer by 61%. The temperature nearest the MCU dropped 6.5°C with the improved thermal management solution and users reported a 25% reduction in perspiration during prolonged use.

Keywords

Wearable, thermal management, consumer products, Virtual Reality, comfort measurements, free convection, heat transfer

Introduction

Wearable technology is a growing market, projected to generate $2.78 billion in revenue by 2024 [1]. Over the next 5 years, it is projected that 117 million headsets will be shipped globally [2]. Individually each headset consumes an insignificant amount of energy, but as a whole these headsets represent 63.2 GWh of energy consumption [3], equivalent to the annual draw of 6,000 American households [4]. With such low energy consumption, the objective of this study is not to reduce power draw but to increase comfort and improve the user experience through improved thermal pathways within the headset package. With wearable electronics, diverting heat away from the user reduces perspiration during operation. Ease of use and ergonomics were cited as major factors in the decision to purchase a VR headset by a recent consumer survey [3]. While software approaches are being developed to improve energy consumption directly [5], this study shows a reduction in perspiration through the diversion of heat to what were non-participating convective heat transfer surfaces. Maximizing the utility of all the surface area on the headset is the key factor to increasing the convective heat transfer by as much as 61%, as will be shown with the data collected in the current study.

Integrated cooling solutions to wearables have already been done, but a qualitative assessment of the user experience before and after implementation of the improved thermal solution is not available in the literature. Google introduced a heat pipe within the VR headset, which more uniformly spread the heat generated from the power-hungry display to the front convective surface area [6]. To reduce perspiration in headphones, thermoelectric devices were installed in the ear cups by HP [7]. Using folded copper plates as shown in Figure 1, the current study seeks to quantify the impact of maximizing the convective heat transfer area on a headset through temperature measurements but also qualitatively assessing the user experience once the improved thermal management solution has been implemented. This study shows how quantitative improvements to heat transfer can be related to

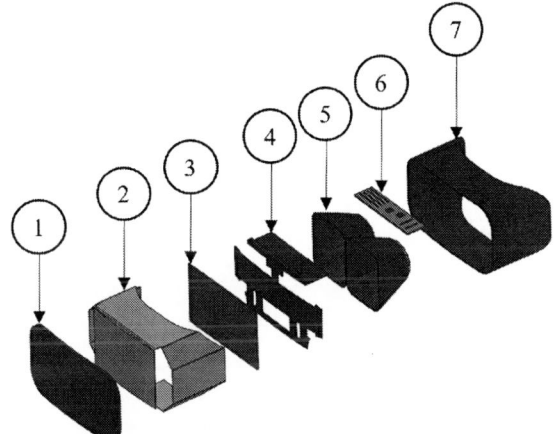

Find Number (FN)	Description
1	Headset Face Plate
2	Folded Copper Plate
3	Display
4	Internal Mounting Structure
5	Eye Cups
6	Primary Circuit Board
7	Headset Cover

Figure 1 – Exploded view of headset assembly with individual components identified as find numbers.

qualitative benefits in user experience which in turn translate to improved market share in the competitive world of wearable products.

Experimental Facility and Testing Procedure

The headset was instrumented with six thermocouples as shown in Figure 2. These thermocouples were calibrated against a NIST-traceable thermistor accurate to within ±0.2°C. The thermocouples were strategically placed to estimate the convective heat transfer around the periphery of the headset (Figure 1, FN 7) and the faceplate (Figure 1, FN 1). Using an 8-channel TC reader, the temperature at these locations was measured at a frequency of 1 Hz. For thermal resistance calculations, the power delivered to the headset is measured using a Monsoon Solution's mobile device power monitor USB-compatible transient power measurement device. Power data throughout all tests were consistent. The resulting wave function can be described by a curve with a central tendency of 3.52W at a sampling frequency of 5KHz at a standard deviation of 97 mW. Ambient temperature in the laboratory is measured

using a hand-held thermocouple reader. The temperature in the lab fluctuates ±0.3°C over the course of any testing. Therefore, it was deemed more important to have continuous temperature measurement on the 6 TCs placed on the headset rather than usurping the limited I/O on the 8-channel reader for transient measurement of the ambient room temperature.

Temperature Increase 5-Minutes After Removal	
Forehead Average	1.1°C
Top	5.6°C
Front	1.4°C
Bottom	2.9°C

Table 1 – Temperature increases after the headset was removed from the user during the initial test.

The first test was completed without the internal copper spreader (Figure 1, FN 2) and indicated that the system was designed for the user's forehead to act as a heat sink. This was further substantiated upon breakdown of the headset and it was found that the MCU was cooled purely by free convection and radiation as no heat spreader was attached from the manufacturer. To test the existing thermal solution, the user operated the headset for 30 minutes. Then, the user removed the headset and placed it on a desk for five minutes while the system was still running. During this five-minute period, all of the temperatures increased by the values shown in Table 1. Headset temperature was observed for an additional 30 minutes after the rest period with the headset off the user but still turned on. The temperature increase at the top (the surface nearest the MCU and the primary convective surface for any heat rising from the internal display) was the largest and shows that the system was relying on the user as a low-temperature thermal reservoir. All the other temperatures increased slightly showing that the excess heat was flowing to other surfaces on the headset. When the user would have been the heat rejection point otherwise. The temperature increase is maintained for over 30 minutes after headset removal due to the lack of a

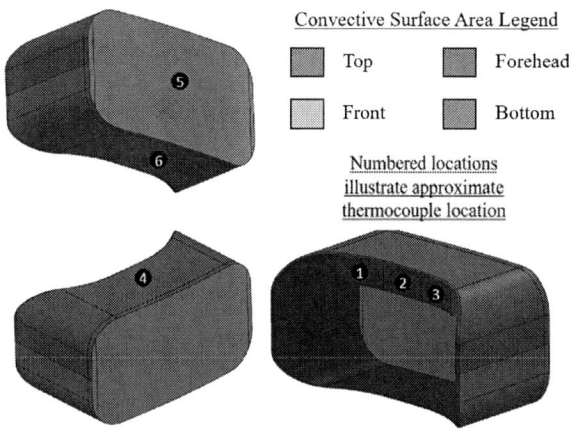

Figure 2 – Data legend showing the area boundaries for convection heat transfer analyses and thermocouple placement for the results in this study.

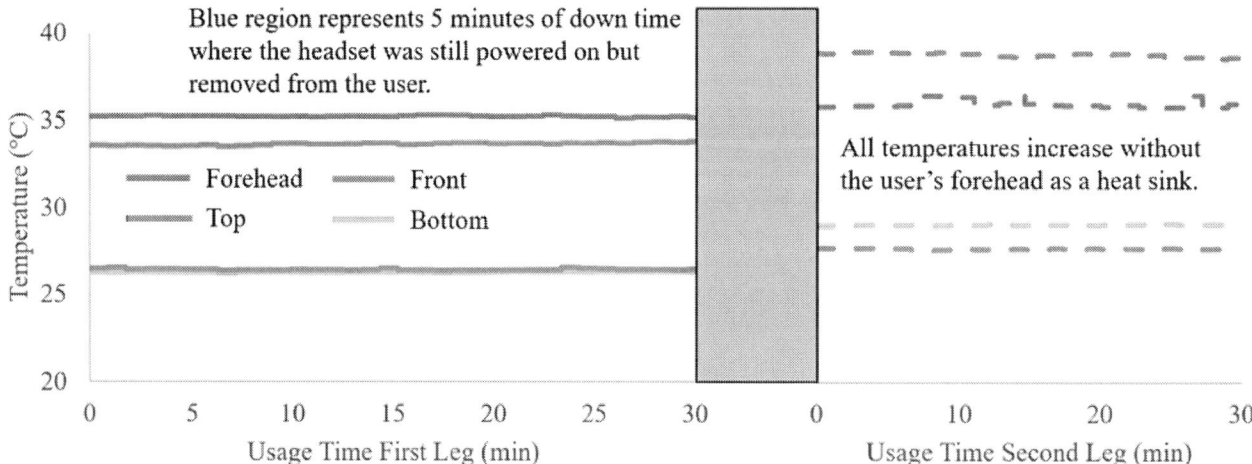

Figure 3 – Transient data of the first test where temperatures increased sharply upon removal of the headset from the user. The device was still powered on once it was removed and at rest for five minutes.

thermal sink via the users head. The steady state data during this initial test is provided in Figure 3, which shows the 30 minutes of game play and the temperature difference after the 5-minute rest period, which is indicated by the shaded blue box.

The first test (Figure 3) drove the definition of the second test. Eight participants were asked to play the same game on the VR headset for 100 minutes. The system was removed and then powered off for a 20-minute cool down period. The number of participants was chosen to reduce the random error in the measurements, and the duration of time was selected to allow the system to reach a steady-state condition prior to headset removal. This test was conducted for both headsets: the one with the integrated heat spreader (Figure 1, FN 2) and the one without. Users were then given a survey to ascertain the degree to which perspiration was experienced, if any, by integration of the heat spreading thermal management approach. Users were aware of what thermal solution was being used during each testing session. Due to scheduling conflicts users differed between the group with and without the heat spreader. Users had significant diversity in head size, intensity of play, hair length, and hair type. Hair was moved to allow forehead TCs to come in direct contact with forehead. All users played the same virtual reality game involving moderate physical activity. Ambient temperature was monitored and fluctuated by ±0.3 °C during testing.

Data Analysis and Results

To analyze the total heat loss of the headset, a control volume was placed around the device. It was first assumed that the net loss would be primarily due to convection and radiation. It was assumed that radiation only occurred from surfaces 4, 5, and 6 as shown on Figure 2. Taking the surface's respective transient temperature measurements and assuming that the surface of the headset was a blackbody, diffuse emitter, and gray, the net radiation loss to the surroundings was calculated. It was found that the heat loss due to radiation was less than 1% of the total loss. With the limited spatial resolution to be corrected in future work, it was assumed that the net radiation heat loss was negligible in comparison to convection and will not be considered for any of the reported analysis.

The temperature results from the second test (where there is no heat spreader within the headset, i.e. out-of-the-box or original thermal design) are shown in Figure 4. Noted on Figure 4 are the points at which the headset was removed, the short-lived (in regard to overall test time) but measurable temperature increase upon headset removal, and the initial region where energy is still being stored within the thermal mass of the headset. Thermocouples located within each color-coded region of Figure 2 were used in conjunction with free convection correlations (Equations 1-3) to determine the heat transfer from those surfaces. For example, TC 4 was used with Equation 1 to determine the heat transfer leaving the headset from the orange surface highlighted in Figure 2. The free convection from the top, bottom, and front surfaces were calculated using Equation 1, Equation 2, and Equation 3, respectively [8]. These equations calculate the dimensionless average Nusselt number; where Ra and Pr are the Rayleigh number for buoyancy driven flow and Prandtl number of the working fluid, air, respectively. The results of the convective heat transfer analysis are shown in Figure 5. In future work, perspiration measurement will be used to quantify the amount of latent heat transfer there is throughout each test. A sensor which measures the volume of sweat over time will be implemented. Volumetric flow rate of sweat generation multiplied by the sweat liquid density and latent heat of vaporization will yield the latent heat transfer with respect to time. The total heat transfer is the sum of the sensible energy storage, represented by the initial exponential temperature increase of the plots shown, plus that transmitted via perspiration which does not effect this initial exponential

$$\overline{Nu_L} = 0.54 Ra_L^{1/4} \tag{1}$$

$$\overline{Nu_L} = 0.52 Ra_L^{1/5} \tag{2}$$

$$\overline{Nu_L} = 0.68 + \frac{0.670 Ra_L^{1/4}}{[1 + \left(\frac{0.492}{Pr}\right)^{\frac{9}{16}}]^{4/9}} \tag{3}$$

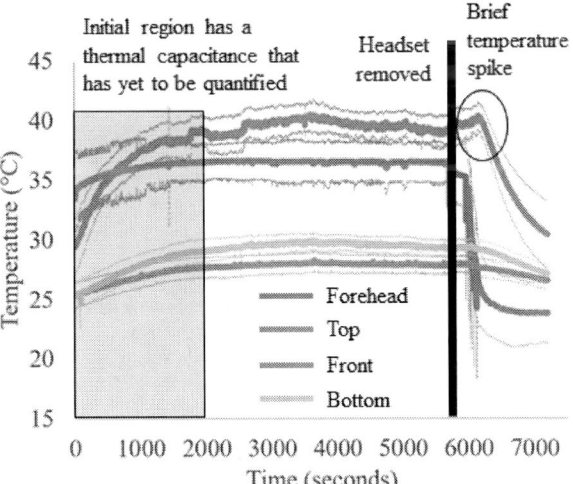

Figure 4 – Average temperature data (solid lines) along with uncertainty (dashed lines) for eight testing replications without the integrated heat spreader.

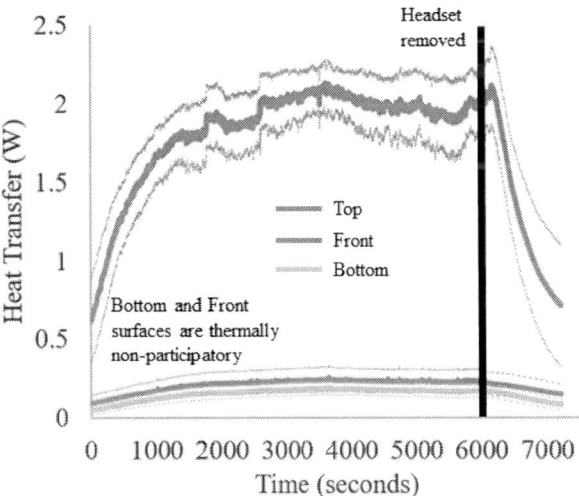

Figure 5 – Convection heat transfer losses (solid lines) along with calculated uncertainty ranges (dashed lines) for eight testing replications.

increase since it is not sensible. Therefore, in order to extract the thermal mass from these curves via regression analysis, the sensible heat gain must be isolated from that transmitted by perspiration.

The forehead temperature stabilized at the nominal skin temperature of 36 °C due to the thermal path way between the TC's and user's forehead. This nominal skin temperature was verified after each test as the user's forehead temperature was measured upon headset removal. The top temperature is shown as the hottest as this is region that is most utilized for convection heat transfer release. This top surface is not only the closest to the MCU of the headset, but it is also the surface that accepts heated air which is buoyantly driven upward from the internal display. The dotted lines in Figure 4 represent the high-band and low-band temperature values from the calculated uncertainty. Uncertainty was calculated by taking the standard deviation of the samples, dividing by the square root of the number of samples and multiplying that result by the appropriate t-student for eight trials. Instrument error accounted for less than 1% of the random error among the eight test replications conducted. The uncertainty ranges from approximately 2-3 °C for the forehead and top temperature measurements while that of the less thermally involved front and bottom faces is approximately 1°C.

As shown in Figure 5, less than 0.25W (~10%) of the heat transfer occurred over the relatively large front and bottom faces during operation. This lack of free surface convection effectiveness puts undue strain on the user through latent heat transfer (i.e. sweating or perspiration). In future work, a perspiration meter will be integrated under the headset to compare the projected latent losses to actual perspiration generation.

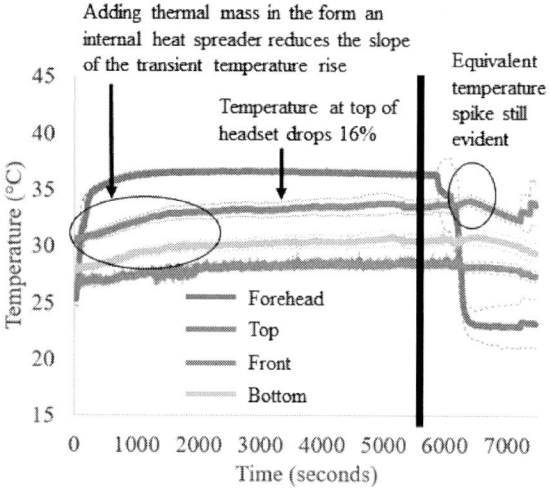

Figure 6 – Average temperature data (solid lines) along with uncertainty (dashed lines) for eight testing replications with the integrated heat spreader.

Figure 6 is the temperature data taken during operational tests where the headset has the integrated thermal management solution/spreader on board. As expected, the forehead temperature remains the same as our bodies work like a thermal reservoir maintained at a constant temperature. However, the top surface, that closest to the MCU and the accepting surface

of all the hot buoyantly-driven air from the internal display, drops its temperature by 6.5 °C compared to the original design.

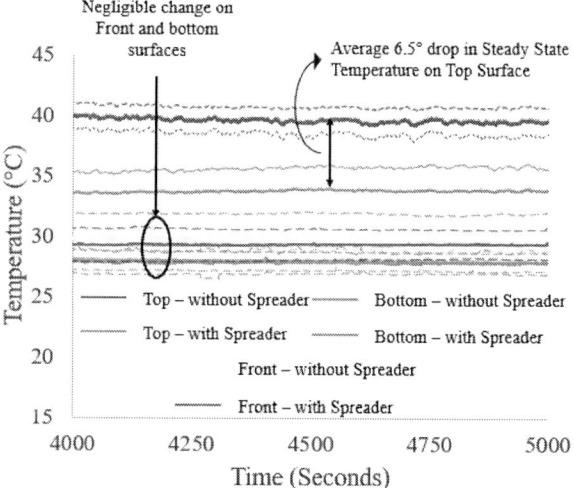

Figure 7 – Steady state measured surface temperature comparison. Average temperature data (solid lines) along with uncertainty (dashed lines).

Figure 7 is an isolated view and comparison of the temperature readings for each separate test. The temperature drop experienced from adding the heat spreader can be seen. The solid lines being measured temperature and dashed lines indicating the maximum and minimum uncertainty. Data showed negligible change in front and bottom headset temperature but a 6.5 °C drop in top surface temperature. Since the transient power data did not change between tests with and without the integrated thermal management solution, this temperature drop indicates that heat is being more effectively spread throughout the headset. The temperature spike upon removal of the headset is consistent between the two headsets at approximately 2.4°C.

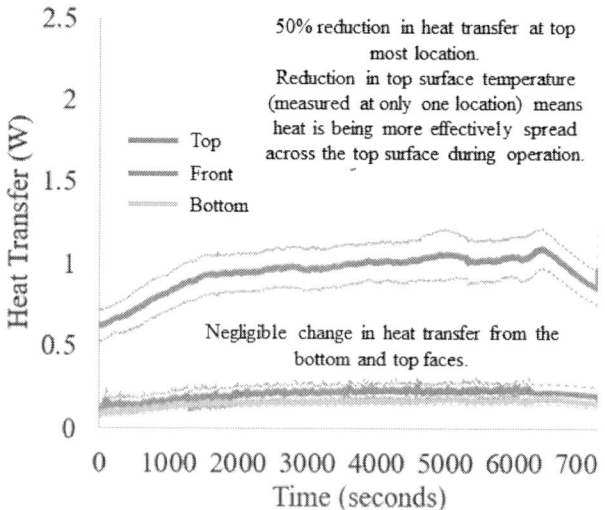

Figure 8 – Convection heat transfer losses (solid lines) along with the calculated uncertainty ranges (dashed lines) for eight testing replications.

Figure 8 shows the resultant free convection heat transfer from all of the surfaces of Figure 2. As expected, the heat transfer from the top surface decreases dramatically, but the expected increases at the front and bottom faces are negligible and within uncertainty. This indicates that the current experimental facility does not contain enough spatial resolution in temperature mapping to accurately capture the breadth of heat spreading around the periphery of the headset. Future work will include mounting an IR camera to the headset during operation to get a temporal and spatially resolved heat transfer profile around the headset.

With adequate temperature data lacking to completely quantify the convective heat transfer along all of the surfaces, the current study surveyed the users about the degree to which they perspire during operation with and without the integrated thermal management solution. With the top surface temperature dropping so dramatically and consequently driving a reduction in the measured heat transfer from that surface, the total amount of power transferred via perspiration actually increased with the integrated spreader solution from an overall energy balance approach. Qualitatively this was not the case, and future work will quantitatively resolve this with the integration of a perspiration measurement device along with time and spatially resolved IR measurement data. The users were asked how much they perspired on a linear scale of 1-10 during use of the system either with or without the additional thermal management equipment. The results are shown in Figure 9, and they indicate that the amount that users that feel that they perspired during use of the headset dropped 25%. The results indicate a relationship between the quality of the thermal management solution and the user experience with wearable technology.

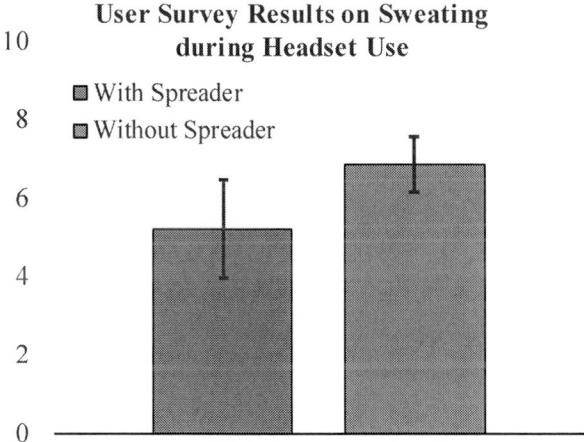

Figure 9 – Qualitative assessment from a user survey illustrating that perspiration discomfort was reduced with the integration of the internal heat spreader into the VR headset.

Conclusions and Future Work

Comfort was increased by using an integrated heat spreader within a commercially available VR headset through a reduction in perspiration over prolonged use. Users reported a 25% reduction in perspiration over an hour and a half of activity. The current study lacks the spatial resolution on the surface of the headset to accurately capture the amount of heat transferred via perspiration from a conventional first law balance. Future work will include a perspiration measurement device installed under the headset during operation along with a mounted IR camera to resolve, both temporally and spatially, the total heat losses from the headset during prolonged use. The internal heat spreader reduced the hottest temperature on the surface of the headset, namely that of the top nearest the primary MCU, by 6.5 °C. This drop, along with the relative stability of the bottom and front temperatures, shows that heat is more effectively spread around the periphery of the headset. However, one thermocouple on each of these three critical faces (top, bottom, and front) was not sufficient to capture the total heat transfer around the device. Through a combination of quantitative assessments proving more effective heat spreading and a survey of the user experience with and without an internal heat spreader, the current study shows a clear relationship between customer satisfaction and thermal management design quality.

References

[1] IMARC Group, 2019, "Industrial Wearable Devices Market: Global Industry Trends, Share, Size, Growth, Opportunity and Forecast." Available: https://www.imarcgroup.com/industrial-wearable-devices-market

[2] International Data Corporation, 2018, "Worldwide Quarterly Augmented and Virtual Reality Headset Tracker," Available: https://www.idc.com/tracker/showproductinfo.jsp?prod_id=1501

[3] International Data Corporation, 2018, "VR Headset Market Rebounds as Standalone Products Gain Traction while AR Headset Market also saw Positive Movement," Available: https://www.idc.com/getdoc.jsp?containerId=prUS44509518

[4] United States Energy Information Administration, 2019, "How Much Electricity does an American Home Use?," Available: https://www.eia.gov/tools/faqs/faq.php?id=97&t=3

[5] Y. Leng, C.-C Chen, Q. Sun, J. Huang, and Y. Zhu, 2019, "Energy-Efficient Video Processing for Virtual Reality, in *Proceedings of the 46th International Symposium on Computer Architecture*, Phoenix, AZ, pp. 91-103.

[6] I. Ali, Heat Pipe Thermal Component for Cooling System, US10416735B2, United States Patent and Trademark Office, Sept. 17, 2019.

[7] HP, "Omen Mindframe Advanced Gaming Headset," Available: https://www8.hp.com/us/en/displays-accessories/gaming/omen-mindframe.html

[8] Bergman, T. L., Lavine, A., & Incropera, F. P. (2019). *Fundamentals of heat and mass transfer*. Hoboken, NJ: John Wiley & Sons, Inc.

Appendix:

A sample calculation of the natural convection for the headset is explained and shown. First the headset needed to be broken up into three different zones for natural convection to occur; the top, front, and bottom surfaces. These zones are color coded in Figure 2. Each plate has its own correlation to find the natural convection heat loss from the surface to the ambient. The Rayleigh number for each surface may be characterized from Equation 4 with each surfaces characteristic length value shown in Table 2. $\beta, \alpha,$ and ν are the expansion coefficient, thermal diffusivity, and specific volume of the air, respectively.

$$Ra_L = \frac{g\beta(T_s - T_\infty)L^3}{\nu\alpha} \tag{4}$$

Surface	Charactersitic Length (mm)
Top	26.7
Bottom	26.7
Front	83.0

Table 2: Characteristic length of each surface used to calculate the corresponding surfaces Rayleigh number

Combining Equations 1,2, and 3 with Equation 5 results in Equations 6, 7, and 8. Each equation was used to calculate the top, bottom, and front instantaneous heat transfer coefficient. The total heat loss of the headset from natural convection is found by summing up each surface's total heat loss from Newtons Law of Cooling, Equation 9. Table 3 shows the values found for the example calculation.

$$\overline{Nu_L} = \frac{\overline{h}l}{k} \tag{5}$$

$$h_{top} = 0.54 Ra_L^{0.25}\frac{k}{L} \tag{6}$$

$$h_{bottom} = 0.52 Ra_L^{0.2}\frac{k}{L} \tag{7}$$

$$h_{front} = \left\{0.68 + \frac{0.670 Ra_L^{0.25}}{[1+(0.492/Pr)\wedge(9/16)\,]\wedge(4/9)}\right\}\frac{k}{L} \tag{8}$$

$$q_{loss} = hA(T_s - T_{amb}) \tag{9}$$

Surface	Rayleigh Number	Heat Transfer Coefficient (W/m2K*K)	Heat Loss (W)
Top	1.46E+04	5.88	1.073
Bottom	6.29E+03	2.96	0.233
Front	2.24E+05	2.01	0.136

Table 3: Values from example calculation

Cross Correlation Method for Images Alignment: Application to 4 Buckets Calculation in Thermoreflectance

Metayrek Youssef[1], Kociniewski Thierry[2], Khatir Zoubir[1]

[1]Institut français des sciences et technologies des transports,
De l'aménagement et des réseaux, IFSTTAR,
25 allée des Marronniers, 78000 Versailles, France
[2]Groupe d'Etude de la Matière Condensée (CNRS and University of Versailles St Quentin),
45 avenue des Etats-Unis, 78035 Versailles cedex, France.
youssef.metayrek@ifsttar.fr, thierry.kociniewski@uvsq.fr, zoubir.khatir@ifsttar.fr

Abstract

Correlation methods are generally used in mechanics to study the deformation in materials. These techniques permit the determination of a moving field by comparing the digital photos before and after deformation. In this article, we have applied correlation techniques to superposed images after displacement due to thermal effect during thermoreflectance measurements on an insulated gate bipolar transistor (IGBT) power module using "4 Buckets" technique at magnification x50. The "4 Buckets" technique uses a simple calculation on light intensities reflected by a component biased with sinusoidal excitation to obtain the variation of reflectivity for each pixel in a picture when the images are superposed perfectly. Then using a calibration, it is possible to obtain a thermal image. We have applied correlation technique to "4 Buckets" images before calculation and studied their effects on thermal images and calibration coefficients. Preliminary results indicate that cross correlation is the best method to superpose each image to a reference and make calculation in order to eliminate any artifacts due to thermal expansion in thermoreflectance measurements.

Keywords

Cross correlation, phase correlation, thermal expansion, thermoreflectance, 4 buckets, calibration.

1. Introduction

Thermoreflectance microscopy is a well-established method for the thermal imaging of power electronic components. The technique provides a submicron spatial resolution (250 nm) coupled with high temperature resolution (less than 1K) that is useful for hot spot detection and failure analysis [1].

Thermoreflectance is a non-contact technique based on the measurement of the relative change in the device surface reflectivity as a function of temperature. As the temperature changes, the reflectivity changes. The relation can be approximated to first order as:

$$\frac{\Delta R}{R} = \frac{1}{R}\frac{\partial R}{\partial T}\Delta T = k\Delta T \qquad (1)$$

Where k is the thermoreflectance coefficient that depends on the sample material, the wavelength of the illuminating light and the composition of the sample [2].

The typical values of thermoreflectance coefficient range from 10^{-6} to 10^{-2} (K^{-1}) which impose a small variation of reflectivity with temperature [1]. Since the variations of reflectivity are small, it is very difficult to measure in continuous regime. Various modulation schemes are used to detect the relative change in reflectivity of the sample surface; one of them is the "4 Buckets" [3]. This method consists in powering the device periodically with sinusoidal or square excitation at a given frequency F, and trigger the camera (CCD) at 4F to take four images I_1-I_4 during each period of the heating. These images (I_1-I_4) correspond to the following four integrals:

$$I_1 = I_0\int_0^{\frac{T}{4}} R(x,y,t)dt, \qquad I_2 = I_0\int_{\frac{T}{4}}^{\frac{T}{2}} R(x,y,t)dt,$$

$$I_3 = I_0\int_{\frac{T}{2}}^{\frac{3T}{4}} R(x,y,t)dt, \qquad I_4 = I_0\int_{\frac{3T}{4}}^{T} R(x,y,t)dt, \qquad (2)$$

A simple calculation based on these four integrals allows us to extract the relative change in reflectivity ($\Delta R/R$) of each pixel as described in eq. 3 by the normalization of the modulated part (ac) and the continuous part (dc).

$$\frac{\Delta R}{R} = \frac{\pi}{\sqrt{2}}\frac{\sqrt{(I_1-I_3)^2+(I_2-I_4)^2}}{I_1+I_2+I_3+I_4} \qquad (3)$$

It is important to recognize that the temperature variation on IGBT surface due to the "4 Buckets" technique results in thermal expansion, which leads to motion from an image to another. These offsets between images generate an error in the thermal measurements because of the sample areas mismatch on the CCD camera pixel matrix and the calculation done on each pixel does not correspond on same pattern. To perform the thermal measurement without any artifacts due to these parasitic movements, it is imperative to perform image recording and post processing registration.

Alignment is a fundamental task in image processing domain. It was used in remote sensing, medical imaging, and microscopy biology [4]. This technique consists in matching two or more images taken at different times, from different sensors or from different points of view. Depending on the nature of the alignment, two types of approaches are possible: global or local. We define the global transformation when all parts of the image move in the same way, whereas local transformation is defined when one of these parts differs in movement from the other parts of the image [5].

Among the global method, the most used are the phase correlation method and cross correlation method. Only a few authors claim to use registration for their thermal measurements by thermoreflectance without going into details about its action on the final results [6-8].

In this paper, we present the use of phase correlation and cross correlation methods for reflectivity variation calculation by thermoreflectance. The influence of alignment on reflectivity variation image and thermoreflectance coefficient calculation are presented.

2. Experiments

We have carried out thermal measurements by thermoreflectance under homodyne conditions on emitter metallization elementary cells structure of the IGBT chip packaged in a module. In order to visualize the cells structure of the IGBT chip, the measurements were carried out using a microscope equipped with an objective x50, yielding a spatial resolution of approximately 0.2 μm (figure 1). A 632 nm wavelength red light was used.

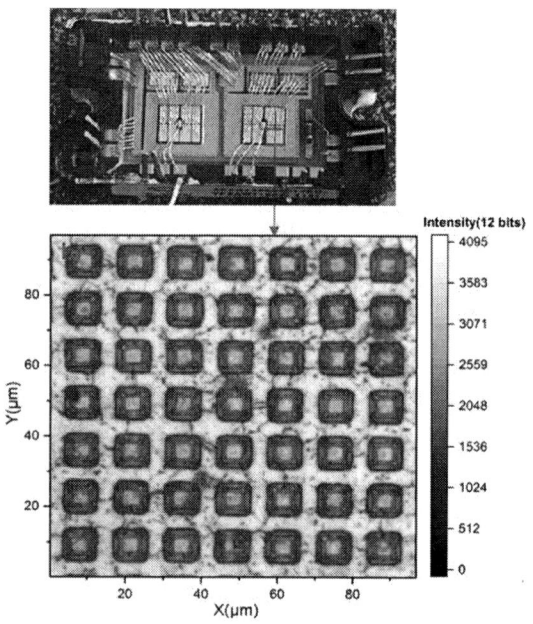

Figure 1: a) Optical general image of an IGBT power module, b) Optical image at x50 magnification obtained on IGBT emitter metallization with photon focus 12 bits camera.

The basic setup of a homodyne high spatial-resolution CCD-based thermoreflectance is presented in figure 2. This homodyne setup supplies a sinusoidal electrical signal at frequency F between collector and emitter connector of the IGBT while continuous illumination is made with light emitting diode (LED) source focused onto the sample by a microscope objective [9]. The light is reflected back to the CCD in response to sinusoidal modulation of the sample and then collected by a computer in images at 4F frequency. Images are registered in order to calculate the relative change in reflectivity given by eq. 3.

The frequency of electrical excitation influences the shape of the thermal signal. For calculation of reflectivity variation, a sinusoidal variation of the temperature is needed. The excitation frequency has been varied in the range of 0.625 to 10 Hz. For frequencies equal to 0.625 Hz, the thermal excitation approximately matches the form of sinus function by nearly 97%.

This frequency is directly linked to the module cooling time. Its value is equal to half the period of electrical excitation, which is around 0.8 s for this module. This time allows to obtain a thermal diffusion length larger than the thickness layer behind the electrical active area and thus to keep a sinusoidal temperature variation which follows the sinusoidal electrical excitation. Note that IGBT modules exhibit high thermal time constant due to the layers stacking of different materials for electrical insulation and mechanical adaptation [10].

The IGBT was excited by different sinusoidal amplitude excitations of collector voltage (Vce) at frequency F=0.625 Hz and Vge=8.5 Volts (gate voltage) fixed in order to have a temperature variation which follows electrical excitation. This excitation was made with an offset sine wave:

$$Vce = \frac{Vcemax}{2}(1 + cos(2\pi Ft)) \qquad (4)$$

The reflected signals were measured by CCD camera at a frequency 2.5 Hz and stored by a computer. Since averaging series of four images increases the dynamics of the measurements [11], we have averaged the signals over 500 periods. Figure 3 presents the " 4 Buckets" images of powered IGBT emitter metallization at different amplitudes (Vce) obtained without alignment. Corresponding powers injected are also precised.

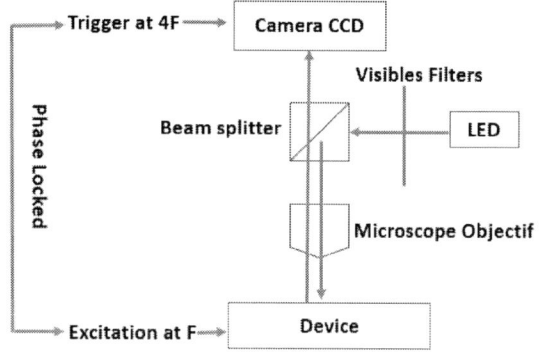

Figure 2: Schematic depiction of a homodyne thermoreflectance microscopy setup.

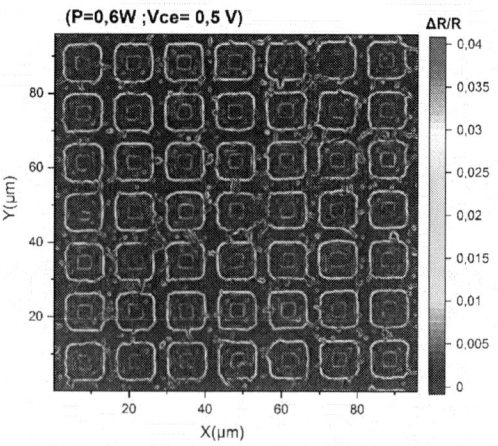

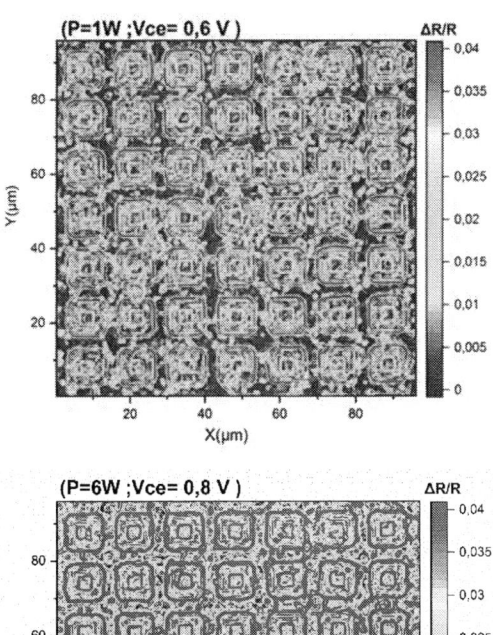

Figure 3 : Normalized thermoreflectance images (ΔR/R) obtained on IGBT emitter metallization at x50 magnification without alignment at different sinusoidal voltages (Vce $_{max}$= 0.5-0.6-0.8 Volts) and Vge=8.5 Volts fixed.

During measurements, optical images have presented a periodic displacement synchronized with electrical excitation of the device. As the amplitude of electrical excitation increased the displacement amplitude increased. These displacements were caused by inhomogeneous thermal expansion. If we compare the "4 Buckets" images at different Vce, we can see that the larger the parasitic motions are, a broader allele appears on the strong contrast gradient zones of the optical image (around elementary cells). The marked sharpness of the edges cells contours increase as we have increased the voltage Vce. This is due to the fact that for some pixels of the CCD camera a dark area of the emitter surface then a lighter area may coincide increasing the difference in "4 Buckets" calculation method.

This movement causes artifacts in thermoreflectance images (ΔR/R) which makes the thermal measurement impossible since the calculation was not made always on the same area but rather on other surrounding areas.

To overcome this problem, an image-processing algorithm must be applied to images in order to align the images to a reference image before calculations.

3. Choice of alignment method for thermal measurements

Image alignment is necessary to align the pixels of translated image to corresponded pixels of the reference image. To perform this alignment, it is important to find the parameters of the geometrical transformation. We distinguish transformations as rotation, translation and elastic [12]. In our case, we limited ourselves only to translations.

Mathematically, this is expressed by:

$$J(x,y) = I(x + t_x , y + t_y) \tag{5}$$

With J is the shifted image corresponding to I, and tx, ty the following translations x and y.

We have observed that all the images move in the same direction. We orient our choice towards the global methods and limit them to translations. Among the global methods, phase correlation and cross correlation were used.

3.1. Phase-Correlation method

Phase correlation is a well-known image matching technique in the frequency domain. This method estimates the relative translation offset between two similar images by using their Fourier transforms. This technique calculates the phase of the cross-power spectrum (CPS):

$$CPS(u,v) = \frac{\alpha(u,v) * \beta(u,v)^*}{|\alpha(u,v) * \beta(u,v)^*|} = e^{-i(u*t_x + v*t_y)}$$

$$(6)$$

Where α(u,v) and β(u,v) are the Fourier transforms of the two images in the frequency domain.

Phase correlation uses the inverse Fourier transform of normalized cross-power spectrum as shown in eq.7 in order to provide a distinct sharp peak (Dirac peak) at the point where the two images are the most similar [13].

$$TF^{-1}(CPS(u,v)) = \delta(x - t_x, y - t_y) \tag{7}$$

In order to assess the performance of phase correlation in our case, we chose an image as a reference and applied the phase correlation algorithm to the others images to determine their displacement field. The algorithm of translation vectors calculation was developed on Matlab and automatized to all the images. Figure 4 represents the typical graph giving the coordinates of the image translation vector by the method of phase correlation. The image chosen for the calculation is the one that showed the largest offset taken in the Vce $_{max}$ = 0.8 Volts series.

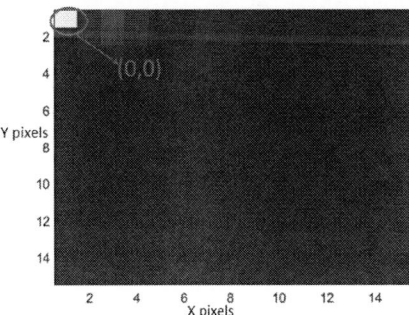

Figure 4: Coordinates of the translation vector for alignment of the two translated images given by phase correlation (Phase correlation image).

139

The calculation failed because the translation vector was found at zero (figure 4) and does not correspond to reality. Same observations were found for all the images. In order to check the validity of our program, we have voluntarily translated an image of 6 pixels in x direction and 6 pixels in y direction and applied the phase correlation. Figure 5 shows the results obtained. We observe that the program has found the correct vector of translation at (6, 6). We notice that blotches of non-uniform color intensity appear in figure 4 since the phase correlation was made between two different images unlike figure 5 where the correlation was made between two identical images only translated.

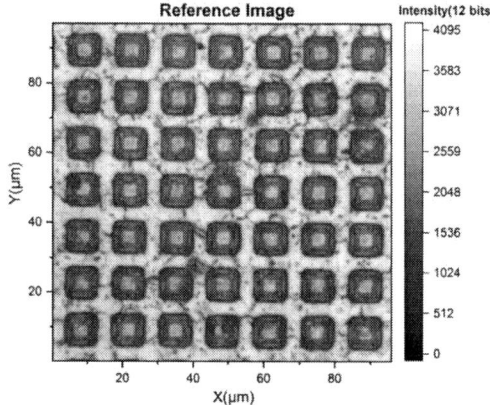

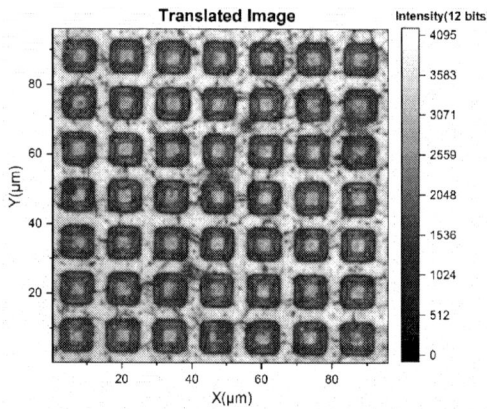

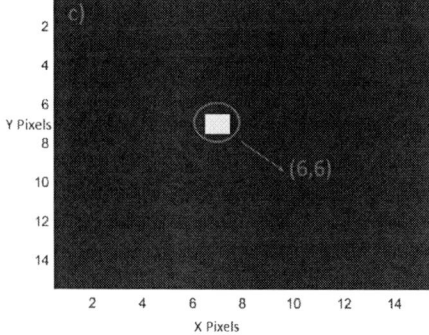

Figure 5: a) Reference image b) Reference image voluntary translated image tx=ty=6 pixels c) Phase correlation image.

The phase correlation technique is not an adapted technique for the alignment of images in the context of measurements made by "4 Buckets" homodyne technique. This could be due

to the change of intensity since the intensity is changed due to reflectivity variation. Therefore, we then considered the cross correlation method.

3.2. Cross-Correlation method

Cross correlation consists in measure of the degree of similarity between two images. This technique is based on the calculation of the correlation coefficient in order to find the offsets that exist between two pair of images obtained before and after surface deformation [14].

Mathematically, the correlation coefficient is given by:

$$r(t_x, t_y) = I(x,y) * J(x,y) = \sum_{x,y} I(x,y)J^*(x - t_x, y - t_y) =$$

$$\sum_{u,v} \alpha(u,v)\beta(u,v)^* \exp[i2\pi(\frac{u*t_x}{M} + \frac{v*t_y}{N}) \qquad (8)$$

Where M and N are the image dimensions. The precision of alignment can also be estimated using the normalized root-mean-square error (NRMSE) (E) [15]:

$$E = \sqrt{1 - \frac{\max\limits_{t_x,t_y} |r(t_x,t_y)|^2}{\sum_{x,y}|I(x,y)|^2 \sum_{x,y}|J(x,y)|^2}} \qquad (9)$$

The correlation coefficient ranges between 0 to 1. If the value is close to 1, the 2 images are very similar while if it is near to 0, the 2 images are very different.

A program was developed using Matlab to automate the calculation of the vectors of translation between the images and a reference image by cross correlation method. This program can align 2000 images in about one minute. We used this program to align the same images used for phase correlation calculations (measurement series at Vce max = 0.8 Volts .The correlation coefficient before alignment was 0.80, and its value increased to 0.99 after alignment.

In order to determine the effect of the alignment on the images by the cross correlation technique, we have followed the intensity of a fixed pixel on the image series realized with a Vce max = 0.8 Volts before and after registration. The results is presented in figure 6. It shows that in the case without alignment, the intensity vary a lot during the experiment with $\Delta I = 400$ (around 15 %), which is too high to compare to the physical effect of temperature on reflectivity. It appears that we have not followed the evolution intensity of the same pixel, but also of pixels that are around due to the displacement of the surface. In other case with alignment, the intensity evolution of the pixel decreases with $\Delta I = 100$ (around 3 %). This intensity variation deconvoluted from the translation effect represents the direct effect of the temperature and the optical noise on the reflectivity coefficient. Note that the average intensity of the images has been calculated with and without registration. The same intensity has been found in two cases. Therefore, the alignment does not affect the intensity. Thus, cross correlation appears to be a good registration method for "4 Buckets" calculation.

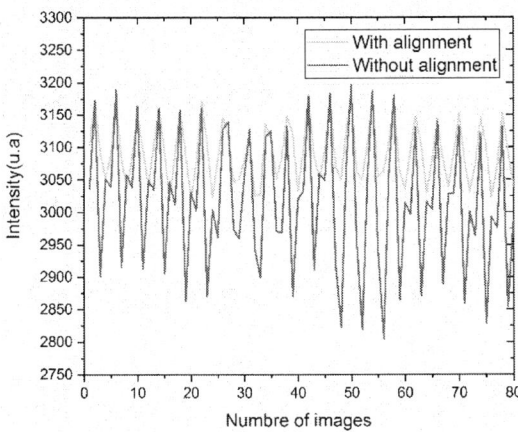

Figure 6: Intensity curves of a pixel during "4 Buckets" acquisition with and without alignment.

4. Influence of registration on calibration

We have calculated the "4 Buckets" images after cross correlation registration. Figure 7 presents the "4 Buckets" images of IGBT emitter metallization obtained with cross correlation alignment at different amplitudes Vce. These figures show that the marked sharpness of contours at the edges of the cells disappeared after performing the alignment. The alignment has eliminated the artifacts in calculation.

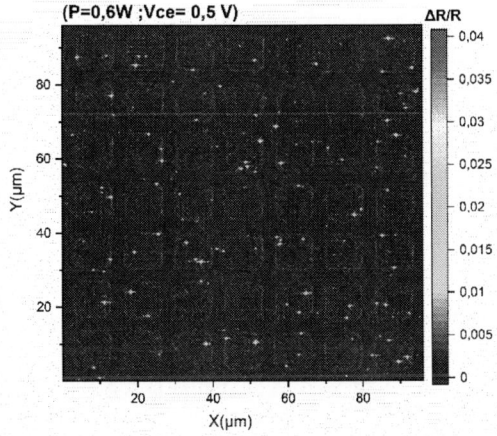

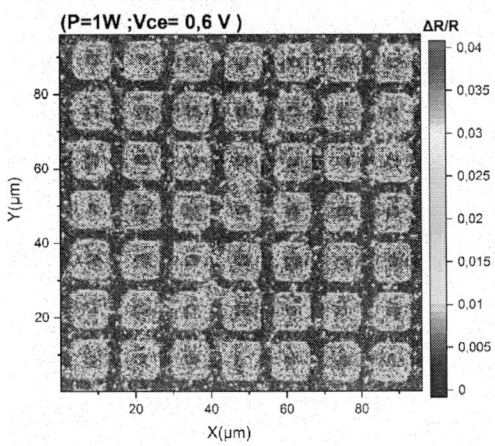

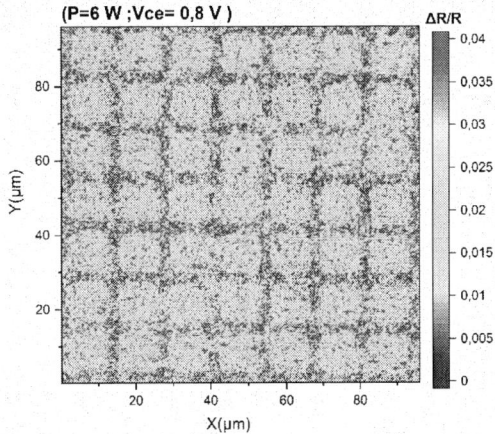

Figure 7: "4 Buckets" images obtained on IGBT emitter metallization at x50 magnification with alignment at different voltages.

In order to obtain a quantitative temperature of the IGBT chip surface, it is important to determine the thermoreflectance coefficients for the measured material. Our calibration method was based on "4 Buckets" measurements correlated to a local temperature measurement on the chip surface using an optical fiber sensor as described in [16].

By placing the GaAs crystal on a chosen region of the device emitter metallization, the temperature variations (ΔT) can be directly measured in the same sampled area by "4 Buckets" measurements. The amplitude of the oscillations gives us the variation of the average temperature on the crystal surface for a Vce $_{max}$.

The increment in temperature variation was created by increasing the Vce$_{max}$ sinusoidal amplitude excitation (Vce $_{max}$= $0 - 0.5 - 0.6 - 0.8$V). By repeating the measurements for different amplitudes, one can get the dependence of the mean value of $\Delta R/R$ with ΔT. Figure 8 presents the thermal signal measured with the GaAs crystal on the emitter metallization as a function of temperature variation with and without registration.

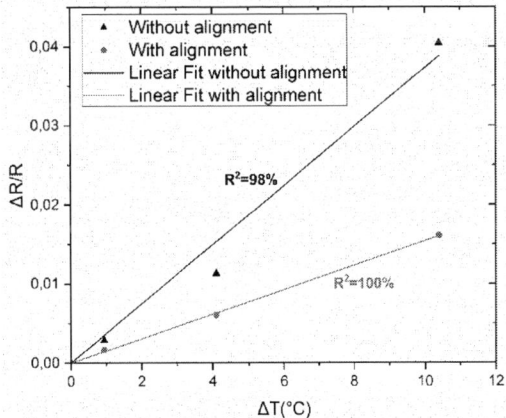

Figure 8: Mean $\Delta R/R$ as a function of ΔT measured by GaAs crystal in two cases: with and without registration.

Linear fits are also presented with their correlation rate. In the case of registration, the thermal signal matches perfectly to a linear function, which is in perfect agreement with the first order relation presented in eq.1. Without registration, a lower correlation rate around 98% is obtained. Such a high correlation rate could validate the fit if registration case was not studied.

$K_{mean} \pm \Delta K_{mean}$ (K^{-1}) with alignment	$K_{mean} \pm \Delta K_{mean}$ (K^{-1}) without alignment
1.55x10^{-3}±1.120 x10^{-5}	3.80x10^{-3}±1.310 x10^{-4}

Table 1: Mean Thermoreflectance coefficients (K_{mean}) of IGBT emitter metallization obtained with and without alignment.

For both cases, the mean thermoreflectance coefficients were extracted by calculating the slope of the linear adjustment of each curve. Results with are summarized in Table 1. The quadratic deviations of the fits with the experimental points were also presented. Thermoreflectance coefficients is higher in the case without alignment than that with alignment. This change in thermoreflectance coefficient is clearly due to the high value of reflectivity variation around the contours at the edges of the cells due to the error in "4 Buckets" calculation. This conclusion is supported by the value of the quadratic deviation of each fit. Cross correlation, alignment is necessary to obtain ΔR/R images and calibrated images without any artefacts due to parasitic movement and thermal expansion even for small displacement when the image contrast is strong.

5. Conclusions

This paper presents the study of two correlation methods for thermal measurements by thermoreflectance in homodyne regime. Experimental results demonstrate that the cross-correlation method is the suited method to align perfectly the images before "4 Buckets" calculation for emitter IGBT metallization. Calibration curves show that with alignment the relation between ΔR/R and ΔT is perfectly linear which is in good argument with the first order theoretical relation. The ΔR/R images without alignment show artifacts in "4 Buckets" images and these displacements induce also error in calibration coefficient estimation.

These results testify the importance of the registration in the thermic maps obtained by thermoreflectance. In our case, measurement artifacts were clearly be highlighted on image areas with a strong intensity gradient (high contrast) but one can ask the question of the effects on homogeneous surfaces in intensity where the artifact would be less marked and where the direct effect would not be appreciable.

References

1. M. Farzaneh, K. Maize, D. Luerßen, J.A. Summers, P.M. Mayer, P.E. Raad, K.P. Pipe, A. Shakouri, R.J. Ram, J. and A. Hudgings, "CCD-based thermoretlectance microscopy: principles and applications " ,J. Phys. D: Appl. Phys., vol. 42, 143001 (20pages), 2009.
2. G. Tessier, S. Holé, and D. Fournier,"Ultraviolet illumination thermoreflectance for the temperature mapping of integrated circuits", Opt. Lett., vol. 28, no. 11, pp. 875-877, 2003.
3. P. M. Mayer, D. Lüerßen, R. J. Ram, and J. A. Hudgings , " Theoretical and experimental investigation of the thermal resolution and dynamic range of CCD-based thermoreflectance imaging ", J. Opt. Soc. Am. A, vol. 24, no. 4, pp. 1156-1163, 2007.
4. Y. Hongshi , L.Jian Guo , " Robust Phase Correlation based feature matching for image co-registration and dem generation",The International Archives of the Photogrammetry, Remote Sensing and Spatial Information Sciences, 2008.
5. M. G. Burzo et al , " Pixel-by-pixel calibration of a CCD camera based thermoreflectance thermography system with nanometer resolution ", 15th International Workshop on Thermal Investigations of ICs and Systems, THERMINIC pp. 130-135,2009
6. M. Bardoux et al,"Thermoreflectance imaging of laser diodes and VCSELs along and perpendicular to the emission direction" , Optics and Lasers in Engineering, vol. 47, pp. 473-476,2009.
7. R. Bhojani et al, " Observation of Current Filaments in IGBTs with Thermoreflectance Microscopy",30th International Symposium on Power Semiconductor Devices and ICs (ISPSD),2018.
8. K. Yazawa et al, "Short Pulse Thermal Response of HBTs",Proceedings of the 11th European Microwave Integrated Circuits Conference, pp. 45-48,2016.
9. S. Grauby and B. C. Forget, " High resolution photothermal imaging of high frequency phenomena using a visible charge coupled device camera associated with a multichannel lock-in scheme" , Review of Scientific Instruments,vol.70,pp. 3603,1999.
10. C. Yun, P. Malberti, M. Ciappa, and W. Fichtner, " Thermal component model for electrothermal analysis of IGBT module systems " , Trans. on . Advan. Pack. vol.24 , pp.401-406, 2001 .
11. G. Tessier, M. Bardoux, C. Filloy, C. Boué, and D. Fournier,"High resolution thermal imaging inside integrated circuits, Sensor Review" , vol.27, pp. 291-297, 2001.
12. E.Gladin,R.Elis, " On the role of spatial phase and phase correlation in vision, illusion, and cognition" , Frontiers in Computational Neuroscience ,vol.9, pp.1-14,2015.
13. S.Freng et al, " A Subpixel Registration Algorithm for Low PSNR Images" , Advanced Computational Intelligence (ICACI), 2012 IEEE Fifth International Conference on (2012).
14. AMR.Sousa,M.A.Pvaz,V.Filipe,J.Xavier ,J.J.L .Morais, " Measurement of displacement fields with sub-pixel accuracy by combining cross-correlation and optical flow" ,ISSN (2012)
15. M.Guizar, S.Thurman, J.Fienup, " Efficient subpixel image registration algorithms " ,Optics Lett. ,vol. 33, no. 2 , 2008.
16. G. Tessier, S. Holé, and D. Fournier, " Quantitative thermal imaging by synchronous thermoreflectance with optimized illumination wavelengths" , Appl. Phys. Lett., vol. 78, no.16, pp. 2267-2269, 2001.

Experimental Measurement and Finite Element Analysis of the Thermal Conductivity of Alumina / Silicone Polymer Composites

Masakazu HATTORI *, Kazuaki SANADA ** and Yasushi KAJITA***
* Fuji Polymer Industries Co., LTD. Aichi, 470-0533, Japan
** Toyama Prefectural University, Toyama, 939-0398, Japan
***Nagoya Municipal Industrial Research Institute, Aichi, 456-0058, Japan
Corresponding author: Masakazu HATTORI,
E-mail: hattori_masakazu@fujipoly.co.jp

Abstract

Polymer-based, high thermal conductivity composite materials are commonly used for transferring heat from heat source to ambient in electronics and electrical systems. Thermal conductivity is a critical property of thermally conductive composite materials. Thermal conductivity is dependent on the volume fraction of the thermally conductive fillers. Although the thermal conductivity is an intrinsic specific value, different values can be reported depending on the thermal conductivity test method used. In this study, samples are prepared with four levels of volume fraction of alumina and silicone polymer. Composite materials are measured for thermal conductivity using transient method (hot disk method, laser flash method) and steady-state method. Moreover, Finite element analysis was conducted on the heat conduction characteristic using the representative volume element (RVE) model. The Digimat software was used to create two and three-dimensional RVE containing randomly distributed thermally conductive fillers. The RVE was modeled to show transient response during laser flash testing using the ANSYS finite element program. Comparison of actual test values are compared to the finite element analysis results.

Keywords

Thermal Conductivity, Composite Material, Thermal Interface Material, Transient Method, Steady-State Method, Finite element analysis, RVE Model

Nomenclature

TIM   Thermal Interface Material
RVE   Representative Volume Element

1. Introduction

Recently, high density, high output, and miniaturization of electronic devices have progressed, and heat countermeasures have become important for improving device reliability. Thermal conductive composite materials called Thermal Interface Material (hereinafter referred to as TIM) are generally used to transfer heat from heat-generating components such as IC to heat sinks and housings that are dissipate heat to ambient. At the interface between the heat-generating component and the radiator, fine irregularities and voids interrupt heat flow. TIM's are designed to fills the voids and lower contact resistance between the heat source and heat sink.

TIM's are generally characterized based on a numerical index of heat transfer, either thermal resistance or thermal conductivity. The thermal conductivity is a numerical value specific to the material, so it must match regardless of the measurement method and conditions. However, different values can be reported depending on measurement method. This is especially common between steady-state method that gives a steady temperature gradient and a constant heat flow, and the transient method that gives the transient heat flow energy. In this study, we verified the difference in measured values between the steady-state and the transient methods using actual measurement and finite element analysis.

2. Experimental/Numerical Method

2.1 Sample preparation

The matrix polymer was a liquid curable silicone polymer, and the thermally conductive filler was spherical alumina with an average particle size of 35μm.

The composite material containing thermally conductive filler was prepared by the following method. First, silicone polymer and spherical alumina were mixed in a planetary mixer so that they were uniformly dispersed while vacuum degassing. Next, it was sandwiched between 100 μm thick polyethylene terephthalate (PET) resin film that had been release-treated on the surface, adjusted to the desired thickness between the rolls with a two-roll mill, and molded into a sheet. Then, in order to prevent sedimentation of the filler, it was quickly cured and cross-linked in a heating chamber. Four levels of heat conductive filler volume fractions of 20, 33, 50, and 60 vol% were prepared.

2.2 Measurement

In order to verify the difference in thermal conductivity between the transient method and the steady-state method, the transient method was measured using laser flash method (manufactured by Netch Co., Ltd.) [1] and hot disk method (manufactured by Kyoto Electronics Industry). The steady-state method is measured with DynTIM (manufactured by Siemens) [2]. The DynTIM steady-state method is compliant with ASTM D5470. A second steady-state thermal conductivity measurement device in accordance ASTM E1530 (using a heat flow sensor) provided by Nagoya City Industrial Research Institute (NMIRI) did.

In the laser flash method, the thermal conductivity λ [W / mK] was calculated by the following equation using the measured thermal diffusivity α [m² / s] and the density and specific heat obtained by the composite law.

$$\lambda = \alpha C_p \rho \quad (1)$$

Cp is the specific heat capacity [J / kgK] and ρ is the density [kg / m³].

2.3 Numerical analysis

Finite element analysis was performed using RVE models. As shown in Figures 1 and 2, the 2D RVE and 3D RVE models were constructed using the composite material modeling software Digimat-FE. ANSYS was used for finite element analysis. The particle size of the thermally conductive filler was set to 35 μm, the same as the average particle size of the model sample. In the RVE model, a filler volume fraction of 50 and 60 vol% was used the size reduction function for particle size due to the difficulty of filling with a single particle. In addition, there was no contact between the fillers, and the distance between particles was controlled.

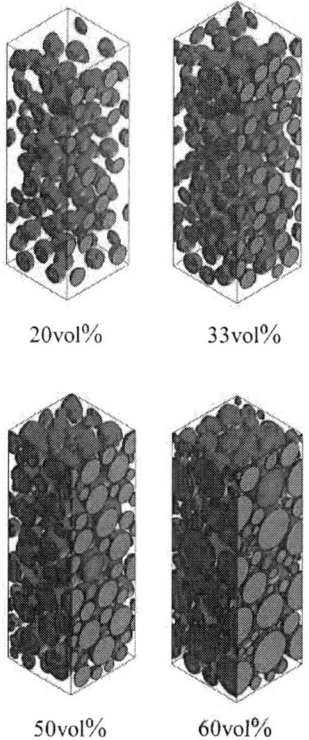

20vol%   33vol%

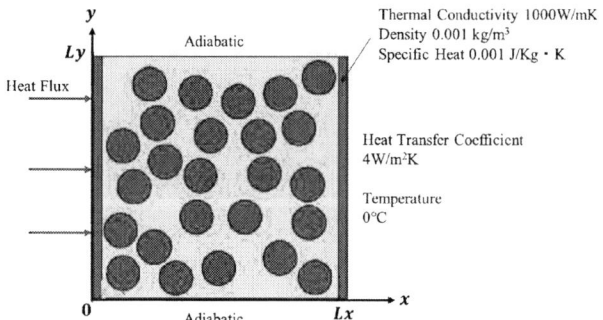

50vol%   60vol%

Figure 2. 3D RVE models

In the laser flash analysis using 2D RVE models, as shown in Figure 3, the boundary conditions were as follows.

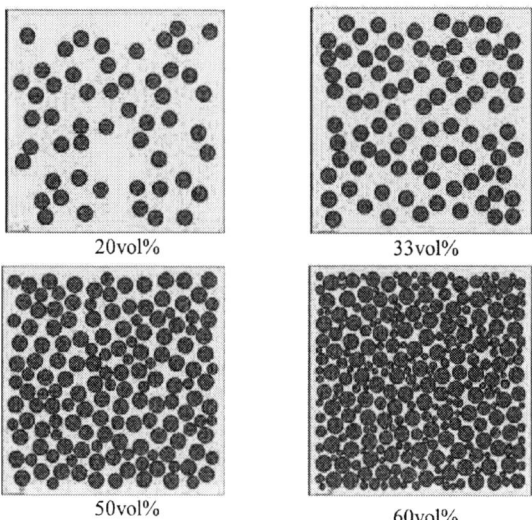

20vol%   33vol%

50vol%   60vol%

Figure 1. 2D RVE models

Figure 3. 2D RVE model of composites and boundary condition

On the $x = 0$ plane, the peak heat amount was 10.9 kW at $t = 0.55$ ms, and a triangular pulse heat flux was given to complete laser irradiation at $t = 1.1$ ms as shown in Figure 4,

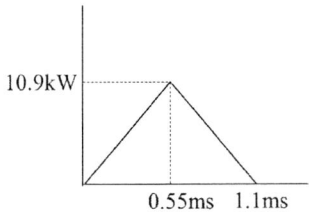

Figure 4. Applied pattern of heat flux

In addition, the ideal member with $x = 0$, Lx surface with thermal conductivity of 1000 [W / mK], density of 0.001 [kg / m^3], and specific heat of 0.001 [J / kgK] to average the in-plane temperature. Assuming natural convection as the boundary condition, the heat transfer coefficient was set to 4 [W / m^{2}K] and the ambient temperature was set to 0 ° C. The y = 0 plane and the $y = Ly$ plane were adiabatic in order to consider periodic symmetry.

In the steady-state analysis, a constant heat flux was given to the $x = 0$ plane, and the temperature difference between the $x = 0$ plane and the $x = Lx$ plane was calculated and thermal conductivity was calculated from the Fourier formula shown as below.

$$\lambda = \Delta T \cdot A/Q \cdot Lx \quad (2)$$

Q is the heat transfer [W], λ is the thermal conductivity [W / mK], ΔT is the temperature difference, A is the area [m^2], and Lx is the distance [m].

3D RVE model of composites and boundary conditions were also shown in Figure 5,

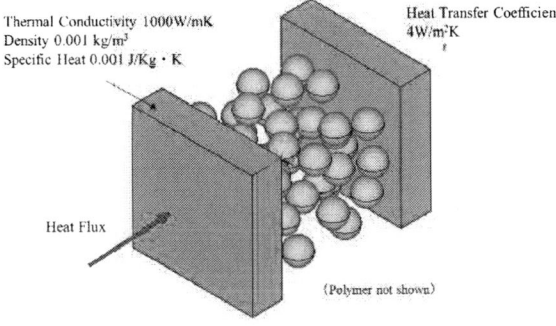

Figure 5. 3D RVE model of composites and boundary Condition

Table 1 shows the physical properties used in the analysis.

Table 1 Material properties

Material	Silicone Polymer	Alumina Filler
Density [kg/m^3]	970	3890
Thermal conductivity [W/mK]	0.2	29
Specific heat [J/kgK]	1600	779

2.4 Thermal conductivity evaluation based on half-time

Evaluation of thermal conductivity using the laser flash method was performed using the time variation of the average temperature of the temperature measurement surface ($x = Lx$ surface) when the laser was applied.

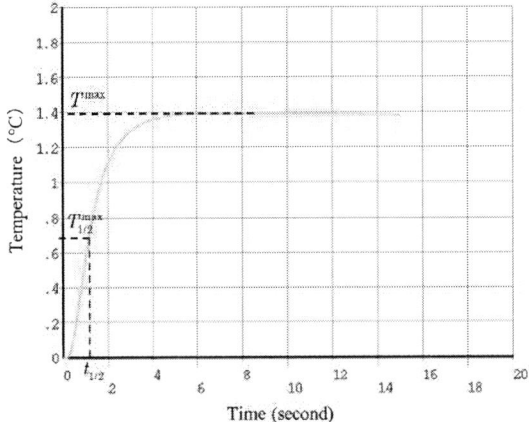

Figure 6. Half-time evaluation using a temperature response curve

Figure 6 shows the relationship between the average temperature of the temperature measurement surface and time.

Using the time $t_{1/2}$ to reach half of the maximum temperature T^{max} (half time), the thermal diffusivity a of the composite material can be obtained from the following equation.

$$\alpha = 0.1388 \frac{Lx^2}{t_{1/2}} \quad (3)$$

The thermal conductivity of the composite material was obtained from equation (1) as in the actual measurement.

3. Results

Table 2 shows the measured results for the thermal conductivity of the composite material. In the case of the same measurement method, the measured thermal conductivity shows almost the same value, but when comparing the results of different measurement methods, the thermal conductivity measured by the steady-state method is about 20% higher than that measured by the transient method.

Table 2 Measured thermal conductivity [W/mK]

Test Method		Volume fraction of alumina [vol%]			
		20	33	50	60
Transient Method	Hot Disk method	0.37	0.56	1.05	1.62
	Laser flash method	0.33	0.52	0.99	1.51
Steady-state Method	ASTM D5470 (DynTIM)	0.33	0.62	1.18	2.04
	ASTM E1530(NMIRI)	0.42	0.58	1.12	1.92

Table 3 shows the results for finite element analysis of the thermal conductivity of the composite material.

In the 2D model, the particles are expressed in a cylindrical shape on the x-y plane, and there is no structural change in the z direction, whereas in the 3D model, the particles are expressed in a spherical shape, so the particles are also distributed in x, y and z direction. Therefore, it is considered that the heat conductivity is higher than that of the 2D model because the heat transfer path is formed in all directions in the 3D model even with the same volume fraction.

Table 3 Predicted thermal conductivity [W/mK]

Test Method		Volume fraction of alumina [vol%]			
		20	33	50	60
Transient Method	2D	0.32	0.46	0.77	1.09
	3D	0.38	0.54	0.89	1.33
Steady-state Method	2D	0.30	0.42	0.65	0.87
	3D	0.36	0.51	0.83	1.20

Figure 7 shows the ratio of the thermal conductivity obtained from finite element analysis using 3D RVE models and the thermal conductivity obtained from measurement using the steady-state method and the laser flash method. In the laser flash method, the ratio did not change greatly even when the filler volume fraction changed. In the steady-state method, the ratio increased as the filler volume fraction increased. This is thought to be due to the contact between the fillers (percolation) when the fillers were highly filled. In the analysis, the model is not a model in which fillers are brought into contact with each other, but a model in which matrix polymer is interposed between fillers.

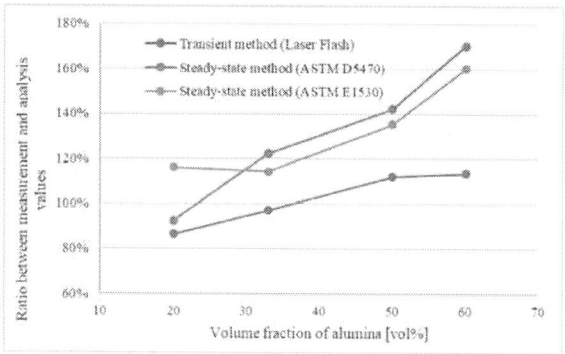

Figure 7. Ratio between measurement and analysis values

4. Conclusions

Measurements and finite element analysis using model samples were performed for the steady-state and transient thermal conductivity of composite materials. From the results,

the difference in thermal conductivity was confirmed between the steady-state and transient measurement methods in the actual measurement. In the comparison between the measured value and the analyzed value, there was a difference depending on the filler volume fraction. Percolation, which is the contact between fillers, is considered to be a factor.

Acknowledgments

This research was carried out as one of the research themes of Chubu Electronics Association 3rd subcommittee.

References

[1] ASTM E1461-92, "Standard Test Method for Thermal Diffusivity of Solids by the Flash Method "ASTM International, (2001)

[2] ASTM D5470-06 (2006) ASTM International

IEEE
445 Hoes Lane
Piscataway, NJ 08854-4141

ISBN 978-1-7281-9588-9